MW00843441

OCCUPATIONAL
NEUROTOXICOLOGY

Edited by

Lucio G. Costa
Luigi Manzo

CRC Press

Boca Raton Boston London New York Washington, D.C.

Acquiring Editor:	Norina Frabotta
Project Editor:	Albert W. Starkweather
Cover design:	Dawn Boyd

Library of Congress Cataloging-in-Publication Data

Occupational neurotoxicology / Lucio G. Costa and Luigi Manzo, editors
    p. cm.
Includes bibliographical references and index.
ISBN 0-8493-9231-4
1. Neurotoxicology. 2. Occupational diseases. I. Costa, Lucio G., 1954–. II. Manzo, L.
RC347.5.O266 1998
616.8—dc21

              98-5973
              CIP

No claim to original U.S. Government works
International Standard Book Number 0-8493-9231-4
Library of Congress Card Number 98-5973
Printed in the United States of America 1 2 3 4 5 6 7 8 9 0
Printed on acid-free paper

The Editors

Lucio G. Costa, Ph.D., is a Professor of Toxicology and Director of the Toxicology Program in the Department of Environmental Health at the University of Washington in Seattle. Dr. Costa received his doctoral degree at the University of Milano, Italy, and after post doctoral training at the University of Texas Medical School in Houston, joined the faculty of the University of Washington in 1983. He holds Adjunct Faculty positions at the University of Milano and Pavia, Italy.

Dr. Costa is a member of the Society of Toxicology, Society for Neuroscience and the Italian Society of Toxicology and of Pharmacology, and was elected Fellow of the American Association for the Advancement of Science. He is also a Fellow of the Academy of Toxicological Sciences and has served as President of the International Neurotoxicology Association. He received the Achievement Award and the Zeneca Lectureship Award from the Society of Toxicology. Dr. Costa is the author of more than 200 publications in the field of neurotoxicology and has served on several national and international panels and review committees.

Luigi Manzo, M.D., is a Professor in the Department of Internal Medicine and Therapeutics and Director of the post-graduate school of Medical Toxicology at the University of Pavia, Italy. He is also involved in graduate and postgraduate training in Toxicology, Occupational Medicine, Hematology, Clinical Neurophysiology, and Psychiatry at the Universities of Pavia and Milan, Italy. Dr. Manzo serves as the Director of the Toxicology Division and the National Toxicology Information Centre at the Salvatore Maugeri Foundation, Pavia Medical Institute. Aside from being the author of more than 100 articles in peer reviewed journals and several book chapters, he has edited 10 books in the areas of toxicology and envirnomental medicine.

Dr. Manzo has been the coordinator of international research programs and has served on peer review panels and the editorial boards of several scientific journals. He currently is a member of the Executive Committee of the Italian Society of Toxicology and the coordinator of the Medical Toxicology Section. He is also an active member of the Society of Toxicology (SOT), the Neurotoxicology Specialty Section, and the International Neurotoxicology Association. Dr. Manzo obtained his training at the University of Pavia, receiving the degrees in Medicine and Surgery and in Pharmacy.

Contributors

Stefano M. Candura
Department of Preventive, Occupational,
  and Community Medicine
University of Pavia
Pavia, Italy

Anna F. Castoldi
Department of Internal Medicine
University of Pavia
Pavia, Italy

Harvey Checkoway
Department of Environmental Health
University of Washington
Seattle, WA

Vittorio Cosi
Institute of Neurological Sciences
University of Pavia
  and C. Mondino Foundation
Pavia, Italy

Lucio G. Costa
University of Washington
Department of Environmental Health
Seattle, WA

Mark R. Cullen
Occupational and Environmental
  Medicine Program
Yale University School of Medicine
New Haven, CT

Donald A. Fox
College of Optometry and
  Department of Biochemical
  and Biophysical Sciences
University of Houston
Houston, TX

A. H. Hall
Rocky Mountain Poison
  and Drug Center
  and University of Colorado Health
  Sciences Center
Denver, CO

Fengsheng He
Institute of Occupational Medicine
Chinese Academy of Preventive
  Medicine
Beijing, China

Perrine Hoet
Industrial Toxicology
  and Occupational Medicine Unit
Catholic University of Louvain
Brussels, Belgium

Anders Iregren
Division of Psychophysiology
Swedish National Institute
  of Occupational Health
Solna, Sweden

Robert R. Lauwerys
Industrial Toxicology
  and Occupational Medicine Unit
Catholic University of Louvain
Brussels, Belgium

Sarah R. Lloyd
Poisindex Information System
Micromedex, South Way
Englewood, CO

Luigi Manzo
Department of Internal Medicine
University of Pavia and Salvatore
  Maugeri Foundation Medical Center
Pavia, Italy

Barry H. Rumack
Rocky Mountain Poison
  and Drug Center
and University of Colorado Health
  Sciences Center
Denver, CO

Anna Maria Seppäläinen
Division of Clinical Neurophysiology
Department of Neurology
University of Helsinki
Helsinki, Finland

William Slikker, Jr.
Division of Neurotoxicology
National Center
  for Toxicological Research
U.S. Food and Drug Administration
Jefferson, AR

G. Triebig
Institute for Occupational
  and Social Medicine
University of Heidelberg
Heidelberg, Germany

Preface

The field of neurotoxicology studies the adverse effects of chemical, biological and physical agents on the function and/or structure of the nervous system. In recent years, there has been an increased concern about the effects of exposures to chemicals that may damage the nervous system, causing sensory, motor, and cognitive effects. The objective of this book is to concisely present information on neurotoxicity resulting from exposure to neurotoxicants in the workplace. Occupational health and safety is a field that involves many different disciplines, and interaction among these fields is essential to deal with the multitude of health problems encountered in the work setting. As such, the recognition and treatment of neurotoxicity requires a multidisciplinary approach which would include professionals from the fields of occupational medicine, toxicology, epidemiology, neurology, industrial hygiene, and psychology.

The first two chapters of the book address the issues of the signs and symptoms of neurotoxicity and of the factors affecting neurotoxic effects. Two additional chapters focus on biological monitoring and on the use of biomarkers, and their application to occupational neurotoxicology. Methodological appearances to occupational neurotoxicity are discussed in a series of chapters dealing with epidemiological and clinical approaches; the latter includes analysis of behavioral, electrophysiological and imaging techniques in the diagnosis of neurotoxicity in occupational illness. Chapter 7 covers the most relevant issue of the potential contribution of occupational neurotoxicants to neurodegenerative or psychiatric disorders. The final chapters discuss two current relevant issues, i.e., the incidence of occupational neurotoxic diseases in developing countries, which remains a major problem to be addressed, and the process of assessing the risks associated with occupational exposure to neurotoxicants, in order to ensure the safety of existing exposure limits in the work place. The editors hope that this volume will promote a greater understanding of the issues of neurotoxicity in the occupational setting, and will be useful to those scientists in industry, government, and academia who are concerned with the prevention of nervous system disorders potentially resulting from occupational exposures.

Lucio G. Costa
Luigi Manzo

Contents

1 Manifestations of Neurotoxicity in Occupational Diseases

L. Manzo and L. G. Costa

CONTENTS

1   INTRODUCTION

Many chemical compounds present in work settings can produce a spectrum of impairments in the human nervous system, ranging from subtle neurological and behavioral disturbances to frank encephalopathy and peripheral nerve disease.

Although recognition of neurotoxic injuries dates to antiquity with descriptions of diseases in lead workers, it has only been since the 1970s that occupational neurotoxicology has grown as a specialty due to the dramatic growth of industrial and agricultural chemicals and the number of situations in which neurotoxicity has been recognized in exposed workers.

This chapter presents an introductory overview of the neurotoxic agents which are commonly encountered in the workplace.

2   OCCUPATIONAL NEUROTOXICANTS

The use of substances with neurotoxic potential is not uncommon in both the industrial setting and agriculture.[1] Thirty percent of the workplace chemicals for which the American Conference of Governmental Industrial Hygienists have recommended maximum exposure concentrations have been so listed in part because of their neurotoxic potential. Many other common chemicals may have neurotoxic properties that have not yet been recognized.[2]

Toxic chemicals that have been involved as etiological factors in occupational nervous system diseases are listed in Table 1. They are divided arbitrarily into five general groups: solvents; pesticides, metals and metalloids, gases, and miscellaneous compounds, including common monomers. The compounds considered in this classification are primarily toxic to the central nervous system or the peripheral nervous system. Many agents are toxic to both. The mode of action is generally multifactorial and the effects are not predictably reversible. Table 1 also includes agents causing expressions of neurotoxicity that are secondary to systemic effects such as hypoxia, respiratory depression, hypotension, or dysrhythmias.

2.1   SOLVENTS

Organic solvents are used widespread for a variety of industrial applications. Although chemically heterogeneous, these compounds are often discussed as a group because of their relatively similar toxicology and the high frequency of exposure to solvent combinations in the workplace. Significant pollution by these substances often occurs in the indoor environment.

Halogenated hydrocarbons which are most commonly encountered in industry include trichloroethylene, chloroform, 1,1,1-trichloroethane, perchloroethylene and methylene chloride (dichloromethane). Research during the past decade has provided a clear understanding of the somatic, behavioral and cognitive changes that result from exposure to these agents. Toxicity may be acute or chronic with changes involving both the central and the peripheral nervous system function.[3]

n-Hexane represents the prototypical industrial solvent causing peripheral neuropathy as the major toxic manifestation.[4] This agent is used primarily as a glue and adhesive solvent, although small amounts may be found in gasoline. Workers at risk include shoemakers, furniture makers, adhesive manufacturers and off-set printers.[5,6] Hexane neuropathy has been described in occupational exposure but also occurs as a consequence of "glue sniffing," i.e., deliberate inhalation of vapors from lacquers, glue or glue thinners containing n-hexane.[7]

Among hexacarbons, methyl n-butyl ketone also causes damage to peripheral nerves in exposed workers.[8] A common metabolite shared by n-hexane and methyl

TABLE 1
Neurotoxicity of Chemicals Described in Human Occupational Disease

Chemical	Effect
Solvents	
carbon disulfide	mild peripheral neuropathy, cranial neuropathy, ataxia, impairment of psychomotor function, Parkinson's syndrome, psychosis, emotional instability, memory impairment
n-hexane	mixed sensimotor neuropathy, spasticity, ataxia, impaired visual acuity
methanol	impaired visual acuity, blindness, headache
methyl *n*-butyl ketone	mixed sensimotor neuropathy, ataxia
perchloroethylene	encephalopathy, impaired psychomotor function, neurasthenia
styrene	neurobehavioral and neuroendocrine alterations
toluene	ataxia, emotional lability
trichloroethylene	cranial neuropathy, memory impairment
1,1,1-trichloroethane	sensory neuropathy
Pesticides	
carbamates	mild to moderate acute cholinergic crisis
chlordecone	tremor, ataxia, opsoclonus, headache, visual disturbances, muscle weakness, loss of recent memory
chlorophenoxy compounds	headache, dizziness, myotonia, neuropathy, fatigue
cyclodienes (chlordane, aldrin)	ataxia, seizures, EEG pattern changes, chronic motor disorders, psychological disorders
dithiocarbamates	muscle weakness, dizziness, incoordination
organophosphates	acute cholinergic crisis, central and autonomic NS toxicity, cranial nerve palsies, delayed peripheral neuropathy, spasticity, impaired psychomotor function
pyrethroids	tremor, choreoathetosis
triphenyl tin	mild encephalopathy
Gases	
carbon monoxide	encephalopathy, delayed syndrome, dystonia
ethylene oxide	headache, neuropathy, seizures
cyanide	ataxic neuropathy, visual disturbances (chronic)
hydrogen sulfide	encephalopathy
methyl bromide	visual and speech impairment, convulsions (acute); neuropathy, cerebellar symptoms (chronic)
methyl chloride	cerebellar dysfunction, ataxia, tremor, blurred vision, loss of recent memory
nitrous oxide	neuropathy
waste anesthetic gases	headache, memory impairment
Metals	
aluminum	encephalopathy
arsenic	mixed sensimotor neuropahy, encephalopathy
lead	motor neuropathy, memory impairment, impaired psychomotor function, neurosthenia

TABLE 1 (continued)
Neurotoxicity of Chemicals Described in Human Occupational Disease

Chemical	Effect
manganese	ataxia, Parkinsonism, dystonia, cognitive impairment, emotional instability, psychosis
mercury	tremor, fatigue, emotional disturbances, neurosthenia, cognitive impairment (elemental Hg)
thallium	peripheral and autonomic neuropathy, ataxia, emotional disturbances, psychosis
trimethyl tin	hyperexcitability, memory impairment, seizures
Miscellaneous	
allyl chloride	sensimotor neuropathy
acrylamide	sensimotor neuropathy, ataxia, neurosthenia
dimethylaminopropionitrile	sacral neuropathy, urinary dysfunction, sexual dysfunction, paresthesias
methyl methacrylate	sensimotor neuropathy
naphthalene	neuropathy, optic neuritis
trinitrotoluene	neuropathy

References: Anthony et al.[61] Baker et al. [92] Beckett et al.[102] Bos et al.[98] Chu et al.[100] Ecobichon,[14] He et al.[99] Kulig,[2] Landrigan and Baker,[1] Ngim et al.[97] O'Donoghue,[93] Ohnishi and Murai, [95] Rosenberg,[88] Spencer and Schaumburg,[96] Squier et al.[94] Yang et al.[101]

n-butyl ketone; 2,5-hexanedione, is apparently responsible for the neurotoxic properties shared by each. By contrast, other compounds quite similar to *n*-hexane in chemical structure such as *n*-pentane, *n*-heptane, and branched hexanes do not have significant peripheral neurotoxicity.[4]

2.2 AGRICULTURAL CHEMICALS

A large proportion of toxic incidents involving pesticides occur in occupational exposure.[9] In developing countries, in particular, occupational pesticide poisoning has been described as a significant social problem.[10] (See also Chapter 14). However, pesticide poisoning is not uncommon in industrialized countries[9,11] and there is concern that long-term exposure to relatively low levels of widely used pesticides may result in chronic neurological *sequelae*.[12,13]

Organophosphate esters are potent neurotoxins acting as irreversible inhibitors of the enzyme acetylcholinesterase in the central, autonomic, and peripheral nervous systems. Carbamic acid esters such as carbaryl also act as cholinesterase inhibitors and cause neurotoxic effects similar to organophosphates. However, their action is less intense and rapidly reversible because of the low stability of the carbamate-cholinesterase complex. Nevertheless, poisoning may result from extreme dermal or inhalation exposure.[14]

For other classes of pesticides, the neurotoxic potential in humans is not adequately defined. Intoxication by chlorophenoxy herbicides has been reviewed by Hayes.[15] There is limited evidence of neurotoxicity associated with exposure to these agents in humans with the exception of one report describing alterations of peripheral nerve conduction velocities in workers employed in manufacturing 2,4-D (2,4-dichlorophenoxyacetic acid) and 2,4,5-T (trichlorophenoxyacetic acid).[16] The results of experimental studies with these agents have been controversial.[14] Paraquat may induce central nervous system effects after exposure to massive doses.[17]

2.3 METALS AND METALLOIDS

The most commonly encountered workplace substance with clearly recognized neurotoxic effects is lead. Sources of lead exposure also exist in the home (leaded paint, ceramics, house dust) and in the general environment (contaminated soil, water, air, and dust). Occupational lead neurotoxicity is most likely to be from chronic exposure, although acute syndromes have been described.[18]

Neurotoxic metals posing health hazards in the workplace also include mercury, arsenic, thallium, manganese and certain organic tin compounds. Elemental mercury is a liquid with sufficient vapor pressure at room temperature to produce concentrations of vapor that can cause adverse health effects when inhaled.[19] Industrial workers at risk for elemental mercury exposure are those employed in chloralkali plants and in battery making and recycling. In addition, mercury levels in dental offices are at times elevated by the use of mercury along with silver in production of amalgam. Among the organic compounds of mercury, the short chain alkyl mercurials, such as methyl mercury, present the greatest hazards to health due to their nervous system toxicity.[20]

Arsenic is a common element and can be found in association with lead (and cadmium) in many commercially mined ores and therefore is of common concern in areas contaminated by smelter smoke and dust. Occupational exposure to arsenic has occurred mainly in the smelting industry and in the manufacture and use of arsenic-based pesticides.[21] Arsenic, like lead, is notable for causing both central nervous system toxicity and peripheral neuropathy. However, the capacity of the human body to tolerate and rapidly excrete arsenic is much greater than for lead. The neurotoxicity of arsenic is likely to be secondary to a very large exposure, such as might be seen in a suicide attempt or non-intentional ingestion, and not to low-level environmental exposure.

Thallium poisoning most commonly occurs following accidental ingestion of thallium containing rodenticides. Chronic occupational intoxication has been described in subjects recovering thallium from flue dust of sulfuric acid waters or in plants handling thallium compounds in the production of optic glass, dyes, and pesticides.[22]

Many organic tin compounds are used as biocides, catalysts and stabilizers in the production of polymers. The neurotoxicity of the short chain alkyl tin derivatives, such as trimethyl-, triethyl- and triphenyltin has been documented in clinical reports of accidental intoxication[23] and has been investigated extensively in laboratory animals.[24]

Manganese is used in the production of metal alloys, dry battery cells, paints, varnish enamel and as an antiknock agent in lead-free petrol. Manganese poisoning has been described in hundreds of cases of miners and industrial workers throughout the world.[25-27,103] Recent studies have also focused on subtle neurobehavioral effects associated with chronic exposure to low levels of manganese in foundry workers[28] and in subjects employed in manganese alloy production.[29]

2.4 GASES

Carbon monoxide and ethylene oxide are neurotoxic gases of considerable importance in occupational toxicology. Carbon monoxide is a widespread occupational hazard which annually causes more deaths than any other toxin or drug overdose.[2] Working environments often polluted by hazardous concentrations of carbon monoxide include chemical, petrochemical, and metallurgic industries, and the ambient air in underground garages.

Ethylene oxide is a known mutagen and reproductive toxin also causing severe irritation to skin and mucous membranes. Its neurotoxicity involves both the central and peripheral nervous systems.[95] Ethylene oxide is one of the most commonly manufactured chemicals in the industrialized countries and the vast majority of exposed workers to this agent are employed in industry. It is also used in hospitals as a sterilizer and represents a significant health hazard in health care workers.[30]

3 EFFECTS

Toxic chemicals can affect virtually any part of the neuraxis and the resulting alterations can be classified in terms of the anatomical site affected and the clinical presentation. These observed changes may be best described as syndromes involving the nervous system rather than as specific disease states, because the pathophysiological bases for the disease are usually unknown.

3.1 CENTRAL NERVOUS SYSTEM

Exposure to neurotoxicants may result in a variety of disorders ranging from severe acute toxic encephalopathy, seizures, headache, and focal brain alterations to mild, generally reversible neuropsychological dysfunction.

Acute poisoning in the workplace usually occurs by inhalation exposure. However, with certain compounds, substantial absorption can occur through the skin, particularly if the product is used in a lipophilic vehicle or the skin is inflamed or abraded.

Coma in acute intoxication may be due to a direct "pharmacological" action of the substance altering synaptic transmission or oxidative metabolism in brain cells or to indirect mechanisms such as metabolic acidosis or impairment of oxygen transport as seen in poisoning by methanol and carbon monoxide, respectively. Both direct and indirect mechanisms may be implicated in severe cases.

Organic (aliphatic and aromatic) solvents are potent anesthetics and acute narcosis follows exposure to high concentrations.[31] Massive poisoning has been associated with CNS damage via respiratory depression and resultant hypoxia. Fatalities due to sudden cardiac dysrhythmias have also been reported in cases involving chlorinated hydrocarbons. At lower exposure levels, solvents may cause a transient syndrome similar to that produced by ethanol consumption. Intoxication has often resulted from deliberate solvent abuse.[32]

Organophosphate poisoning can affect the nervous system after either acute or significant chronic exposure. The inhibition of acetylcholinesterase results in accumulation of acetylcholine at cholinergic synapses leading to overstimulation of cholinergic receptors. In massive exposures, immediate signs and symptoms are those reflecting cholinesterase inhibition in the brain (confusion, dizziness, ataxia, respiratory depression, coma, and seizures), or excessive stimulation of peripheral acetylcholine receptors, either nicotinic (muscle weakness and fasciculations, hypertension, and tachycardia), or muscarinic (salivation, urination, diarrhea, abdominal cramps, nausea, vomiting, bradycardia, bronchorrhea, miosis, sweating and wheezing).[14] Recovery of normal enzyme activity in organophosphate poisoning is slow since reactivation occurs only through synthesis of additional cholinesterase molecules. Repeated insecticide exposure may similarly result in cumulative depression of cholinesterase stores.

Signs and symptoms of carbamate poisoning are less lasting and persist for 6 to 8 hours in most cases. They include nausea, vomiting, mild abdominal cramping, and headache. Disturbed vision, dizziness, weakness and difficulty in breathing may occur.[14]

Convulsions may develop in cases of massive overdosage due to systemic toxicity, for example anoxia in carbon monoxide or hypocalcemia induced by ethylene glycol. In addition, epileptic fits can be observed as specific features of chemical toxicity without other evidence of brain injury as shown in poisoning by chlordane and other organochlorine insecticides.[14]

Tremor in chlordecone (Kepone) poisoning is another example of effect occurring in the absence of brain lesions. Clinical features of chlordecone-induced tremor are characteristic: tremor is irregular, non purposive, and most severe when the limb is static but unsupported against gravity. It differs from the tremor seen in mercury poisoning which is fine and affects the eyelids, tongue, and outstretched hands. Chlordecone also causes an abnormality of eye movement (opsoclonus) consisting of irregular bursts of involuntary, abrupt, rapid jerks of both eyes simultaneously that usually are horizontal but are multidirectional in severely affected individuals.[96]

Another common finding in occupational neurotoxicity is headache which is typically observed as the initial toxic manifestation in workers exposed to carbon monoxide, waste anesthetic gases, n-hexane, toluene, and halogenated hydrocarbons. Headaches associated with nausea and flushing are provoked by carbon di-sulfide in workers also consuming alcoholic beverages. This was considered to be a "disulfiram-like" syndrome secondary to carbon disulfide-induced alteration of ethanol metabolism.[33] A similar reaction has been reported in subjects occupationally exposed to TMTD (tetramethylthiuram disulfide), calcium cyanamide, n-butyl aldoxime and dimethylformamide.[33]

3.1.1 Selected Syndromes

Metals Central nervous system toxicity has often been described in occupational exposure to metals and their compounds.

Organic lead is a well known cause of encephalopathy in adults, usually occurring in the context of deliberate inhalation of leaded gasoline. Hallucinations and other behavioral changes, and extrapyramidal effects, including myoclonus, ataxia, tremor, and chorea, are prominent.[2] Inorganic lead remains an important occupational hazard. Manifestations of lead neurotoxicity, as currently encountered in workers, predominantly involve psychological functions (see below) and thus differ significantly from those reported in the earlier part of this century when more overt peripheral neuropathy occurred because of exposure to very high lead levels.

Typical feature in chronic intoxication by elemental mercury inhalation is hand tremor occurring as one of the earliest and most frequent symptoms. Patients also present with erethism, a condition which includes unusual sensitivity and irritability and may be accompanied by insomnia, poor attention and apathy followed by hallucinations and mania.[34]

Arsenic encephalopathy is a long-lasting or permanent syndrome which has been described in situations including acute massive exposure, prolonged consumption of arsenic-contaminated alcoholic beverages, and chronic exposure in the workplace.[2] Clinical findings are nonspecific and include confusion, disorientation, agitation, and hallucinations. Headache, drowsiness, memory loss, diplopia, optic neuritis, and seizures are also observed.[2]

Central nervous system effects in thallium poisoning include sleep disorders, behavioral alterations and Parkinsonism, but peripheral nerve changes accompanied by hair loss are the prominent findings in most cases.[35]

Signs and symptoms of chronic manganese intoxication can be divided into three phases. The prodromal phase, also termed "manganese psychosis," may become clinically apparent after 12–24 months of exposure. Manifestations include memory impairment, mental excitement, compulsive behavior and emotional instability often accompanied by sexual dysfunction. These symptoms usually last about one to three months. Disturbances reflecting lesion in the basal ganglia subsequently appear with tremor, speech disorders, clumsiness of movements, and altered balance and gait. The final stage exhibits muscular rigidity, and fine tremors usually involving the upper limbs.[26]

Gases   In carbon monoxide poisoning, the initial symptoms of headache, nausea, malaise, and lethargy may progress rapidly to loss of consciousness. This is commonly associated with high carboxyhemoglobin levels (>40% of the total hemoglobin). Carbon monoxide poisoning is one of the few examples of acute poisoning that may cause significant central nervous system damage in a delayed fashion.[36] Clinically, delayed neurological sequelae may develop three days to three weeks after recovery from the initial exposure has appeared to be complete. Findings include mental deterioration, personality changes, speech disturbance, confusion, urinary/fecal incontinence, gait disturbance, weakness, and Parkinsonism. The pathogenesis of this particular syndrome is a matter of current controversy. During the acute stage, the effects of carbon monoxide in the brain have generally been related to functional hypoxia from impaired oxygen transport. However, patients who have never been significantly hypoxic may still develop delayed sequelae and, on the other hand, delayed neurotoxicity apparently does not develop in some patients despite their persistent initial coma. Postulated mechanisms of the delayed neurological sequelae consider (a) hypoxic-ischemic stress related to defective oxygen transport to the cells, (b) inhibition of cytochromes leading to impairment of mitochondrial electron transport in neural cells, (c) brain lipid peroxidation resulting

TABLE 2
Neurobehavioral Effects of Solvents[40]

Syndrome	Severity	Other Denominations
Organic effective syndrome	Minimal	Neurasthenic syndrome
Mild chronic encephalopathy	Moderate	Psycoorganic syndrome
Severe chronic encephalopathy	Severe	Solvent dementia

from reperfusion injury, and (d) overstimulation of excitatory amino acid receptors.[36] Other reported complications in carbon monoxide poisoning include hearing loss, retinal hemorrhages, and basal ganglion infarcts.

Acute, massive exposure to ethylene oxide can induce irritation of skin and mucous membranes, nausea, headache, and then obtundation and seizures.[37] Peripheral nerve damage may also occur as discussed elsewhere in this book. More chronic and moderate exposure levels have been reported to cause headache, fatigue, dizziness, irritability, depression, and significant, long-lasting cognitive dysfunction.

Intoxication by methylene chloride is manifested by headache, dizziness and neuropsychiatric changes. However, at the concentrations generally found in workplaces, the main hazard from methylene chloride is related to its metabolic conversion to carbon monoxide that can be deleterious in subjects with coronary heart diseases.[3]

3.2 PSYCHOLOGICAL DYSFUNCTION

Although exposure to industrial toxins has long been known to affect behavior, only recent studies using quantitative methods have demonstrated distinctive patterns of neurobehavioral aberrations following long term exposure to numerous substances which are commonly encountered in the workplace.[29,38,39] A critical description of computer-assisted methods in neurobehavioral toxicology is given by Iregren in this book (See Chapter 11).

3.2.1 Organic Solvents

Long term exposure to organic solvents has been implicated in neurotoxic syndromes primarily involving psychological and cognitive functions. Various methods have been proposed to classify these impairments. The scheme proposed by the World Health Organization[40] is shown in Table 2.

In a different classification,[41] the mildest form of changes (Type 1 or organic affective syndrome) is characterized by subjective complaints of headache, irritability, malaise, fatigue, loss of interest in daily events, and anxiety not accompanied by deficits on neuropsychological testing. This type of effect typically is reversible. The second form (Type 2, mild chronic toxic encephalopathy) is characterized by objective abnormalities on neurobehavioral testing and by symptoms that are similar to organic affective syndrome but more pronounced. Sustained personality and mood disturbances (Type 2a effect) or impairment in intellectual function with memory loss, and difficulty in learning and concentration (Type 2b) may be seen at this level of effect, singly or in combination. Workers exposed to solvents may exhibit either of these syndromes, depending on the intensity and duration of their exposure. The most severe form (Type 3) is associated with progressive and usually irreversible neuropsychological, intellectual, and emotional decline.

Neuropsychological dysfunction has been determined according to these criteria in workers employed in a number of activities implicating exposure to solvents and solvent mixtures.[3,38,39,42-45]

In certain countries, the issue of solvent-related syndromes has become of great importance in worker's compensation. However, several controversies remain in this area and the available scientific evidence is not univocal.[38,46] Causation issues are difficult because toxicity usually becomes evident only after many years in subjects with chronic exposure, and most affected patients have been exposed to numerous chemicals including alcohol, tobacco, and street drugs over that time period. Unsolved questions also remain about criteria for specifying the relative contributions of various sensory, cognitive and motor functions to the observed patterns of impairments.[38] The classifications themselves are subjective, and psychological assays used to determine the classification of a given patient are often not standardized.

3.2.2   Metals

Subtle central nervous system alterations have been described in workers with long-term exposures to mercury, lead and manganese. Exposure to mercury vapor was associated with a spectrum of functional and subjective disturbances, including poor psychomotor performance, deterioration of intellectual capacities, and alteration of the emotional state.[47-49,97]

Mergler and coworkers[29] administered a battery of neurofunctional tests to workers employed in manganese alloy production. The results indicated a spectrum of neurobehavioral changes involving the emotional state, motor function, cognitive flexibility and olfactory perception threshold. Average whole blood manganese levels in the workers were 1.03 µg/100 mL (vs. 0.68 µg/100 mL in control) and the ambient air concentration of respirable manganese was 0.04 mg/m^3.

Central nervous system function impairment evidenced by psychological and behavioral deficits was described in lead workers.[50-52] In a study by Pasternak and coworkers,[51] the reported symptoms were fatigue, irritability, difficulty in concentrating, and inability to perform tasks requiring sustained concentration. These findings were associated with abnormalities on standardized neuropsychological testing suggestive of impairment of verbal intelligence, memory, and perceptual speed.

3.3   SENSORY FUNCTION ALTERATIONS

Various industrial chemicals were shown to induce damage to the visual and auditory systems in humans.

Visual toxicity has been described in cases of methanol, thallium and carbon disulfide intoxication. Methanol poisoning is often manifested by loss of visual acuity to a varying degree. The observed alterations in the electroretinogram suggest the involvement of different retinal cell types.[53] Typically, optic nerves are affected by methanol in the absence of peripheral nerve damage, differing from thallium or carbon disulfide intoxication in which both optic neuropathy and peripheral neuropathy occur.[54]

Abnormalities involving central vision, color vision, contrast sensitivity and visual-evoked potentials have been reported in subjects with chronic low dose exposure to organic-solvents.[55,56] (See also Chapter 8).

There are several studies suggesting auditory system alterations in subjects exposed to carbon disulfide, styrene, trichloroethylene and toluene.[57]

In combined exposure, ototoxins may interact causing changes that are not necessarily predicted on the basis of knowledge or the effect of the individual agents.[58] Common hazards that are encountered simultaneously in the workplace are noise and exposure to organic solvents. There are data suggesting that overexposure to noise interacts synergistically with carbon disulfide, toluene[59] and other organic solvents.[57] The damage incurred by the agents acting together may exceed the simple summation of the damage each agent produces alone.[60] The therapeutic use of ototoxic drugs, such as aminoglycoside antibiotics and certain diuretics, may be an additional factor exacerbating auditory dysfunction in workers at risk.

3.4 Peripheral Neuropathies

The range of chemicals implicated in these syndromes is extensive. Only agents which are commonly involved in occupational diseases are emphasized here and the reader is referred to reference books[96] and original reviews[61,62] for a more comprehensive discussion of the literature.

Two basic forms of peripheral nerve damage have been associated with excessive exposure to neurotoxic agents in the workplace: axonopathy and segmental demyelination. The former can result from metabolic derangement of the neuron and is manifested by degeneration of the distal portion of the nerve fiber in a process that was originally termed "dying-back axonopathy."[63] Toxicants can provoke by various mechanisms a "chemical transection" of the axon at some point along its length leading to axonal degeneration distal to the transection.[61] Myelin sheath degeneration may occur secondarily. Nerve conduction rates are usually normal until the condition is relatively far advanced. Distal muscles show changes of denervation. Recovery may occur by axonal regeneration but is slow and often incomplete.

Segmental demyelination results from primary destruction of the neuronal myelin sheath and intramyelinic edema, with relative sparing of the axons. This process often begins at the nodes of Ranvier and slows nerve conduction. Segmental demyelination may also result from direct toxicity to the myelinating cell. Characteristically, there is no evidence of muscle denervation, although disuse atrophy may occur if paralysis is prolonged. Recovery associated with remyelination is rapid and usually complete in mild to moderate neuropathies.[61] In many instances, axonal degeneration and segmental demyelination coexist and the clinical manifestations of neuropathy in exposed individuals may represent a combination of both pathological processes.

3.4.1   Clinical Syndromes

The list of occupational agents causing a dying-back type of axonal degeneration includes *n*-hexane, methyl-*n*-butyl ketone, acrylamide, carbon disulfide, thallium, trivalent arsenic and certain organophosphate compounds (Table 1).[61,64] Peripheral neuropathy and sensory disturbances similar to those caused by acrylamide have been described in dental technicians and orthopedic surgeons occupationally exposed to methyl methacrylate, an agent which is widely used in the manufacture of acrylic plastics for prosthetic dentistry and reconstructive orthopedic surgery.[65,66] Sensory disturbances predominate in the relatively rare neuropathy seen with 1,1,1-trichloroethane.[67,68]

Many, but not all, organophosphates have been found to cause a syndrome of delayed neurotoxicity involving the peripheral nerve.[69,70] Clinically, the syndrome begins 8–14 days following exposure with numbness and fatigue of the lower extremities. This may progress to the proximal muscles over the next several weeks. Deep tendon reflexes are frequently depressed. Sensory loss is minimal. The cholinergic findings of acute poisoning may have been initially absent and therefore the diagnosis may be elusive. The disease may progress for one to three months, and recovery is typically slow and often incomplete. Residual motor conduction velocities and abnormal sensory-evoked potential amplitudes may be permanent.

Recently, an "intermediate syndrome" has been described as an allegedly distinct entity in several case series and reports of organophosphate intoxication.[71,72] The features that distinguish it from the delayed neuropathy include the time of onset (one–four days vs. two–three weeks), effects primarily involving the proximal musculature and cranial nerves, the tetanic fade vs. denervation seen at electromyography, and the more rapid recovery time (4–18 days vs. 6–12 months).

The peripheral neurotoxicity of ethylene oxide generally begins as limb weakness and fatigability, which may or may not be associated with central nervous system effects.[37] On examination, distal limb weakness is obvious, and vibratory and position sense is decreased. Ataxia, wide-based gait, clumsy alternating hand movements, and decreased deep tendon reflexes are also common. Nerve conduction studies demonstrate abnormal slowing, and electromyograms show scatted positive sharp waves and fibrillation potentials with denervation potentials. These findings may improve over time, but improvement does not appear to be predictable.[2]

The clinical features of metal-induced neuropathies are probably the best documented among work-related neurotoxic diseases. The classic description of lead neuropathy is of an asymmetrical weakness of the wrists (wrist drop), with no loss of sensation. The extensors of the fingers and thumb may be affected before the wrist. Peripheral nerve conduction velocities are usually slowed, even in workers who have not yet become symptomatic from lead intoxication. Such abnormalities tend to develop within two to three months of the onset of work in individuals presenting blood lead levels in the range of 40 to 60 µg/dl and become more apparent as the blood lead level rises. After removal of the patient from exposure, symptoms and abnormalities resolve slowly over weeks to months.[62,73]

Far more typical in their presentation are the neuropathies induced by arsenic and thallium intoxication.[64] Symptoms may develop after only a single acute exposure and usually begin with numbness, tingling, and burning paresthesias of the feet and later the hands in a stocking and glove distribution. The paresthesias are quite painful and are exacerbated by a pressure stimulus. Vibration, position, and temperature sensation are also decreased. Sensory abnormalities spread proximally in a symmetrical pattern. Within a few days of the onset of paresthesias, distal muscle weakness begins and spreads proximally. Foot drop and wrist drop are prominent and impressive wasting of the distal muscles occurs in proportion to the degree of other signs. Muscle tenderness, cramps, and fasciculations occasionally occur. Deep tendon reflexes are decreased and then disappear. Sensory nerve action potentials and motor nerve conduction are severely affected and may remain so for years. Even after rapid diagnosis and treatment, the peripheral neuropathy in both thalium[35] and arsenic[74] poisoning may be permanent.

3.5 CRANIAL NEUROPATHY

Cranial nerves can be preferentially involved in certain toxic neuropathies. In trichloroethylene poisoning, a typical finding has been facial numbness and neuralgia reflecting a lesion in the fifth cranial nerve.[75] It is not clear if trigeminal neuropathy in trichloroethylene intoxication is caused by the compound itself or by its metabolite dichloroacetylene (DCA). Both trichloroethylene and DCA have neurotoxic effects but inhalation of DCA was shown to produce more obvious trigeminal nerve lesions than exposure to pure trichloroethylene.[76]

3.6 AUTONOMIC NEUROPATHY

Autonomic neuropathy is relatively uncommon but may occasionally occur in occupational intoxication giving rise to bowel and bladder disturbance and postural hypotension. Symptoms of autonomic nervous system dysfunction have been described in workers exposed to organic solvents[77,78] and pesticides.[79]

A catalyst containing dimethylaminopropionitrile, an agent structurally similar to acrylamide, was apparently the causative factor in two epidemics of neurotoxicity primarily affecting the urinary bladder among workers in polyurethane manufacturing plants.[80] The clinical picture in these patients included urinary retention accompanied by paresthesia, weakness, insomnia and sexual dysfunction.

3.7 BRAIN TUMORS

Little is known about relations between brain tumors and occupational chemical exposures. Epidemiological studies (reviewed by Thomas and Waxweiler[81]) suggest that excess brain and central nervous system cancer risk may be associated with employment in certain occupations (e.g., artists, laboratory professionals, veterinarians, embalmers) or industries (rubber, oil refinery or chemical plant workers). The most convincing evidence in this area was provided by cohort studies examining the relationship between occupational exposure to polyvinyl chloride (PVC) and risk of cancer. Among individuals who had worked for at least one year in a job involving exposure to vinyl chloride, the total number of brain tumors observed was significantly greater than expected.[81] Most of the worker groups examined in these studies were potentially exposed to multiple chemicals. However, they presumably had some exposure patterns in common; for example, exposure to organic solvents, acrylonitrile, formaldehyde, polycyclic aromatic hydrocarbons and phenolic compounds. Some of these substances, in particular vinyl chloride, acrylonitrile and certain polycyclic aromatic hydrocarbons, have been shown to cause brain tumors in laboratory animals.

4 MECHANISMS

Nervous system components are apparently more vulnerable to some substances than to others depending on the mode of action of the toxicant and the molecular target involved. Thus, neurotoxic agents can be classified according to their sites of action and their primary cellular and subcellular targets in neural tissues.[61,64]

Unfortunately, mechanisms of toxicity and primary cellular and subcellular targets have been elucidated for only a limited number of neurotoxicants. A number of processes are evidently implicated. Changes in the neuronal, axonal, myelin, or muscle membrane structure can impair excitability and impede impulse transmission. Alterations in protein, fluid content, or ionic exchange capability of the membrane may provoke swelling of neurons and astrocytes, damage to capillary endothelium, or increase extracellular fluid. Toxic effects on specific neurotransmitter mechanisms block neurotransmitter access to post-synaptic receptors, produce false neurotransmitter effects, or alter the synthesis, storage, release, re-uptake, or enzymatic inactivation of natural neurotransmitters.[61]

Chemicals have variable ability to cross the blood-brain and blood-neural barriers. Compounds that are nonpolar are more lipid soluble and therefore have greater access to nervous tissue than do highly polar and less lipid-soluble ones. In addition, the choroid plexus is able to sequester xenobiotics and act as a sink to exclude their uptake into the brain.[82] However, an alternative route of entry into the central nervous system has been demonstrated for airborne metals (e.g., manganese), whereby the toxic agent is taken up by olfactory neurons in the nose and is then retrogradely transported into the brain.[83] Localized direct transport through the olfactory nerves may serve to by-pass the blood-brain barrier and lead to accumulation of the metal in the brain.[84]

5 VARIATION IN TOXIC RESPONSE

Whether a chemical elicits neurotoxic effects depends not only on the inherent characteristics and site specificity of the compound but also on a number of physiological and environmental factors.[85] The individual rate of metabolism can influence the response to chemicals which have to be activated to toxic metabolites after absorption in the body.[86] Similarly, chemical and physico-chemical interactions may alter the nature and/or the magnitude of toxic responses after exposure to combinations of pollutants. For example, methyl ethyl ketone, a compound devoid of neurotoxic potential, can potentiate the peripheral nerve damage caused by n-hexane.[87,98] Also, physical hazards such as noise which is commonly encountered in worksites, may exacerbate susceptibility to toxins acting on the auditory system.[57]

Understanding the difference between effects elicited by acute (high-dose) and chronic (moderate or low-dose) exposures is of great importance in neurotoxic risk assessment. There are several examples of toxins which produce pure pictures of damage to one portion of the nervous system or are capable of inducing combinations of damage at various nervous system sites depending on the level and duration of exposure.[85] It is also readily apparent that a single overdosage of certain chemicals is more likely to result in neurotoxicity than the prolonged absorption of the same compound at lower doses. In some instances, the reduced susceptibility in chronic exposure was shown to reflect the enormous adaptive capacity of the nervous system which can mitigate or even prevent a toxic insult.

There is always a certain amount of variability in the response to toxic chemicals among individuals, although variability in neurotoxic effects appears to be less pronounced compared to effects involving other organ systems.[85] Notably, while extremes of age or health may result in individual variations of effects observed at the same level of exposure, this does not generally occur in the occupational setting where the range of ages is narrower and most individuals are in good health.[88]

Other important aspects of neurotoxic disease are those related to the latency and reversibility of effects, and recovery from insult. In most cases, maximum symptoms are seen with maximum exposure, with little delay in the onset of symptoms, and improvement rapidly occurs with either removal of the toxic substance or removal of the individual from the place of exposure. However, many exceptions are known.[85] As already noted, delayed nervous system disorders may develop in individuals exposed to a single dose of carbon monoxide[36] or certain organophosphates.[69,70] There is also some evidence suggesting that extremely low levels of pollutants that are insufficient to produce neurological symptoms may actually predispose susceptible individuals to the later development of a progressive nervous system disorder. The possible role of remote chemical exposures in the pathophysiology of neurodegenerative disorders is of great current concern. This topic is discussed in Chapter 7 and in recent literature articles.[88-91]

6   REFERENCES

1. Landrigan, P. J. and Baker D. B., "The recognition and control of occupational disease." *JAMA,* 266, 676, 1991.
2. Kulig, K., "Clinical neurotoxicology of industrial and agricultural chemicals," in *Neurotoxicology: Approaches and Methods*, Chang, L. W. and Slikker, W., Jr., Eds., Academic Press, Orlando, FL, 1995, Chapter 40.
3. Seppalainen, A. M., "Changing picture of solvent neurotoxicity," in Advances in *Modern Environmental Toxicology*, Vol. 23, Mehlman, M. A. and Upton, A., Eds, Princeton Scientific Publ., Princeton, NJ, 1994, p. 223.
4. Spencer, P. S., Couri, D., and Schaumburg, H. H., "*n*-Hexane and methyl *n*-butyl ketone," in *Experimental and Clinical Neurotoxicology*, Spencer, P. S. and Schaumburg, H. H., Eds, Williams & Wilkins, Baltimore, MD, 1980, Chapter 32.
5. Abbritti, G., Siracusa, A., Cianchetti, C., Coli, C. A., Curradi, G., Perticoni, G. G., and De Rosa, F., "Shoe-maker's polyneuropathy in Italy: the aetiological problem," *Br. J. Ind. Med.*, 33, 92, 1976.

6. Chang, C.-M., Yu, C.-W., and Fong, K.-Y., "*n*-Hexane neuropathy in offset printers," *J. Neurol. Neurosurg. Psychiat.*, 56, 538, 1993.

7. Altenkirch, H., Mager, J., Stoltenburg, G., and Helmbrecht, J., "Toxic polyneuropathies after sniffing a glue thinner," *J. Neurol.*, 214, 152, 1977.

8. Allen, N., Mendel, J. R., Billmaier, D. J., Fontaine, R. E., and O'Neill, J., "Toxic polyneuropathy due to methyl *n*-butyl ketone," *Arch. Neurol.*, 32, 209, 1975.

9. Thompson, J. P., Casey, P. B., and Vale, J. A., "Pesticide incidents reported to the Health and Safety Executive 1989/90-1991–92," *Hum. Exp. Toxicol.*, 14, 630, 1995.

10. Anon, "Pesticides in the third world," *Lancet*, 336, 1437, 1990.

11. Rosenstock, L., Keifer, M., Daniell, W. E., McConnel, R., and Claypoole, K., "Chronic central nervous system effects of acute organophosphate pesticide intoxication," *Lancet*, 338, 223, 1991.

12. Weiss, B., "Behavior as an early indicator of pesticide toxicity," *Toxicol. Ind. Hlth.*, 4, 351, 1988.

13. Jonkman E. J., De Weerd, A. W., Poortvliet, D. C. J., Veldhuizen, R. J., and Emmen, H., "Electroencephalographic studies in workers exposed to solvents or pesticides," *Electroencephalogr. Clin. Neurophysiol.*, 82, 438, 1992.

14. Ecobichon D. J. "Toxic effects of pesticides," in *Casarett & Doull's Toxicology*, Klaassen, C. D., Ed., 1996, p. 643.

15. Hayes W. J., *Pesticides Studies in Man*, Williams & Wilkins, Baltimore, MD, 1982.

16. Singer, R., Moses, M., and Valciukas, J., "Nerve conduction velocity studies of workers employed in the manufacture of phenoxy erbicides," *Environ. Res.*, 29, 297, 1982.

17. World Health Organization, "Paraquat and Diquat. Environmental Health Critera 78," Geneva, Switzerland, WHO, 1988.

18. Silbergeld, E. K., "Neurotoxicology of lead," in *Neurotoxicology*, Blum, K. and Manzo L., Eds., Marcel Dekker, New York, 1985, Chapter 14.

19. Foà, V., "Neurotoxicity of elemental mercury: Occupational aspects," in *Neurotoxicology*, Blum, K. and Manzo, L., Eds., Marcel Dekker, New York, 1985, Chapter 15.

20. Watanabe, C. and Sato, H., "Evolution of our understanding of methylmercury as a health threat," *Environ. Hlth. Perspect.*, 104(Suppl. 2), 367, 1996.

21. Feldman, R. G., Niles, C. A., Kelly-Hayes, M., Sax, D., Dixon, W. J., Thompson, D. J., and Landaw, E., "Peripheral neuropathy in arsenic smelter workers," *Neurology*, 29, 939, 1979.

22. Manzo, L., Minoia, C., and Sabbioni, E., "Toxicity of detrimental metal ions: Thallium," in *Handbook of Metal-Ligand Interactions in Biological Fluids*, Vol. 2., Berthon, G., Ed., Marcel Dekker, New York, 1995, p. 766.

23. Manzo, L., Richelmi, P., Sabbioni, E., Pietra, R., Bono, F., and Guardia, L., "Poisoning by triphenyl tin acetate," *Clin. Toxicol.*, 18, 1343, 1981.

24. Walsh, T. J. and DeHaven, D. L., "Neurotoxicity of alkyltins" in *Metal Neurotoxicity*, Bondy, S. C. and Prasad, K. N., Eds., CRC Press, Boca Raton, FL, 1988, Chapter 6.

25. Mena I., Marin, O., Fuenzalida, S., and Cotzias, G., "Chronic manganese poisoning," *Neurology*, 17, 128, 1987.

26. Seth, P. K. and Chandra, S. V., "Neurotoxic effects of manganese," in *Metal Neurotoxicity*, Bondy, S. C. and Prasad, K. N., Eds., CRC Press, Boca Raton, FL, 1988, Chapter 2.

27. Huang, C.-C., Chu, N.-S., Lu, J.-D., Wang, J.-D., Tsai, J.-L., Tzeng, J.-C., Wolters, E. C., and Calne, D. B., "Chronic manganese intoxication," *Arch. Neurol.*, 46, 1104, 1989.

28. Iregren, A., "Psychological test performance in foundry workers exposed to low levels of manganese," *Neurotoxicol. Teratol.*, 12, 673, 1990.

29. Mergler, D., Huel, G., Bowler, R., Iregren, A., Bélanger, S., Baldwin, M., and Tardif, R.: "Nervous system dysfunction among workers with long-term exposure to manganese." *Environ. Res.* 64, 151, 1994.

30. Brashear, A., Unverzagt, F. M., Farber, M. O., Bonnin, J. M., Garcia, J. G. N., and Grober, E., Ethylene oxide neurotoxicity: "A cluster of 12 nurses with peripheral and central nervous system toxicity," *Neurology*, 46, 992, 1996.

31. McCarthy, T. B. and Jones, R. D., "Industrial gasing poisoning due to trichloroethane, perchloroethylene, and 1,1,1- trichloroethane, 1961–1980," *Br. J. Ind. Med.*, 40, 450, 1983.

32. Blum, K., *Handbook of Abusable Drugs*, Gardner Press, New York, 1984, Chapter 10.

33. Freundt, K. J., "Industrial chemicals and alcohol: Interactions and work-site risk," in *Advances in Neurotoxicology*, Manzo, L., Ed., Pergamon Press, Oxford, UK, 1980, p. 151.

34. Marsh, D. O., "The neurotoxicity of mercury and lead," in *Neurotoxicity of Industrial and Commercial Chemicals*, Vol. 1, O'Donoghue, J. L., Ed, CRC Press, Boca Raton, FL, 1985, Chapter 5.

35. Manzo, L. and Sabbioni, E., "Thallium toxicity and the nervous system," in *Metal Neurotoxicity*, Bondy, S. C. and Prasad, K. N., Eds., CRC Press, Boca Raton, FL, 1988, Chapter 3.

36. Butera, R., Candura, S. M., Locatelli, C., Varango, C., Li, B., and Manzo, L., "Neurological sequelae of carbon monoxide poisoning. Role of hyperbaric oxygen." *Indoor Environ.*, 4, 134, 1995.

37. Gross, J. A., Haas, M. L. and Swift, T. R., "Ethylene oxide neurotoxicity: report of four cases and review of the literature," *Neurology*, 29, 978, 1979.

38. Stollery, B. T., "Long-term cognitive sequelae of solvent intoxication," *Neurotoxicol. Teratol.*, 18, 471, 1996.

39. White R. F., Proctor S. P., Echeverria, D., Schweikert, J., and Feldman, R. G., "Neurobehavioral effects of acute and chronic exposure in the screen printing industry," *Am. J. Ind. Med.*, 28, 221, 1995.

40. World Health Organization, "Principles and methods for the assessment of neurotoxicity associated with exposure to chemicals. Environmental Health Criteria Document 60," WHO, Geneva, Switzerland, 1986.

41. Baker E. L. and Seppalainen, A. M., "Human aspects of solvent neurobehavioral effects," *Neurotoxicology*, 7, 43, 1986.

42. Kumar, P., Gupta, B. N., Pandya, K. C., and Clerk, S. H., "Behavioral studies in petrol pump workers," *Int. Arch. Occup. Environ. Hlth.*, 61, 35, 1988.

43. Knave, B., Persson, H. E., and Goldberg, M., "Long-term exposure to jet fuel. An investigation of occupationally exposed workers with special reference to the nervous system," *Scand. J. Work Environ. Hlth.*, 2, 152, 1976.

44. Lebrèche F. P., Cherry N. M., and McDonald J. C., "Psychiatric disorders and occupational exposure to solvents." *Br. J. Ind. Med.* 49: 820, 1992.

45. Nelson N. A., Robins T. G., White R. F., and Garrison R. P., "A case-control study of chronic neuropsychiatric disease and organic solvent exposure in automobile assembly plant workers." *Occup. Environ. Med.*, 51, 302, 1994.

46. Hakkola, M., Honkasalo, M. L., and Pulkkinen, P., Neuropsychological symptoms among tanker drivers exposed to gasoline," *Occup. Med.*, 46, 125, 1996.

47. Kishi, R., Doi, R., Fukuchi, Y., Satoh, H., Satoh, T., Ono, A., Morikawa, F., Tashiro, K., and Takahata, N., "Subjective symptoms and neurobehavioral performance of ex-mercury miners at an average of 18 years after the cessation of chronic exposure to mercury vapor," *Environ. Res.*, 62, 289, 1993.

48. Piikivi, L. and Hanninen, H., "Subjective symptoms and psychological performance of chlorine-alkali workers," *Scand. J. Work Environ. Hlth.*, 15, 69, 1989.
49. Soleo, L., Urbano, M. L., Petrera, V. and Ambrosi, L., "Effects of low exposure to inorganic mercury on psychological performance," *Br. J. Ind. Med.*, 47, 105, 1990.
50. Ehle, A. L. and McKee, D. C., "Neuropsychological effect of lead in occupationally exposed workers: a critical review," *Crit. Rev. Toxicol.*, 20, 237, 1990.
51. Pasternak, G., Becker, C. E., Lash, A., Bowler, R., Estrin, W. J., and Law, D., "Cross-sectional neurotoxicology study of lead-exposed cohort," *Clin. Toxicol.*, 27, 37, 1989.
52. Tang, H. W., Liang, Y-X., Hu, X-U, and Yang, H-G., "Alterations of monoamine metabolites and neurobehavioral function in lead exposed workers," *Biomed. Environ. Sci.*, 8, 23, 1995.
53. Ruedemann, A. D., "The electroretinogram in chronic methyl alcohol poisoning in human beings," *Am. J. Ophthalmol.*, 54, 34, 1960.
54. Grant, W. M., "The peripheral visual system as a target," in *Experimental and Clinical Neurotoxicology*, Spencer, P. S. and Schaumburg, H. H., Eds., Williams & Wilkins, Baltimore, MD, 1980, Chapter 6.
55. Coe, J. E. and Douglas, R. B., "Ocular responses to chemical and physical injury," in *Occupational Medicine*, 3rd Ed., Zenz, C, Dickerson, O.B., and Horvarth, E. P., Eds., Mosby, St. Louis, MO, 1994, Chapter 8.
56. Donoghue, A. M., Dryson, E. W., and Wynn-Williams, G., "Contrast sensitivity in organic-solvent-induced chronic toxic encephalopathy," *J. Occup. Environ. Med.*, 37, 1357, 1995.
57. Morata, T. C., Dunn, D. E., and Sieber, W. K., "Occupational exposure to noise and ototoxic solvents," *Arch. Environ. Hlth.*, 49, 359, 1994.
58. Prosen, C. A. and Stebbins, W. C., "Ototoxicity," in *Experimental and Clinical Neurotoxicology*, Williams & Wilkins, Baltimore, MD, 1980, Chapter 5.
59. Morata, T. C., Nylen, P., Johnson, A. C., and Dunn, D. E., "Auditory and vestibular functions after single or combined exposure to toluene: a review," *Arch. Toxicol.*, 69, 431, 1995.
60. Morata, T. C., Dunn, D. E., Kretschmer, L.W., Lemaster, G. K. and Keith, R. W., "Effects of occupational exposure to organic solvents and noise hearing," *Scand. J. Work Environ. Hlth.*, 19, 245, 1993.
61. Anthony, D. C., Montine T. J., and Graham D. G., "Toxic responses of the nervous system." In *Casarett & Doull's Toxicology*, C. D. Klaassen, Ed., 1996, Chapter 16.
62. Ludolph, A. C. and Spencer, P. S., "Toxic neuropathies and their treatment," in *Clinical Neurology*, Hartung, H. P., Ed., Baillière Tindall, London, 1995, p. 505.
63. Cavanagh, J. B., "Peripheral nervous system toxicity: a morphological approach," in *Neurotoxicology*, Blum, K. and Manzo, L., Eds., Marcel Dekker, New York, 1985, Chapter 1.
64. Cavanagh, J. B., "Neurotoxic effects of metals and their interactions," in *Recent Advances in Nervous System Toxicology*, Galli, C. L., Manzo, L., and Spencer, P. S., Plenum Press, New York, 1988, p. 177.
65. Rajaniemi, R., "Clinical evaluation of occupational toxicity of methyl methacrylate monomer in dental technicians," *J. Soc. Occup. Med.*, 36, 56, 1986.
66. Donaghy, M., Rushworth, G., and Jacobs, J. M., "Generalized peripheral neuropathy in a dental technician exposed to methyl methacrylate monomer," *Neurology*, 41, 1112, 1991.
67. House R. A., Liss G. M., and Wills M., "Peripheral sensory neuropathy associated with 1,1,1-trichloroethane." *Arch. Environ. Hlth.*, 49, 196, 1994.

68. House R. A., Liss G. M., Wills M., and Holness, D. L., "Paresthesias and peripheral neuropathy due to 1,1,1-trichloroethane," *J. Occup. Environ. Med.*, 38, 123, 1996.

69. Lotti, M., "The pathogenesis of organophosphorous polyneuropathies in humans: perspectives for biomonitoring," *Trends Phamacol. Sci.*, 8, 175, 1987.

70. Abou-Donia, M. B., "Organophosphorous ester-induced delayed neurotoxicity," *Ann. Rev. Pharmacol. Toxicol.*, 21, 511, 1981.

71. Senanayake, N. and Karalliede, L., "Neurotoxic effects of organophosphorous insecticide: an intermediate syndrome." *N. Engl. J. Med.*, 316, 761, 1987.

72. De Blecker, J., Van Den Neucker, K., and Williams, J., "The intermediate syndrome in organophosphate poisoning: presentation of a case and review of the literature," *Clin. Toxicol.*, 30, 321, 1992.

73. Schwartz, J., Landrigan, P. J., and Feldman, R. G., "Threshold effect in lead-induced peripheral neuropathy," *J. Pediat.*, 112, 12, 1988.

74. Le Quesne, P. M. and McLeod, J. G., "Peripheral neuropathy following a single exposure to arsenic," *J. Neurol. Sci.*, 32, 437, 1977.

75. Seppalainen, A. M., "Halogenated hydrocarbons," in *Neurotoxicology*, Blum, K. and Manzo, L., Eds., Marcel Dekker, New York, 1985, Chapter 22.

76. Feldman, R. G., "Occupational exposure to trichloroethylene: controversies concerning neurotoxicity," in *Advances in Modern Environmental Toxicology*, Vol. 23, Mehlman, M. A. and Upton, A., Eds, Princeton Scientific Publ., Princeton, NJ, 1994, p. 235.

77. Juntunen, J., Hupli, V., Hernberg, S., and Luisto, M., "Neurological picture of organic solvent poisoning in industry: a retrospective clinical study in 37 patients," *Int. Arch. Occup. Environ. Hlth.*, 46, 219, 1980.

78. Murata K., Araki S., Yokoyama K., and Maeda, K., "Autonomic and peripheral nervous system dysfunction in workers exposed to mixed organic solvents," *Int. Arch. Occup. Environ. Hlth.*, 63, 335, 1991.

79. Ruijten M. W., Sallé H. J. A., Verberk M. M., and Smink M., "Effect of chronic mixed pesticide exposure on peripheral and autonomic nerve function." *Arch. Environ. Hlth.*, 49, 188, 1994.

80. Keogh, J. P., Pestronk, A., Wertheimer, D., and Moreland, R., "An epidemic of urinary retention caused by dimethylaminopropionitrile," *J. AMA*, 243, 746, 1980.

81. Thomas, T. L. and Waxweiler, R. J., Brain tumors and occupational risk factors. *Scand. J. Work Environ. Hlth.*, 12, 1, 1986.

82. Zheng, W., Perry, D. F., Nelson, D. L., and Aphosian, H. V., "Choroid plexus protects cerebrospinal fluid against toxic metals," *FASEB J.*, 5, 2188, 1991.

83. Hastings, L. and Evans, J. E., "Olfactory primary neurons as a route of entry for toxic agents into the CNS," *Neurotoxicology*, 12, 707, 1991.

84. Gianutsos, G., Morrow, G. R., and Morris, J. B., "Accumulation of manganese in rat brain following intranasal administration," *Fundam. Appl. Toxicol.*, 37, 102, 1997.

85. Schaumburg, H. H. and Spencer, P. S., "Recognizing neurotoxic disease," *Neurology*, 37, 276, 1987.

86. Perucca, E. and Manzo, L., "Metabolic activation of neurotoxicants," in *Recent Advances in Nervous System Toxicology*, Galli, C. L., Manzo, L. and Spencer, P. S., Plenum Press, New York, 1988, p. 67.

87. Ralston, W. H., Hilderbrand, R. L., Uddin, D. E., Andersen, M. E., and Gardier, R. W., "Potentiation of 2,5-hexandione neurotoxicity by methyl ethyl ketone," *Toxicol. Appl. Pharmacol.*, 81, 319, 1985.

88. Rosenberg, N. L., "Basic principles of clinical neurotoxicology." In *Neurotoxicology: Approaches and Methods*, Chang, L.W. a,nd Slikker, W. Jr, Eds., Academic Press, Orlando, FL, 1995, Chapter 39.

89. Hawkes, C. H., Cavanagh, J. B., and Fox, A. J., "Motoneuron disease: a disorder secondary to solvent exposure?" *Lancet.* 1989; i: 73-76.

90. Reuhl K. R., "Delayed expression of neurotoxicity: the problem of silent damage," *Neurotoxicology,* 12: 341, 1991.

91. Weiss B., "Unique dimensions of neurotoxic risk assessment." In *Neurotoxicology: Approaches and Methods,* L.W. Chang and Slikker, W., Jr., Eds., Academic Press, Orlando, FL, 1995, pp. 815.

92. Baker E. L., "Environmentally related disorders of the nervous system." *Med. Clin. North Am.* 74: 325, 1990.

93. O'Donoghue J. L., *Neurotoxicity of Industrial and Commercial Chemicals,* Vol. 1,2, CRC Press, Boca Raton, FL, 1985.

94. Squier M. V., Thompson J., Rajgopalan B., "Neuropathology of methyl bromide poisoning." *Neuropathol. Appl. Neurobiol.,* 18: 579, 1992.

95. Ohnishi A. and Murai Y., "Polyneuropathy due to ethylene oxide, propylene oxide, and butylene oxide." *Environ. Res.* 60, 242, 1993.

96. Spencer, P. S. and Schaumburg, H. H., *Experimental and Clinical Neurotoxicology,* Williams & Wilkins, Baltimore, MD, 1980.

97. Ngim C. H., Foo S. C., Boey K. W., and Jeyaratnam J., "Chronic neurobehavioural effects of elemental mercury in dentists." *Br. J. Ind. Med.,* 49, 782, 1992.

98. Bos, M. J., De Mik, G., and Bragt, P. C., "Critical review of the toxicity of methyl *n*-butyl ketone: risk from occupational exposure," *Am. J. Ind. Med.,* 20, 175, 1991.

99. He, F. and Zang, S., "Effects of allyl chloride on occupationally exposed workers," *Scand. J. Work Environ. Hlth.,* 11 S4, 43, 1985.

100. Chu C.-C., Huang C.-C., Chen R.-S., and Shih T.-S., "Polyneuropathy induced by carbon disulphide in viscose rayon workers." *Occup. Environ. Med.* 52, 404, 1995.

101. Yang, Y-J., Huang, C-C., Shih, T.-S., and Yang, S.-S., "Chronic elemental mercury intoxication: clinical and field studies in lampsocket manufactures," *Occup. Environ. Med.,* 51, 267, 1994.

102. Beckett, W. S., Moore, J. L., Keogh, J. P. and Bleecker, M. L., "Acute encephalopathy due to occupational exposure to arsenic," *Br. J. Ind. Med.,* 43, 66, 1986.

103. Wolters E. C., "Manganese-induced Parkinsonism," in *Mineral and Metal Neurotoxicology,* Yasui, M., Strong, M. J., Ota, K., and Verity, A. M., CRC Press, Boca Raton, FL, 1997, 319.

2 Metabolism and Toxicity of Occupational Neurotoxicants: Genetic, Physiological, and Environmental Determinants

Stefano M. Candura, Luigi Manzo,
Anna F. Castoldi, and Lucio G. Costa

CONTENTS

0-8493-9231-4/98/$0.00+$.50

1   INTRODUCTION

1.1   Metabolism as a Preliminary Step to Excretion

As all xenobiotics, the majority of occupational neurotoxicants absorbed into the body are biotransformed to some degree, yielding one or more metabolites that are usually more polar and water-soluble than the parent compounds. This enhanced water solubility decreases the ability of the chemical to partition into biological membranes, thus reducing the distribution of the metabolite(s) to the various tissues and organs, including the nervous system. Additionally, it decreases the renal tubular and intestinal readsorption of the toxicant, ultimately promoting its excretion by the urinary and biliary fecal routes.[1-4]

1.2 PHASE I AND PHASE II REACTIONS

Biotransformation processes are usually enzymatic in nature. They are conventionally divided into two distinct phases. Phase I reactions include oxidation, reduction, and hydrolysis. They result in functionalization, i.e., the addition or uncovering of specific functional groups (e.g., –OH, –SH, –NH$_2$, –COOH) that are required for subsequent metabolism by Phase II enzymes. Phase II reactions are biosynthetic: the functional group is linked to an endogenous substrate (e.g., D-glucuronate, acetate, sulfate, glutathione), resulting in the formation of water-soluble conjugates. In some cases, the added chemical group is recognized by specific carrier proteins involved in facilitated diffusion or in active transport, thus increasing the cell's ability to remove the xenobiotic. The conjugation products may undergo further metabolism, as in the case of glutathione conjugates ultimately excreted as mercapturic acids. Phase I and II reactions are generally coordinated, with the product of one reaction becoming the substrate of the other. However, metabolism may be restricted to Phase I or Phase II reactions only. In the latter case, the parental molecule must possess functional groups suitable for conjugation.[1-4]

The metabolic fate of a particular chemical is determined by its physical/chemical properties. Volatile organic compounds (e.g., industrial solvents) or gases (e.g., carbon monoxide) may be eliminated, at least in part, through the lungs with no biotransformation. Compounds with functional groups may be conjugated directly, while others require Phase I reactions before conjugation.[1]

1.3 METABOLIC ACTIVATION

Although xenobiotic metabolism generally represents a detoxication process, there are many exceptions to this rule and in several cases biotransformation products contribute in full or in part to the toxic activity of the parent compound. Active metabolites are more frequently produced by Phase I than by Phase II reactions, the P450-dependent microsomal oxidations being the best documented activation pathways. Biosynthetic reactions usually lead to detoxication, even though conjugations leading to metabolic activation (e.g., glutathione conjugation of dihaloalkanes resulting in the formation of sulfur mustards) have also been documented.[4,5]

1.4 TISSUE LOCALIZATION OF BIOTRANSFORMATION

The liver plays the major role in the metabolic transformation of chemicals, including those endowed with neurotoxic properties.[2] However, a significant degree of enzymatic activity may also be detected in body fluids such as blood, and in other organs such as kidney, intestine, lung, and skin, i.e., the main sites of chemical absorption or excretion. To a minor extent, xenobiotic metabolism also occurs in the adrenals, testis, placenta and brain. Experimentally, almost every tissue tested has shown metabolic activity toward some foreign compounds. However, the rate of extrahepatic metabolism is not as high as in liver, and the total capacity is usually lower.[1,3]

1.5 SUBCELLULAR LOCALIZATION OF BIOTRANSFORMING ENZYMES

The enzyme systems implicated in the metabolism of xenobiotics have a particular subcellular localization. Phase I enzymes are predominantly located in the smooth endoplasmic reticulum (microsomes), while Phase II enzymes are primarily contained in the soluble fraction of the cell (cytosol). Few enzymes (e.g., monoamino oxidases, rhodanase) are also found in other organelles such as mitochondria.[1,4]

1.6 XENOBIOTIC METABOLISM IN BRAIN

A variety of xenobiotics is metabolized in neurons, glial cells and cerebral microvessels. Such biotransformations may lead to *in situ* bioactivation or detoxication, thus conceivably contributing to local modulation of pharmacological and/or toxicological responses in certain brain areas or in specific cells within these areas.[6]

As in other sites, nervous tissue biotransformation of foreign compounds is predominantly mediated through specific drug-metabolizing enzymatic systems, involving both Phase I and Phase II reactions. However, certain chemicals with close resemblance to endogenous substrates are often modified by brain enzymes associated with metabolism of endogenous compounds. Such is the case of 1-methyl-4-phenyl-1,2,3,6-tetrahydropyridine (MPTP), a chemical causing a Parkinsonian syndrome which is clinically undistinguishable from idiopathic Parkinson's disease. MPTP is metabolically activated to 1-methyl-4-phenylpyridinium ion (MPP+) by monoamino oxidase B (MAO-B). This reaction is likely to occur within glial cells. MPP+ is then released from glial cells and actively accumulated into dopaminergic neurons (the ultimate target of MPTP/MPP+ neurotoxicity) by the catecholamine uptake system.[6,7]

1.7 INTESTINAL MICROBIAL BIOTRANSFORMATION

Intestinal microflora plays an important role in xenobiotic metabolism. The more than 400 bacterial species known to exist in the intestinal tract possess a potential for biotransformation equivalent to, or greater than, the liver. They can alter xenobiotic bioavailability by metabolizing ingested chemicals to molecules that may be absorbed to a greater or lesser extent, or modify metabolic products which are secreted into the gut directly from the blood or through the bile, saliva, or by swallowing respiratory tract mucus. The presence of bacterial deconjugating enzymes (e.g., β-glucuronidases and arylsulfatases) leads to enterohepatic recycling.[1,4]

1.8 METABOLISM AS A DETERMINANT OF NEUROTOXICITY

Metabolism plays a fundamental role in determining the intensity and duration of most neurotoxicants.[2,5] The enzymes that protect from the neurotoxicity of certain substances may be responsible for the adverse effects of others. Human susceptibility to the neurotoxicity of a given chemical often depends on the delicate balance between detoxication and metabolic activation that exists during exposure to the compound. Since the activity of the metabolizing enzymes is affected by a number of endogenous and exogenous factors, this balance may differ considerably among individuals and even in the same subject.[4] Thus, genetically determined and acquired variations in metabolism might explain inter- and intraindividual variability in neurotoxic responses.

The following paragraphs provide a general overview of the main metabolic reactions and enzymes involved in the metabolism of foreign compounds, including neurotoxicants, and of the genetic, physiologic and environmental factors which influence them. A number of examples particularly relevant to occupational medicine are also presented.

2 PHASE I REACTIONS

2.1 THE CYTOCHROME P450 SYSTEM

2.1.1 Components and Isoforms

Most Phase I reactions are catalyzed by an enzyme system which involves a specific heme-containing cytochrome, known as cytochrome P450 since its complex with CO exhibits a spectrophotometric absorption peak at 450 nm. The cytochrome P450 system (also referred to as the polysubstrate monooxygenase system or the mixed-function oxydase [MFO] system) is actually composed of two enzymes embedded in the phospholipid matrix of the endoplasmic reticulum: the cytochrome and NADPH-cytochrome P450 reductase.[1,4] P450s represent a superfamily of enzymes, occurring under multiple forms in animals, plants, yeast and bacteria. Their name is formed by the CYP root followed by an arabic number and upper case letter designating the family and subfamily, respectively. Individual forms are denoted by an arabic number that follows the subfamily letter. To date, eight major mammalian gene families have been identified. Families 1, 2, 3 and 4 are involved in xenobiotic Phase I reactions, while families 11, 17, 19 and 21 are involved in steroid hormone biosynthesis. Family 2 is by far the largest, with five major subfamilies. It is the most active in metabolizing many clinically used drugs, as well as several environmental contaminants. The 2B subfamily includes the major phenobarbital-inducible cytochromes P450, while the 2E subfamily contains the major ethanol-inducible form (CYP2E1). Members of the 2A, 2C and 2D subfamilies are in most cases constitutively expressed.[1,8]

2.1.2 Oxidative Reactions

The P450 system catalyzes many oxidation reactions that may be subdivided into: aliphatic, aromatic, alicyclic and heterocyclic hydroxylation; epoxidation; N-, S- and O-dealkylation; N-oxidation; N-hydroxylation; S-oxidation; desulfuration and deamination.[1,3] The following examples illustrate the importance of these metabolic processes for occupational neurotoxicology.

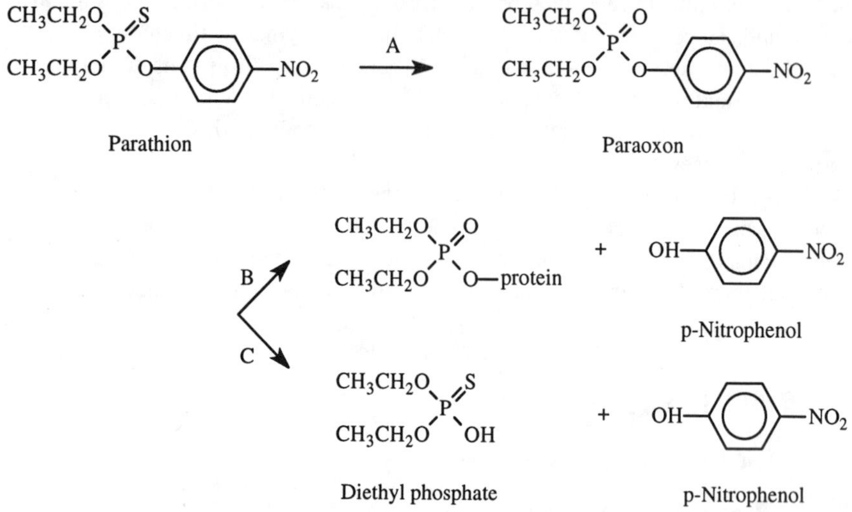

FIGURE 1   Metabolism of the organophosphate parathion. Reaction A represents activation by P450-mediated oxidative desulfuration, reaction B represents phosphorylation of proteins (non-enzymatic hydrolysis), and reaction C represents detoxication by paraoxonase.

2.1.3   Activation of Organophosphorus Compounds

P450-mediated oxidation represents an important activation step of many organo-phosphates. At least four major activative reactions have been identified: oxidative desulfuration (conversion of P = S to P = O) (Figure 1); thioether oxidation (production of a sulfoxide [S = O], followed by the formation of a sulfone [O = S = O]); hydroxylation; and cyclization.[9] The bulk of activation occurs within the liver. However, certain organophosphorus compounds can be metabolically activated in extrahepatic sites, including the nervous system.[10] For example, several phosphorothionate pesticides are bioactivated by the microsomal and mitochondrial fractions of rat brain.[11]

2.1.4   Activation of Neurotoxic Aliphatic Hexacarbons

Aliphatic hydroxylation is typical of industrial solvents such as *n*-hexane and methyl *n*-butyl ketone (MnBK, 2-hexanone). These compounds have been shown to induce remarkably similar patterns of peripheral neuropathy in exposed workers.[12] The basis for this similarity in neurotoxic action resides in their metabolism (Figure 2). *n*-Hexane is successively oxidized to 2-hexanol, 2,5-hexanediol, 5-hydroxy-2-hexanone, and 2,5-hexanedione (2,5-HD), which is the major metabolite of *n*-hexane in humans. MnBK is also metabolized to 2,5-HD, through conversion to 5-hydroxy-2-hexanone, or indirectly, by transformation to 2-hexanol (Figure 2). 2,5-HD is the active agent responsible for both *n*-hexane and MnBK neurotoxicity. In experimental animals, 2,5-HD has been found to produce a polyneuropathy undistinguishable from that caused by *n*-hexane and MnBK under a variety of exposure conditions.[12,13] Moreover, 2,5-HD is several fold more potent than *n*-hexane

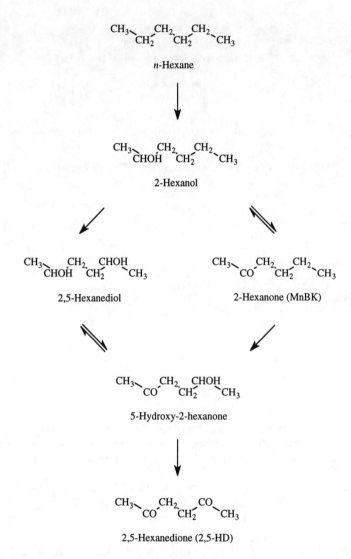

FIGURE 2 Metabolism of *n*-hexane and methyl *n*-butyl ketone (MnBK, 2-hexanone). Both compounds are oxidized by the microsomal P450 system to the neurotoxic metabolite 2,5-hexanedione (2,5-HD).

or MnBK in causing neurotoxicity.[13] It appears that the neurotoxicity of 2,5-HD resides in its γ-diketone structure, since 2,3-2,4-hexanedione and 2,6-heptanedione are inactive, whereas 2,5-heptanedione and 3,6-octanedione and other γ-diketones are neurotoxic.[14]

2.1.5   Activation of Carbon Disulfide

Although the exact mechanism(s) by which carbon disulfide (CS_2) causes peripheral neuropathy and other neurotoxic effects (retinopathy, hearing loss to high-frequency

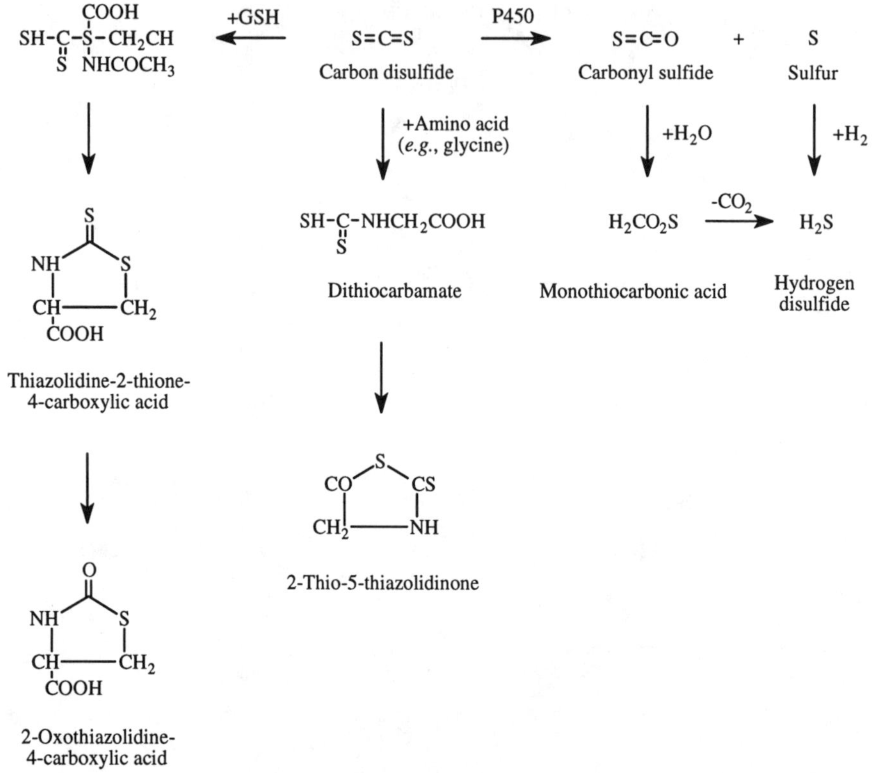

FIGURE 3 Metabolism of carbon disulfide (CS_2). Glutathione conjugation is a detoxication reaction, whereas P450-mediated oxidation and amino acid conjugation yield hydrogen disulfide (H_2S) and dithiocarbamates, respectively, which are responsible for CS_2 neurotoxicity.

tones) are yet to be determined, it is believed that CS_2 toxicity is, at least in part, mediated by metabolic activation.[15] Following exposure to CS_2, very little of the parent compound is excreted unchanged. Most of the absorbed dose is eliminated as sulfur-containing urinary metabolites, some of which are shown in Figure 3. P450-mediated oxidative desulfuration of CS_2 yields elemental sulfur and carbonyl sulfide, which in turn is hydrated to monothiocarbonic acid (Figure 3). The latter decomposes to hydrogen disulfide (H_2S) and CO_2. Additional H_2S is produced by sulfur reduction in the strong reducing environment of the cell. Dithiocarbamates formed as metabolites of CS_2 may contribute to neurotoxicity, possibly by interfering with distribution of essential metals, such as copper and zinc.[15,16]

2.1.6   Reductive Reactions

The cytochrome P450 system also participates in some reductive reactions, such as nitro-, azo-, arene oxide-reduction, and reductive dehalogenation. These reactions may detoxify a xenobiotic, but they often result in more toxic intermediates.[1-4] Carbon tetrachloride and halotane are activated by accepting electrons from P450 resulting in radical anions, which fragment into free radicals upon cleavage of the carbon-halogen bond.[4] Another example is aromatic nitro reduction of the organophosphate EPN (*O*-ethyl *O*-4-nitrophenyl phenylphosphonothionate) to amino EPN.[2]

2.1.7   Brain Cytochromes P450

The constitutive expression of P450 isoforms such as CYP1A1, CYP1A2, CYP2E1, and CYP3A has been demonstrated in several human brain regions.[17] In the nervous tissue, cytochromes P450 are predominantly located in neuronal cells,[17,18] and enzymatic activity is higher in mitochondria than in microsomes.[19] The further characterization of specific isoforms, and their possible capability to metabolize different substrates selectively, would clarify the role of the P450 system in the local modulation of toxicants acting on the central nervous system.

2.2   FAD-CONTAINING MONOOXYGENASES

The microsomal FAD-containing monooxygenase (FMO, mixed-function amine oxidase) competes with the P450 system in the oxidation of amines. Multiple forms of the enzyme are present in several species and tissues, however it is unlikely that there are as many isoforms of FMO as there are of cytochromes P450. The FMO system converts tertiary amines to amine oxides, secondary amines to hydroxyl amines and nitrones, and primary amines to hydroxylamines and oximes. It also oxidizes sulfur compounds (sulfides, thioethers, thiols, thiocarbamates) and organophosphates.[1,4] In particular, FMO activates certain organophosphorus pesticides containing at least one C-P bond (e.g., fonofos), or a thioether linkage (e.g., phorate). Such compounds are substrates for both cytochromes P450 and FMO.[10]

Recently, the presence of FMO has also been demonstrated in the human brain, where the enzyme is preferentially located in the neuronal cell bodies.[20] Brain FMO activity shows considerable interindividual variations,[20] which might partly account for the variability of neurotoxic responses.

2.3   ALCOHOL, ALDEHYDE, KETONE OXIDATION-REDUCTION SYSTEMS

Alcohols, aldehydes, and ketones are functional groups of many occupational neurotoxicants. Moreover, they often occur from oxidative or hydrolytic metabolic reactions. These functional groups may be biotransformed by several soluble enzymes, such as alcohol dehydrogenase (ADH), aldehyde oxidase, aldehyde dehydrogenase (ALDH), and aldehyde/ketone reductases.[1] These enzymatic systems are present in a number of mammalian tissues. Since alcohol, aldehyde, or ketone groups often confer pharmaco/toxicological properties, their oxidation or reduction generally represents a detoxication pathway. There are, however, exceptions. For example, ADH, a cytosolic NAD-dependent enzyme located primarily in the liver, oxidizes ethanol to acetaldehyde (Figure 4A), which contributes to ethanol toxicity. To a minor extent, conversion of ethanol to acetaldehyde is also carried out by the microsomal ethanol oxidizing system (MEOS), which involves CYP2E1, and by cytosolic catalase, which uses hydrogen peroxide (H_2O_2) to perform the oxidation. Acetaldehyde is in turn oxidized to acetic acid by ALDH (Figure 4A), which is

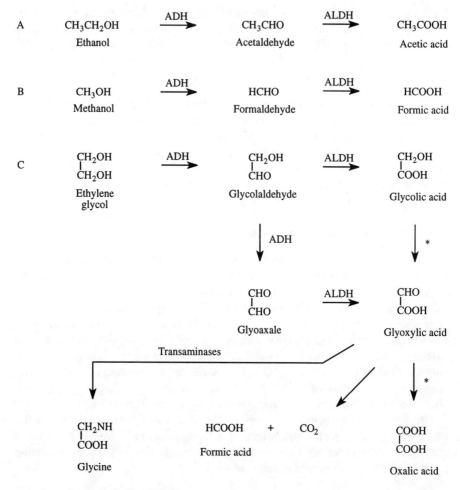

FIGURE 4 Role of alcohol dehydrogenase (ADH) and aldehyde dehydrogenase (ALDH) in the metabolism of ethanol (A), methanol (B) and ethylene glycol (C). Asterisks indicate reactions that may be catalyzed by glycolic acid oxidase or by lactate dehydrogenase. Conversion of glyoxylic acid to formic acid and CO_2 involves coenzyme A and flavin mononucleotides.

another cytosolic (in humans) NAD-requiring enzyme. Finally, acetic acid enters the Krebs cycle (as acetyl-CoA), and is degradated to CO_2 and water.

Other alcohols or polyalcohols may act as substrates for ADH and ALDH. While most of these reactions lead to detoxication of alcoholic substances, exceptions to this rule (i.e., the metabolites are more toxic) have also ben reported. Such is the case of the conversion of methanol to formic acid (Figure 4B), and of ethylene glycol to a series of organic acids (Figure 4C). As a result, metabolic acidosis develops after acute exposure to these compounds. Additionally, formic acid damages the optic nerve, seemingly by inhibiting the mitochondrial enzyme cytochrome oxidase, thus reducing the level of ATP available to nerve fibers. It was also postulated that the intraretinal metabolism of methanol, rather than elevated formic acid levels, is required to initiate methanol-induced retinal toxicity. There is some evidence suggesting that the local oxidation of methanol to formate parallels the depletion of retinal ATP and that the retinal glial cells (Müller cells) may be the initial target in the mechanism of methanol-induced retinal damage.[21] Since metabolic activation is clearly important in the toxicity of both methanol and ethylene glycol, therapeutic treatment of the poisoning involves blocking the first step of their metabolic pathways by using ethanol or 4-methylpyrazole as inhibitors of ADH.[22]

2.4 PEROXIDASE-DEPENDENT COOXIDATION

Peroxidases are present in several tissues and cell types. They catalyze a common generalized reaction in which the reduction of H_2O_2 or lipid hydroperoxides is coupled to the oxidation of a cosubstrate. Many of the metabolites produced are reactive electrophiles which bind to DNA and other macromolecules. Classes of compounds undergoing peroxidase-dependent metabolism include aromatic amines, phenols, hydroquinones, polycyclic aromatic hydrocarbons (PAH) and N-methyl carbamates.[1,2]

2.5 ESTERASES AND AMIDASES

A large number of nonspecific esterases and amidases (both cytosolic and microsomal) are contained in mammalian tissues. Esterases are broadly classified in four classes: arylesterases (A-esterases), carboxylesterases (B-esterases), acetylesterases (C-esterases), and cholinesterases. There is however considerable overlap in substrate specificity.[1] Moreover, esterases also possess amidase activity and vice versa.[3] Serum paraoxonase arylesterase deactivates several organophosphorus compounds (Figure 1).[9,23]

2.6 EPOXIDE HYDROLASE

Epoxides undergo hydration catalyzed by the enzyme epoxide hydrolase, which is located in the smooth endoplasmic reticulum, conveniently near to the P450 system which produces the epoxide. Both aliphatic and aromatic epoxides may be substrates for the enzyme, which is found in a wide variety of sites, including the nervous tissue and cerebral microvessels.[19] The reaction usually represents a detoxication process, since the metabolic products are *trans* diols that are less electrophilic and thus less chemically reactive than the parent compounds.[1,3]

3 PHASE II REACTIONS

3.1 GLUCURONIDATION

Conjugation with D-glucuronic acid represents the major Phase II reaction. It is catalyzed by UDP-glucuronosyltransferases, a microsomal enzyme system independent of, but functionally related to, the cytochrome P450 system. There is evidence for a minimum of 11 distinct forms of UDP-glucuronosyltransferase,[1] of which at least five are expressed in humans.[24] Glucuronidation may occur with compounds containing functions such as the hydroxyl, the carboxyl, the amino, the mercapto and the dithiocarboxyl groups.[3]

3.2   N-ACETYLATION

In N-acetylation reactions, amino groups situated on the original molecule or its Phase I metabolite(s) react with acetyl coenzyme A to form an amide bond.[25] The enzymes that catalyze N-acetylation are designed as acetyl-CoA:amine N-acetyl transferases. They are cytosolic enzymes found in several tissues. Examples of substrates include aromatic primary amines, hydrazines, hydrazydes, sulfonamides, and certain primary aliphatic amines.[3]

3.3   SULFATION

The addition of an inorganic sulfate moiety to a hydroxyl group is a major route of conjugation for both endogenous and foreign compounds. The reaction is catalyzed by cytosolic sulfotransferases, a group of enzymes found primarily in liver, kidney, intestinal tract and lungs, and requires the coenzyme 3'-phosphoadenosine-5'-phosphosulfate (PAPS).[3] There are four classes of sulfotransferases: aryl sulfotransferase, which conjugates phenols, catecholamines, and organic hydroxylamines; hydroxysteroid sulfotransferase, which conjugates hydroxy steroids and certain primary and secondary alcohols; estrone sulfotransferase, which is active on phenolic groups in the aromatic ring of steroids; and bile salt sulfotransferase.[1]

3.4   METHYLATION

Hydroxyl, amino and thiol groups may be methylated by one of a series of methyltransferases mostly located in the cytosol. This reaction requires S-adenosylmethionine as a methyl donor. Methylation differs from most other conjugation reactions in that it actually masks functional groups. It is a common biochemical reaction for the metabolism of endogenous compounds, but is not a quantitatively important pathway for xenobiotic transformation.[1,3]

Hepatic S-methyl transferases detoxify H_2S. Besides being an environmental pollutant, this compound is produced by anaerobic bacteria in the intestinal tract and by CS_2 metabolism (Figure 3). H_2S is methylated to methane thiol (CH_3SH), which is further methylated to dimethylsulfide (CH_3-S-CH_3).[1]

Biomethylations of toxicological importance also occur outside the body. A number of metals, such as mercury, may be methylated by microorganisms in the environment, a reaction which changes the physico-chemical characteristics and, hence, the environmental fate and toxicological properties of the metal. Methylmercury (MeHg) can cross the blood-brain barrier and accumulates in the nervous tissue. Demethylation by intestinal microflora is instead a protective reaction, since it decreases methylmercury retention by interrupting enterohepatic recycling of the compound and increases the fecal excretion of the inorganic mercury thus generated.[26]

Demethylation of MeHg also occurs in the human brain,[27] and long-term MeHg exposure studies on macaque monkeys have demonstrated a slow accumulation of inorganic mercury in several brain areas.[28] *In situ* formation of inorganic mercury has been suggested to play an important role in MeHg neurotoxicity.[28]

3.5 AMINO ACID CONJUGATION

Conjugation with one of a variety of amino acids (glycine, glutamine, taurine, ornithine) involves compounds containing a carboxylic acid group. These reactions result in the formation of an amide bond between the carboxylic acid group of the xenobiotic and the amino group of the amino acid. In humans, glutamine is the most common amino acid used. Substrates for amino acid conjugation include aromatic carboxylic acids, and arylacetic acids.[1] As already stated, amino acid conjugation of CS_2 produces dithiocarbamates (Figure 3), which are partly responsible for CS_2 neurotoxicity.[15,16]

3.6 GLUTATHIONE CONJUGATION

Reduced glutathione (GSH) is a tripeptide (glu-cys-gly) found in many mammalian tissues (including the brain), and especially in the liver. It has a major protective role as it acts as a scavanger for reactive compounds of various types. The sulphydryl group attacks the reactive part of the xenobiotic and/or its metabolite(s). This metabolic pathway has also been shown to cause some toxic reactions (e.g., metabolic activation of haloalkanes).[29] GSH conjugations are catalyzed by a family of enzymes (glutathione-S-transferases) localized in both the cytoplasm and the endoplasmic reticulum, and showing different but overlapping substrate selectivity. Substrates include aromatic, heterocyclic, alicyclic and aliphatic epoxides; aromatic halogen and nitro compounds; alkyl halides; unsaturated aliphatic compounds; and several metals. GSH conjugates usually undergo further metabolism which involves removal of the glutamyl residue, followed by loss of glycine and cysteine acetylation, with the formation of mercapturic acids.[1]

Oxidative stress is a mechanism of neurotoxicity for several chemicals, and therefore GSH conjugation in nerve tissue plays a major protective role. In addition, brain GSH, which is mostly located in glial cells, appears to be involved in other important functions such as protein synthesis and neurotransmission.[30] Although cerebral GSH levels are generally quite stable and GSH is not easily depleted, several studies indicate that modification of GSH homeostasis in the nervous system modulates the neurotoxic effects of chemicals such as cadmium,[31] manganese,[32] methyl iodide,[33] styrene[34] and its Phase I metabolite styrene oxide.[35]

3.7 RHODANASE

Cyanide (CN^-) is metabolized in the liver and skeletal muscle by the mitochondrial enzyme rhodanase (sulfur transferase), through conjugation with thiosulfate ($S_2O_3^{2-}$), that acts as a sulfur donor. The reaction yields thiocyanate (SCN^-), which is far less toxic than CN^-, and sulfite ($S_2O_3^{2-}$).[1]

4 FACTORS AFFECTING NEUROTOXICANT METABOLISM

4.1 GENETIC FACTORS

4.1.1 General Remarks

The importance of genetics in determining interindividual differences in xeno-biotic metabolism has been recognized for more than 40 years.[36,37] The inherited traits influencing foreign compound disposition may be either rare single gene defects, or genetic polymorphisms, defined as Mendelian traits that exist in the population in at least two phenotypes neither of which is rare (*i.e.,* the genotype of lowest frequency is present in more than 1% of individuals).[38] For a gene to be polymorphic, it must encode a product that is not required for development, repro-ductive vigor or a crucial physiological function, otherwise such a high frequency of occurrence would not be expected. In other words, the existence of polymorphism in enzymes responsible for foreign compound metabolism implies that these enzymes are not necessary for survival under normal circumstances. Genetically determined differences in xenobiotic disposition can be associated with marked variability in the response to drug therapy, and may cause increased susceptibility to environmentally or work-related neurotoxic diseases.[39] Moreover, they have a considerable influence on biological monitoring strategies and risk assessment.[40-42]

4.1.2 Cytochromes P450 Polymorphism

Genetic polymorphisms of human cytochromes P450 are currently under exten-sive investigation. Polymorphism has been mostly described for the CYP2C and CYP2D subfamilies. The enzymes of the CYP3A subfamily exhibit a wide range of interindividual variability in their levels in liver, but no genetic component respon-sible for polymorphism in their expression has been confirmed.[8]

The best characterized P450 polymorphism is that of cytochrome CYP2D6 (debrisoquine 4-hydroxylase). Its discovery followed the observation of unusual subject sensitivity to the hypotensive effect of debrisoquine,[43] and of unusual neu-rological side effects in some patients treated with sparteine.[44] It was subsequently clarified that the debrisoquine-sparteine polymorphism is caused by mutations of the CYP2D6 gene, which is part of a gene cluster on chromosome 22.[45] Enzymatic activity is deficient in 7–10% of Caucasian subjects, who therefore show a slower oxidative metabolism.[46,47] The frequency of poor metabolizers appears to be mark-edly lower in non-Caucasian populations.[38] Since many therapeutic agents are oxidized by CYP2D6, the polymorphism has important clinical consequences, particularly in the use of cardiovascular and psychoactive drugs.[8] With regard to neurotoxicology, CYP2D6 detoxifies 1,2,3,4-tetrahydroisoquinoline (TIQ) and MPTP. MPTP and other environmental neurotoxins with similar chemical struc-tures have been implicated in the pathogenesis of Parkinson's disease. Some researchers characterized the genotype that regulates 2D6 expression in Parkinson patients, suggesting that slow metabolizers have a two–three fold increased disease susceptibility.[48,49]

4.1.3 NAT2 Polymorphism

The genetic polymorphism controlling N-acetylation is the earliest described and the most widely recognized. Using one of several probe compounds (e.g., isoniazid, caffeine, sulfamethazine) individuals may be separated into either fast or slow acetylator phenotypes. About half of Caucasian populations are slow acetylators. Compared with fast acetylators, slow acetylators achieve a higher parent blood concentration of xenobiotics which are detoxified by N-acetylation, and are more likely to develop neurotoxic effects (e.g., arylamine-induced peripheral neuropathy).[38] Two enzymes catalyzing N-acetylation (NAT1 and NAT2) are expressed in the human liver. NAT2 exists as two isoforms (NAT2A and NAT2B), and is responsible for the genetic polymorphism.[50] NAT1 and NAT2 are encoded at two genetic loci on chromosome 8.[51] At least 4 alleles, one fast (F1) and 3 slow (S1, S2, S3) may be present at the polymorphic NAT2 locus. The F1 allele is dominant and produces the fast phenotype in both heterozygotes and homozygotes.[52] The most common allele in Caucasians is the S1 which has a frequency of 45%.[52] In the Japanase population, the absence of this allele in association with a 68% frequency of the F1 allele determines a markedly lower incidence of slow acetylation (6.6%).[53]

4.1.4 ADH and ALDH Polymorphism

Genetic factors also determine important intersubject variations in the metabolism and neurotoxicity of ethanol and industrial alcohols.[38,39] ADH is a dimeric enzyme whose subunit chains are determined by at least six genetic loci, of which at least two, ADH2 and ADH3, are polymorphic. The isoenzymes vary markedly in their kinetic properties, and the frequency of the alleles encoding them vary across ethnic group.[54] Both polymorphic ADH loci have been localized to the long arm of chromosome 4.[55] ALDH is present in at least four forms, whose characteristics and activity vary substantially.[56] Acetaldehyde oxidation is mostly catalyzed by ALDH2, encoded by a gene localized to chromosome 12.[57] In Asians and American Indians, a null allele at the ALDH2 locus causes impaired elimination of acetaldehyde and flushing reactions following ingestion or inhalation of relatively small amounts of ethanol. The null allele is dominant, and heterozygotes exhibit ALDH deficiency.[38]

4.1.5 ALA-D Polymorphism

∂-Aminolevulinic acid dehydratase (ALA-D) is a main biochemical target for lead toxicity. This enzyme catalyzes the asymmetric condensation of two molecules of ∂-aminolevulinic acid (ALA) to form porphobilinogen (PBG), the monopyrrole precursor of heme. Inhibition of erythrocyte ALA-D activity is one of the most sensitive biomarkers of lead poisoning.[58] ALA-D inactivation results in a proportional accumulation of ALA in blood and increased secretion in urine.[59]

ALA-D is a polymorphic enzyme encoded by a single gene localized to chromosome 9q34.[60,61] Expression of two alleles (ALA-D^1 and ALA-D^2) results in three distinct isozymes, designated ALA-D 1-1, 1-2 and 2-2. In Caucasian populations, the gene frequencies of the ALA-D^1 and ALA-D^2 alleles are 0.9 and 0.1, respectively.[60] The activities of the erythrocytic ALA-D isozymes are essentially the same.[60] However, among occupationally exposed lead workers or environmentally exposed children, individuals carrying the ALA-D² allele were found to have a median blood lead level 9–11 µg/dl higher than similarly exposed subjects who were homozygous for the ALA-D¹ allele.[59,62] Thus, it appears that the ALA-D2 subunit binds lead more effectively, and hence that individuals with the ALA-D² allele may be more susceptible to lead poisoning.[59]

4.1.6   Paroxonase Polymorphism

Human paraoxonase exhibits a substrate-dependent polymorphism.[22] One iso-form with arginine at position 192 hydrolyzes paraoxon (the active metabolite of parathion [Figure 1]) rapidly, whereas the other with glutamine at position 192 at a slower rate.[63] This polymorphism is also observed with the oxons of methyl par-athion, chlorthion, and EPN, but not with the oxon of chlorpyrifos. In humans, three genotypes have been identified: individuals homozygous for the low activity allele, individuals homozygous for the high activity allele, and heterozygotes.[64] In popu-lations of Northern European origin, 48% of the individuals are homozygous for the low activity allele, 9.6% are homozygous for the high activity allele, and 43% are heterozygotes. A large variation of enzymatic activity also exists within a genetic class. For example, there is at least a 13-fold variation in activity among the indi-viduals homozygous for the low-activity allele.[64] At present it is not clear whether this difference results from differential rates of synthesis among individuals, different rates of breakdown, or a combination of the two. In any case, the combination of the genetic polymorphism and the variability in expression results in a greater than 60-fold difference observed for paraoxon hydrolysis in humans. Since several animal studies provide convincing evidence that serum paraoxonase activity is an important determinant of sensibility to organophosphate poisoning, it is currently speculated that paraoxonase status may be a useful genetic marker for susceptibility to specific organophosphorus compounds.[41]

4.1.7   Genetic Differences in Glucuronidation

UDP-glucuronosyltransferase activity is reduced in a number of inherited diseases known as familial hyperbilirubinemias. Gilbert's syndrome is a mild form occurring in 2–5% of the population. The patients present chronic unconjugated hyperbilirubinemia and an impaired ability to metabolize menthol, which can thus be used as a probe drug to diagnose the disease. Glucuronidation of many other drugs and toxicants is also impaired.[4] The large prevalence of Gilbert's syndrome in the population makes it an important genetic deficiency when considering interindividual variation in neurotoxicant metabolism. Crigler-Najjar syndrome is a less common and more severe pathological condition. Infants often develop neurological damage from bilirubin encephalopathy (kernicterus) due to the immaturity of the blood-brain barrier. A proportion of patients (Type II) respond favorably to enzymatic inducers such as phenobarbital. The remaining patients (Type I) present higher unconjugated bilirubin levels (>20 mg/dl) and are not responsive to barbiturate therapy, suggesting a difference from Type II in the molecular basis of the genetic defect.[4]

4.2 Physiological Factors

4.2.1 Age

Both Phase I and Phase II reactions take place at a slower rate in neonates, especially those born prematurely. Enzyme levels are low during fetal life and tend to increase rapidly after birth. The isozyme patterns of the fetus and the newborn are also qualitatively different from those observed in adults. In children, on the other hand, many metabolic reactions tend to occur at a faster rate when compared with adults.[65] The relevance of these metabolic differences in environmental medicine is limited. More important are instead the variations that occur during aging. Evidence for a decreased metabolic clearance has been provided for a number of oxidized drugs, while for others oxidation rate shows negligible changes. The response to enzyme inducers may also decrease with age, but evidence for this is conflicting. Aging seems to affect conjugating capacity much less than oxidation capacity.[66,67] As in younger subjects, in the elderly the net result of neurotoxic action depends on the balance between activation and detoxication produced by biotransforming enzymes. If metabolism results in water-soluble, less toxic products, the metabolic impairment occurring during aging will decrease detoxication and excretion of the compound, thereby increasing its neurotoxicity. On the other hand, if biotransformation yields a more toxic metabolite, the neurotoxicity of a compound will be decreased.

4.2.2 Sex

Marked differences between male and female experimental animals in the pharmacologic and toxicologic responses to a number of xenobiotics have been noted. For example, parathion is approximately twice more toxic to female than to male rats. The major sensitivity of females results from a reduced hepatic capacity to detoxify the pesticide as well as other chemicals. Once again, if a metabolite produces the toxic response, then male rats will usually show the greater effects. This gender variability mostly reflects differences in the pattern of liver P450 isoenzymes. Sex differences in extrahepatic metabolism also occur.[1] Despite the large sex-linked variations observed in rats and certain strains of mice, in humans these differences are usually relatively minor and of less importance. However, the biotransformation ability of women can vary during menstrual cycle and by use of oral contraceptives, and differences in pharmacokinetics and biotransformation capacity for a variety of drugs have been reported between men and women.[1] Thus, sex differences in the biological response should always be considered when exposing workers to neurotoxic chemicals.

4.2.3 Pregnancy

The rate of neurotoxicant metabolism may be affected by several factors during pregnancy. These include hormonal changes, a physiological rise in body temperature, the metabolic contribution of the feto-placentar unit, variations in xenobiotic binding to plasma proteins and hemodynamic changes influencing the rate of delivery of the xenobiotic to the metabolic site.[65] Due to liver immaturity, the fetus is severely limited in its ability to biotransform xenobiotics, and the developing brain is exquisitely sensitive to neurotoxic insults.[68] This explains why industrial hygienists usually recommend to avoid any occupational exposure to potential neurotoxic chemicals during gestation.

4.2.4   Nutritional Status and Diet

Animal experiments clearly indicate that nutritional factors can affect the metabolism of foreign compounds by altering the biotransforming enzymes. Mineral (Ca, Cu, Fe, Mg, Zn) or vitamin (C, E, and B complex) deficiencies reduce the rates of xenobiotic biotransformation, mostly by decreasing P450 activities. Moreover, vitamin deficiencies can alter the redox state of the cell, thus hampering the synthesis of the high-energy cofactors required for Phase II reactions. Low-protein diets increase the toxicity of several xenobiotics which are active in their parental form and reduce the toxicity of those that require metabolic activation to express their toxic actions.[1] In rodents, food deprivation can reduce the tissue levels of conjugating agents and coenzymes. An overnight fast reduces hepatic GSH content by approximately 50%, thus potentiating the neurotoxicity of the compounds which are detoxified by this route. Dietary constituents that inhibit certain biotransformation reactions are also known.[1]

In humans, short-term fasting exerts relatively minor effects on xenobiotic metabolism, while malnutrition and protein deficiency prolong the half-lives of several drugs.[69] On the other hand, increasing the ratio of protein to carbohydrate or fat in the diet stimulates xenobiotic metabolism.[70] Certain dietary constituents influence oxidation processes. For example, flavonoids (flavone, tangeretin, nobiletin) found in edible plants stimulate hepatic hydroxylation.[70] Both oxidative and conjugation reactions are stimulated by a diet containing cruciferous vegetables such as Brussel sprouts and cabbage. Such variations involve enzyme induction by naturally-occurring indoles (indole-3-acetonitrile, indole-3-carbinol, 3',3'-diindolylmethane).[69,70] Eating charcoal-broiled meat for several days also accelerates oxidative metabolism but does not affect Phase II processes. This effect is probably due to PAH formed during food combustion, which are known inducers of the P450 system.[69,70] Food also influences the disposition of ingested chemicals whose clearance depends on hepatic blood flow as well as on enzyme activity. This probably occurs as a result of a transient post-prandial increase in hepatic blood flow, shunting the xenobiotic past the liver, thus allowing a greater proportion of the dose to escape metabolism.[71]

4.2.5   Physical Activity

Physical exercise and fitness are host factors with impact on hepatic xenobiotic metabolism, contributing to the individual variation in the biological response. Inhalation exposure during physical activity (a condition often encountered in the workplace) enhances the pulmonary uptake of neurotoxicants by increasing the ventilatory rate.[72] This is not necessarily paralleled by an increase in the rate of their metabolism. Indeed, pharmacokinetic studies indicate that moderate to heavy physical exercise for a few hours reduces liver blood flow leading to a decreased disposition of xenobiotics exhibiting flow-limited metabolism (high clearance drugs).[72,73] Hepatic disposition of xenobiotics exhibiting capacity-limited metabolism (low clearance drugs) is not affected.[72] Improved physical fitness is accompanied by accelerated hepatic biotransformation of low clearance drugs, while the effects on the metabolism of high clearance drugs are variable, depending on the compound examined.[73]

4.2.6 Biological Rhythms

It is well established that in experimental animals the rate of xenobiotic biotransformation differs depending on the time of day at which the exposure to xenobiotics takes place. This variability is generally correlated with variations in endocrine functions as influenced by the light-dark cycle to which the animal is exposed, and is thought to mostly depend on variation in P450s activities. In rodents, concentrations of hepatic GSH also exhibit circadian rhythmicity, with levels being highest at the end of the dark cycle. Thus, during the night (*i.e.,* the feeding period) GSH accumulates, but during the light period it declines.[1] This fluctuation can modify the neurotoxicity of chemicals that are metabolized by GSH conjugation.

Pharmacokinetic studies indicate that, also in humans, several parameters characterizing drug disposition vary according to the time of the day at which treatment is given.[74,75] For the most part, these investigations have been carried out on healthy volunteers utilizing commonly prescribed drugs (e.g., aspirin), although some studies of drugs involving a neurotoxic risk (e.g., lithium, cisplatin) have also been performed on patients requiring specific treatment.[74] The relationship between chronobiology and occupational medicine has attracted the attention of scientists only recently.[76] The potential relevance for shift workers of cyrcadian changes in neurotoxicant metabolism is a fertile ground for future investigations.

4.3 ENVIRONMENTAL FACTORS

4.3.1 General Remarks

Several drugs, industrial contaminants, food additives and recreational substances possess the ability to modify the rate or the capacity of neurotoxicant metabolism. In most cases, these variations are brought about by stimulation (induction) or inhibition of the biotransforming enzymes, and result in an increase or decrease, respectively, in the basal level of enzyme activity. Enzyme induction and inhibition are not mutually excluding and may occur at the same time, thus providing an explanation for the inconsistent and apparently contradictory nature of certain toxicant interactions. Additionally, environmental factors do not exist in isolation, and very little is known on how they interact with each other and with hereditary influences. Thus, they account in large part for the unpredictability of the responses to occupational neurotoxicants in the individual workers.

4.3.2   Enzyme Induction

Several chemicals stimulate the xenobiotic biotransforming enzymes in the liver and/or other metabolic sites. This stimulatory action involves primarily the P450 system, but can also affect other metabolic pathways (e.g., glucuronidation, GSH conjugation). Enzyme induction is reversible: cessation of exposure to the inducing agent results in a return to basal enzymatic activity.[1-4]

Cytochromes P450 induction is characterized by the following features: (a) it can be demonstrated only *in vivo*, with some delay following exposure to the inducing agent; (b) biochemically, it requires *de novo* synthesis of proteins; (c) morphologically, it is associated with proliferation of the smooth endoplasmic reticulum and, in some species, hepatic hypertrophy; (d) different inducers act on different isoenzymes and therefore may have variable effects on the metabolism and toxicity of various substrates.[77,78]

The most widely studied inducing agents are phenobarbital and PAH. The induction caused by phenobarbital is generally more pronounced in the liver than in extrahepatic tissues. The cluster of induced enzymes includes, among others, the subfamilies CYP2B, CYP2C and CYP3A. The spectrum of affected substrates is not limited to xenobiotics, but extends to endogenous compounds, such as steroid hormones and vitamin D3.[78] Despite intensive study, the molecular mechanism(s) responsible for the inducing effect of phenobarbital has not been fully clarified. It appears however that the induction is regulated at the transcriptional level and involves a dramatic increase in the mRNAs encoding specific P450 isoenzymes.[79]

PAH such as benzo(a)pirene and 3-methylcholantrene are environmental pollutants formed by incomplete combustion of organic matter. Besides being formed in food during cooking, they can be found in the exhaust gases generated by vehicle engines, industrial plants, domestic heating appliances, and tobacco smoke.[80] PAH appear to induce selectively few specific forms of cytochromes P450, namely CYP1A1 and CYP1A2. The molecular mechanism responsible for CYP1A1 induction involves PAH interaction with a cytosolic receptor, denoted as the Ah receptor. The PAH-receptor complex then is translocated into the nucleus, where it causes selective transcriptional gene activation resulting in *de novo* synthesis of enzymes. In addition to benzo(a)pirene and 3-methylcholantrene, chemicals which bind to the Ah receptor include 2,3,7,8-tetrachloro-dibenzo-*p*-dioxin (TCDD), benzo(a)anthracene and β-naphtoflavone.[77,78]

Other major classes of inducing agents include steroids and related compounds (testosterone, spironolattone); antibacterials (rifampicin); anticonvulsants (phenytoin, carbamazepine); polychlorinated and polybrominated biphenyls; and halogenated pesticides (DDT, aldrin, hexachlorobenzene, lindane, chlordane). Some of these produce a spectrum of induction similar to phenobarbital or PAH, while others stimulate the synthesis of other P450 isoforms.[1,69]

The solvent methyl ethyl ketone (MEK) is occasionally present in mixtures with *n*-hexane and MnBK. MEK is itself devoid of neurotoxic activity but accelerates the development of hexacarbon toxicity in animal models. This action appears related to the ability of MEK to induce the enzymes involved in the formation of neurotoxic 2,5-HD (Figure 2). It is of interest that phenobarbital not only fails to potentiate *n*-hexane and MnBK neurotoxicity but may actually slow its onset. A possible explanation is that this drug, besides stimulating 2,5-HD production, also increases 2,5-HD detoxification through the oxidative pathway.[5]

4.3.3   Enzyme Inhibition

Inhibition of the xenobiotic metabolizing enzymes may be caused by a variety of mechanisms, such as decreased biosynthesis or increased breakdown of the enzymes, allosteric changes in enzyme conformation, competition for enzyme active sites or cofactors, and loss of functional tissue (e.g., hepatic necrosis). The most common type of inhibition is competition of two different xenobiotics for the substrate binding site of cytochrome P450, resulting in mutual metabolic inhibition. In this case, the degree of inhibition depends on the relative affinities of the substrates for the binding site. Individual P450 isoforms show different sensitivities to the inhibitory action of selected chemicals.[1] Carbon monoxide competes with molecular oxygen (O_2) for the reduced heme moiety, thus inhibiting P450-mediated oxido-reductive metabolism. Other compounds such as secobarbital, halogenated alkanes (carbon tetrachloride), alkenes (vinyl chloride, trichloroethylene), and acetylene act as "suicide" inhibitors. These chemicals are activated by the cytochrome P450 system forming reactive metabolites which bind covalently to the pyrrole nitrogens present in the heme moiety. This interaction results in heme destruction and loss of enzymic activity.[1]

In the clinical setting, enzyme inhibition is probably more important than induction, since it generally only requires a single dose of drug rather than repeated doses as does induction. Inhibition may also be relevant to the neurotoxic effects of substances encountered in the workplace or in the general environment. For example, the solvent dimethylformamide inhibits the metabolism of ethanol and of other industrial alcohols.[3]

4.3.4   Effects of Ethanol

It is well established that ethanol exerts multiple and complex effects upon the metabolism of numerous xenobiotics.[81] This has several implications for occupational medicine since alcoholic beverages are extensively consumed by the working population, occasionally in the course of the workday (for example during lunch time) and particularly after work. The metabolic effects of alcohol consumption are varied and relate to the nature of the exposure. Enzyme inhibition is usually observed after acute ingestion, whereas induction prevails as a consequence of chronic alcohol intake.[69,81] In addition, chronic alcoholism is almost invariably associated with malnutrition and deficiencies of vitamins (thiamine, folate) and minerals (Ca, Se, Zn). Such altered nutritional status may enhance the susceptibility to neurotoxic effects of ethanol itself or of other neurotoxicants, both as a result of nutrient-related impairment of neural functioning and because of variations in xenobiotic biotransformation and excretion.[82]

The main mechanism involved in the acute effects of ethanol on metabolic processes appears to be competition for binding with cytochromes P450. In some cases, the NADH generated by ethanol oxidation may inhibit Krebs cycle activity, thus depleting intermediates required for the generation of NADPH (*i.e.,* the coenzyme of the P450 system).[81] Acute interactions have been described for a number of commonly prescribed drugs, general anesthetics and industrial solvents.[81] For example, after ethanol intake, blood xylene levels were found to increase by about 1.5- to 2-fold, while urinary methylhippuric acid excretion fell by approximately 50%, suggesting that ethanol decreases the metabolic clearance of xylene by about one half during xylene inhalation.[83] These changes may have important practical consequences. Indeed, a proportion of work-related accidents are associated with alcohol consumption. One may wonder to what extent these events may be caused not only by ethanol itself, but also by an interaction with occupational toxicants exerting a depressant action on the central nervous system.

Chronic ethanol intake stimulates hepatic oxidative metabolism, due to induction of CYP2E1 and other microsomal enzymes, but may inhibit certain Phase II reactions such as glucuronidation, methylation and GSH conjugation. Additionally, in chronic alcoholics ethanol causes liver damage, which leads to lowered levels of xenobiotic-metabolizing enzymes.[81,84] Extrahepatic enzymes and cofactors can also be affected. In the rat, under subchronic exposure conditions that mimicked combined exposure of workers consuming moderate amounts of alcoholic beverages, ethanol and styrene induced 23 and 30% depletion of brain GSH, respectively. Concomitant administration of the toxicants caused a cerebral GSH depletion in the order of 60%.[34]

The complexity of the effects of ethanol upon xenobiotic metabolism is also exemplified by a recent epidemic of methanol poisoning which occurred in Italy due to consumption of methanol-adulterated wine. Some features of the toxic syndrome differed from the typical effects observed in acute poisoning with pure methanol. In particular, the latency of toxic manifestations was prolonged (two–three days) as was the methanol half-life (up to 30–35 hours), probably reflecting inhibition by ethanol of methanol oxidation to formic acid (*i.e.,* the neurotoxic metabolite) (Figure 4).[21] Moreover, a mild or even asymptomatic course was observed in some patients despite consumption of large quantities of methanol-wine. Finally, the toxic response exhibited large individual variability with most fatalities and cases of visual loss and blindness involving chronic alcoholic patients.[82] The severity of the intoxication in alcoholics was probably related to the folate deficiency which is a well known complication of alcohol abuse. As shown in Figure 4, the formic acid produced by oxidative metabolism of methanol is subsequently degradated to CO_2 by a series of enzymatic reactions involving a folate-dependent pathway.[82]

4.3.5   Effects of Smoking

Tobacco smoke is an extremely complex mixture containing hundreds of different chemicals, whose effects on the metabolism of neurotoxicants are largely unknown. In humans, cigarette smoke has been reported to induce the oxidation of several substrates, but not to affect significantly the metabolism of others.[69,84] Inducing effects of marijuana smoke have also been described.[69] Enzyme induction by smoking can be partly ascribed to nicotine and to the PAH generated by tobacco combustion, yet other constituents of smoke probably play an important role. Indeed, in experimental animals PAH accelerate markedly the biotransformation of benzo(a)pirene and other substrates both in the liver and in extrahepatic tissues, whereas the inducing effect of cigarette smoke is much more prominent at extrahepatic sites such as the placenta. Moreover, the spectrum of enzymatic activities induced by cigarette smoking is much wider in the human placenta than in the human liver.[84] These findings illustrate the importance of exploring factors affecting neurotoxicant metabolism at extrahepatic sites.

5 CONCLUSIONS

After absorption into the human organism, solvents, metals, pesticides, and other occupational neurotoxicants are usually metabolized into more hydrosoluble compounds, thus facilitating excretion from the body. Biotransformation processes include oxido-reductive (Phase I) and conjugative (Phase II) reactions, which mostly take place in the liver and kidney. Metabolism of occupational and environmental contaminants plays a crucial role in determining their neurotoxicity, since it may result either in detoxication or in the production of active metabolites. Phase I reactions are mostly catalyzed by the microsomal cytochrome P450 system, which is responsible for the metabolic activation of numerous neurotoxicants such as organophosphates, n-hexane, MnBK, and carbon disulfide. Phase II reactions involve conjugation with endogenous substrates, and usually lead to detoxication. Xenobiotic biotransforming enzymes (e.g., MAO-B, P450s, FMO, epoxide hydrolase, GSH-S-transferases) are also present in the brain, where they modulate the local effects of several neurotoxic chemicals (e.g., MPTP, methylmercury, styrene).

Marked inter- and intraindividual differences exist in the activity of the metabolizing enzymes and, therefore, in the responses to neurotoxic exposures. Such variability is in large part genetically determined (gene defects and polymorphisms), and partly due to physiological and environmental factors. Genetic polymorphisms have been identified for several important enzymatic systems such as the CYP2D subfamily of cytochromes P450, NAT2, ADH, ALDH, ALA-D, and paraoxonase. Subjects with familial hyperbilirubinemias (Gilbert's syndrome, Crigler Najjar syndrome) present an impaired glucuronidation of several substrates. Physiological variables affecting neurotoxicant biotransformation include age, gender, pregnancy, nutritional status, physical activity, and circadian rhythms. Metabolic variations induced by exogenous chemicals (e.g., drugs, ethanol, smoking, environmental pollutants) are mostly related to enzyme induction and/or inhibition phenomena.

Since human susceptibility to the neurotoxicity of a given chemical relies upon the balance between metabolic detoxication and activation, the complex and interacting influences on xenobiotic metabolism must be adequately considered by hygienists, clinicians, and regulators. Knowledge of neurotoxicant metabolic pathways, and their genetic, physiological and environmental determinants, is crucial to understand individual responses, as well as to develop and validate biomarkers of exposure, effect and individual susceptibility.

6 REFERENCES

1. Sipes, I. G., and Gandolfi, A. J., "Biotransformation of toxicants," in *Casarett and Doull's Toxicology. The Basic Science of Poisons, Fourth Edition,* Amdur, M. O., Douli, J. and Klassen, C. D., Eds., Pergamon Press, New York, 1991, Chapter 4.
2. Abou-Donia, M. B., "Disposition, metabolism, and toxicokinetics," in *Neurotoxicology,* Abou-Donia, M. B., Ed., CRC Press, Boca Raton, FL, 1992, Chapter 1.
3. Timbrell, J. A., "Biotransformation of xenobiotics," in *General & Applied Toxicology,* Ballantyne, B., Marrs, T., and Turner, P., Eds., Macmillan Press, Basingstoke, UK, 1993, Vol. I, Chapter 4.
4. deBethizy, J. D., and Hayes, J. R., "Metabolism. A determinant of toxicity," in *Principles and Methods of Toxicology, Third Edition,* Hayes, A. W., Ed., Raven Press, New York, 1994, Chapter 3.
5. Perucca, E., and Manzo, L., "Metabolic activation of neurotoxicants," in *Recent Advances in Nervous System Toxicology,* Galli, C. L., Manzo, L., and Spencer, P. S., Eds., Plenum Press, New York, 1988, 67.
6. Ravindranath, V., and Boyd, M. R., "Xenobiotic metabolism in brain," *Drug Metab. Rev.,* 27, 419, 1995.
7. Di Monte, D. A., "Potential role of glial cells in Parkinson's disease," *Neurotoxicology,* 16, 536, 1995.
8. Gonzales, F. J., and Idle, J. R., "Pharmacogenetic phenotyping and genotyping. Present status and future potential," *Clin. Pharmacokinet.,* 26, 59, 1994.
9. Costa, L. G., "Organophosphorus compounds," in *Recent Advances in Nervous System Toxicology,* Galli, C. L., Manzo, L., and Spencer, P. S., Eds., Plenum Press, New York, 1988, 203.
10. Sultatos, L. G., "Mammalian toxicology of organophosphorus pesticides," *J. Toxicol. Environ. Hlth.,* 43, 271, 1994.
11. Chambers, J. E., "The role of target site activation of phosphorothionates in acute toxicity," in *Organophosphates: Chemistry, Fate and Effects,* Chambers, J. E., and Levi, P. E., Eds., Academic Press, New York, 1992, 229.
12. Couri, D., and Milks, M., "Toxicity and metabolism of the neurotoxic hexacarbons *n*-hexane, 2-hexanone, and 2,5-hexanedione," *Annu. Rev. Pharmacol. Toxicol.,* 22, 145, 1982.
13. Krasavage, W. J., O'Donoghue, J. L., DiVincenzo, G. D., and Terhaar, C. J., "The relative neurotoxicity of methyl *n*-buthyl ketone, *n*-hexane and their metabolites," *Toxicol. Appl. Pharmacol.,* 52, 433, 1980.
14. Spencer, P. S., Bischoff, M. C., and Schaumburg, H. H., "On the specific molecular configuration of neurotoxic aliphatic hexacarbon compounds causing central peripheral distal axonopathy," *Toxicol. Appl. Pharmacol.,* 44, 17, 1978.
15. Bus, J., "The relationship of carbon disulfide metabolism to development of toxicity," *Neurotoxicology,* 4, 73, 1985.
16. Lukás, E., "Eight years of experience with experimental CS_2 polyneuropathy in rats," *G. Ital. Med. Lav.,* 1, 7, 1979.
17. Farin, F. M., and Omiecinski, C. J., "Regiospecific expression of cytochrome P450s and microsomal epoxide hydrolase in human brain tissue," *J. Toxicol. Environ. Hlth.,* 40, 317, 1993.
18. Ravindranath, V., Anandatheerthavarada, H. K., and Shankar, S. K., "Xenobiotic metabolism in human brain. Presence of cytochrome P450 and associated monooxygenase," *Brain Res.,* 496, 331, 1989.

19. Ghersi-Egea, J.-F., Perrin, R., Muller, B. L., Grassiot, M.-C., Jeandel, C., Floquet, J., Cuny, G., Siest, G., and Minn, A., "Subcellular localization of cytochrome P450, and activities of several enzymes responsible for drug metabolism in the human brain," *Biochem. Pharmacol.*, 45, 647, 1993.

20. Bahmre, S., Bhagwat, S. V., Shankar, S. K., Boyd, M. R., and Ravindranath, V., "Flavin-containing monooxygenase mediated metabolism of psychoactive drugs by human brain microsomes," *Brain Res.*, 672, 271, 1995.

21. Garner, C. D., Lee, E. W., Terzo, T. S., and Louis-Ferdinand, R. T., "Role of retinal metabolism in methanol-induced retinal toxicity," *Toxicol. Environ. Hlth.*, 44, 43, 1995.

22. Jacobsen, D., and McMartin, K. E., "Methanol and ethylene glycol poisoning: mechanism of toxicity, clinical course, diagnosis and treatment," *Med. Toxicol.*, 1, 309, 1986.

23. Geldmacher-von Mallinckrodt, M., and Diebgen, T. L., "The human serum paraoxonase. Polymorphism and specificity," *Toxicol. Environ. Chem.*, 18, 79, 1988.

24. Tephley, T. R., "Isolation and purification of UDP-glucuronosyltransferase," *Chem. Res. Toxicol.*, 3, 509, 1990.

25. Price-Evans, D. A., "*N*-acetyltransferase," *Pharmacol. Ther.*, 42, 157, 1984.

26. Manzo, L., Costa, L. G., Tonini, M., Minoia, C., and Sabbioni, E., "Metabolic studies as a basis for the interpretation of metal toxicity," *Toxicol. Lett.*, 64/65, 677, 1992.

27. Davis, L. E., Kornfeld, M., Mooney, H. S., Fiedler, K. J., Haaland, K. Y., Orrison, W. W., Cernichiari, E., and Clarkson, T. W., "Methylmercury poisoning: Long-term clinical, radiological, toxicological, and pathological studies of an affected family," *Ann. Neurol.*, 35, 680, 1994.

28. Vahter, M. E., Mottet, N. K., Friberg, L. T., Lind, S. B., Charleston, J. S., and Burbacher, T. M., "Demethylation of methyl mercury in different brain sites of *Macaca fascicularis* monkeys during long-term subclinical methyl mercury exposure," *Toxicol. Appl. Pharmacol.*, 134, 273, 1995.

29. Van Bladeren, P. J., den Besten, C., Bruggeman, I. M., Mertens, J. J. W. M., van Ommen B., Spenkelink, B., Rutten, A. L. M., Temmink, J. H. M., and Vos, R. M. E., "Glutathione conjugation as a toxication reaction," in *Metabolism of Xenobiotics*, Gorrod, J. W., Oelschlager, H., and Caldwell, J., Eds., Taylor and Francis, London, 1988, 267.

30. Costa, L. G., "Effect of neurotoxicants on brain neurochemistry," in *Neurotoxicology*, Tilson, H., and Mitchell, C., Eds., Raven Press, New York, 1992, Chapter 6.

31. Shukla, G. S., Srivastava, R. S., and Chandra, S. V., "Glutathione status and cadmium neurotoxicity: studies in discrete brain regions of growing rats," *Fundam. Appl. Toxicol.*, 11, 229, 1988.

32. Liccione, J. J., and Maines, M. D., "Selective vulnerability of glutathione metabolism and cellular defense mechanisms in rat striatum to manganese," *J. Pharmacol. Exp. Ther.*, 247, 156, 1988.

33. Bonnefoi, M. S., "Mitochondrial glutathione and methyl iodide-induced neurotoxicity in primary neural cell cultures," *Neurotoxicology*, 13, 401, 1992.

34. Coccini, T., Di Nucci, A., Tonini, M., Maestri, L., Costa, L. G., Liuzzi, M., and Manzo, L., "Effects of ethanol administration on cerebral non-protein sulfhydryl content in rats exposed to styrene vapour," *Toxicology*, 106, 115, 1996.

35. Trenga, C. A., Kunkel, D. D., Eaton, L. D., and Costa, L. G., "Effect of styrene oxide on rat brain glutathione," *Neurotoxicology*, 12, 165, 1991.

36. Daly, A. K., Cholerton, S., Armstrong, M., and Idle, J. R., "Genotyping for polymorphism in xenobiotic metabolism as a predictor of disease susceptibility," *Environ. Hlth. Perspect.*, 102 (suppl. 9), 55, 1994.

37. Motulsky, A. G., "Drug reactions, enzymes and biochemical genetics," *JAMA*, 165, 835, 1957.

38. May, D. G., "Genetic differences in drug disposition," *J. Clin. Pharmacol.*, 34, 881, 1994.

39. Propping, P., "Genetic aspects of neurotoxicity," in *Neurotoxicology*, Blum, K., and Manzo, L., Eds., Marcel Dekker, New York, 1985, Chapter 9.

40. Barret, J. C., Vainio, H., Peakall, D., and Goldstein, B. D., "12th Meeting of the Scientific Group on methodologies for the safety evaluation of chemicals: susceptibility to environmental hazards," *Environ. Hlth. Prospect.*, 105 (suppl. 4), 699, 1997.

41. Costa, L. G., and Manzo, L., "Biochemical markers of neurotoxicity: research strategies and epidemiological applications," *Toxicol. Lett.*, 77, 137, 1995.

42. Bolt, H. M., "Genetic disposition in occupational toxicology," *Toxicol. Lett.*, 78 (suppl. 1), 4, 1995.

43. Mahgoub, A., Dring, L. G., Idle, J. R., Lancaster, R., and Smith, R. L., "Polymorphic hydroxylation of debrisoquine in man," *Lancet*, 1, 584, 1977.

44. Eichelbaum, M., Spannbrucker, N., Steincke, B., and Dengler, H. J., "Defective *N*-oxidation of sparteine in man: a new pharmacogenetic defect," *Eur. J. Clin. Pharmacol.*, 16, 183, 1979.

45. Gonzales, F. J., Vilbois, F., Hardwick, J. P., McBride, O. W., Nebert, D. W., Gelboin, H. V., and Meyer, U. A., "Human debrisoquine 4-hydroxylase (P450IID1):cDNA and deduced amino acid sequence and assignment of the CYP2D locus to chromosome 22," *Genomics*, 2, 174, 1988.

46. Evans, D. A. P., Mahgoub, A., Sloan, T. P., Idle, J. R., and Smith, R. L., "Family and population study of genetic polymorphism of debrisoquine oxidation in a White British population," *J. Med. Genet.*, 17, 102, 1980.

47. Steiner, E., Bertilsson, L., Sawe, J., Bertling, I., and Sjoqvist, F., "Polymorphic debrisoquine hydroxylation in 757 Swedish subjects," *Clin. Pharmacol. Ther.*, 44, 431, 1988.

48. Armstrong, M., Daly, A. K., Cholerton, S., Bateman, D. N., and Idle, J. R., "Mutant debrisoquine hydroxylation genes in Parkinson's disease," *Lancet*, 339, 1017, 1992.

49. Smith, C. A. D., Gough, A. C., Leigh, P. N., Summers, B. A., Harding, A. E., Marangnore, D. N., Sturman, S. G., Schapira, A. H. V., Williams, A. C., Spurr, N. K., and Wolf, C. R., "Debrisoquine hydroxylase gene polymorphism and susceptibility to Parkinson's disease," *Lancet*, 339, 1375, 1992.

50. Grant, D. M., Blum M., Beer, M., and Meyer, U. A., "Monomorphic and polymorphic human arylamine N-acetyltransferase: a comparison of liver isozymes and expressed products of two cloned genes," *Mol. Pharmacol.*, 39, 184, 1990.

51. Blum, M., Grant, D. M., McBride, W., Heim, M., and Meyer, U.A., "Human arylamine *N*-acetyltransferase genes: isolation, chromosomal location and function expression," *DNA Cell Biol.*, 9, 193, 1990.

52. Hickman, D., and Sim, E., "*N*-acetyltransferase polymorphism comparison of phenotype and genotype in humans," *Biochem. Pharmacol.*, 42, 1007, 1991.

53. Horai, Y., and Ishizaki, T., "*N*-Acetylation polymorphism of dapsone in a Japanese population," *Br. J. Clin. Pharmacol.*, 25, 487, 1988.

54. Bosron, W. F., and Li, T.-K., "Catalytic properties of human liver alcohol dehydrogenase isoenzymes," *Enzyme*, 37, 19, 1987.

55. Tsukahara, M., and Yoshida, A., "Chromosomal assignment of the alcohol dehydrogenase cluster locus to human chromosome 4q21-23 by *in situ* hybridization," *Genomics*, 4, 218, 1989.

56. Goedde, H. W., Agarwal, D. P., "Pharmacogenetics of aldehyde dehydrogenase (ALDH)," *Pharmacol. Ther.*, 45, 345, 1990.

57. Hsu, L. C., Yoshida, A., and Mohandas, T., "Chromosomal assignment of the genes for human aldehyde dehydrogenase-1 and aldehyde dehydrogenase-2," *Am. J. Hum. Genet.*, 38, 641, 1986.

58. Masci, O., Sannolo, N., and Castellino, N., "Biological monitoring," in *Inorganic Lead Exposure. Metabolism and Intoxication*, Castellino, N., Castellino, P., and Sannolo, N., Eds., CRC Press, Boca Raton, FL, 1995, Chapter 8.

59. Wetmur, J. G., Lehnert, G., and Desnick, R. J., "The ∂-aminolevulinate dehydratase polymorphism: higher blood lead levels in lead workers and environmentally exposed children with the 1-2 and 2-2 isozymes," *Environ. Res.*, 56, 109, 1991.

60. Battistuzzi, G., Petrucci, R., Silvagni, L., Urbani, F. R., and Caiola, S., '∂-Aminolevulinate dehydrase: a new genetic polymorphism in man," *Ann. Hum. Gen.*, 45, 223, 1981.

61. Potluri, V. R., Astrin, K. H., Wetmur, J. G., Bishop, D. F., and Desnick, R. J., "Human ∂-aminolevulinate dehydratase: chromosomal localization to 9q34 by *in situ* hybridization," *Hum. Genet.*, 76, 236, 1987.

62. Ziemsen, B., Angerer, J., Lehnert, G., Benkmann, H.-G., and Goedde, H. W., "Polymorphism of delta-aminolevulinic acid dehydratase in lead-exposed workers," *Int. Arch. Occup. Environ. Hlth.*, 58, 245, 1986.

63. Humbert, R., Adler, D. A., Disteche, C. M., Hassett, C., Omiecinski, C. J., and Furlong, C. E., "The molecular basis of the human serum paraoxonase activity polymorphism," *Nat. Genet.*, 3, 73, 1993.

64. Furlong, C. E., Richter, R. J., Seidel, S. O., Costa, L. G., and Motulsky, A. G., "Spectrophotometric assay for the enzymatic hydrolysis of the active metabolites of chlorpyrifos and parathion by plasma paraoxonase/arylesterase," *Anal. Biochem.*, 180, 242, 1989.

65. Perucca, E., "Drug metabolism in pregnancy, infancy and childhood," *Pharmacol. Ther.*, 34, 129, 1987.

66. Loi, C.-M., and Vestal, R. E., "Drug metabolism in the elderly," *Pharmacol. Ther.*, 36, 131, 1988.

67. Dawling, S., and Crome, P., "Clinical pharmacokinetic considerations in the elderly: An update," *Clin. Pharmacokin.*, 17, 236, 1989.

68. Nelson, B. K., "Developmental neurotoxicology of environmental and industrial agents," in *Neurotoxicology*, Blum, K., and Manzo L., Eds., Marcel Dekker, New York, 1985, Chapter 8.

69. Mucklow, J. C., "Environmental factors affecting drug metabolism," *Pharmac. Ther.*, 36, 105, 1988.

70. Conney, A. H., Buening, M. K., Pantuck, C. B., Fortner, J. G., Anderson, K. E., and Kappas, A., "Regulation of human drug metabolism by dietary factors," *Ciba Found. Symp.*, 76, 147, 1980.

71. Melander, A., and McLean, A., "Influence of food intake on presystemic clearance of drugs," *Clin. Pharmacokin.*, 8, 286, 1983.

72. Ylitalo, P., "Effect of exercise on pharmacokinetics," *Ann. Med.*, 23, 289, 1991.

73. Døssing, M., "Effect of acute and chronic exercise on hepatic drug metabolism," *Clin. Pharmacokin.*, 10, 426, 1985.

74. Reinberg, A., and Smolensky, M. H., "Circadian changes of drug disposition in man," *Clin. Pharmacokin.*, 7, 401, 1982.

75. Marks, V., English, J., Aherne, W., and Arendt, J., "Chronopharmacology," *Clin. Biochem.*, 18, 154, 1985.

76. LaDou, J., "Rhythmic variations in medical monitoring tests," *Occup. Med.*, 5, 479, 1990.

77. Netter, K. J., "Mechanisms of monooxygenase induction and inhibition," *Pharmacol. Ther.*, 10, 515, 1980.

78. Okey, A. B., "Enzyme induction in the cytochrome P-450 system," *Pharmacol. Ther.*, 1990, 45, 241.

79. Pike, S. F., Shepard, E. A., Rabin, B. R., and Phillips, I. R., "Induction of cytochrome P-450 by phenobarbital is mediated at the level of transcription," *Biochem. Pharmacol.*, 34, 2489, 1985.

80. Manzo, L., and Weetman, D. F., Eds., *Toxicology of Combustion Products*, Fondazione Clinica del Lavoro, Pavia. Italy, 1992.

81. Lieber, C. S., "Mechanisms of ethanol-drug-nutrition interactions," *J. Toxicol. Clin. Toxicol.*, 32, 631, 1994.

82. Manzo, L., Locatelli C., Candura, S. M., and Costa, L. G., "Nutrition and alcohol neurotoxicity," *Neurotoxicology*, 15, 555, 1994.

83. Riihimaki, V., Savolainen K., Pfaffli, P., Pekari, K., Sippel, H. W., Laine, A., "Metabolic interaction between m-xylene and ethanol," *Arch. Toxicol.*, 49, 253, 1982.

84. Pelkonen, O., and Sotaniemi, E. A., "Environmental factors of enzyme induction and inhibition," *Pharmac. Ther.*, 33, 115, 1987.

3 Biological Monitoring of Occupational Neurotoxicants

P. Hoet and R. Lauwerys

CONTENTS

0-8493-9231-4/98/$0.00+$.50
© 1998 by CRC Press LLC

The objective of this chapter is to summarize the principal biological methods currently used for assessing exposure to some neurotoxic industrial chemicals.

1 ALUMINIUM

For the general population, food constitutes the main source of exposure to aluminium. Gastrointestinal absorption is low (<0.1%); however, simultaneous intake of citric acid increases the absorption considerably. In the occupational setting, some absorption may occur following inhalation of high concentrations of aluminium dust[18] but there is no data on the rate of pulmonary absorption of aluminium. Absorbed aluminium is mainly excreted via the kidneys. Kinetics studies suggest the existence of two urinary excretion phases; a relatively fast one (half-life of about eight hours) and a slower one (half-life of several years).[62,89,90] Actually, a fraction of absorbed aluminium is probably retained and stored in several compartments of the body (particularly the skeleton) and excreted from these compartments at differents rates over many years.[27,58] Patients with impaired renal function are at risk of aluminium accumulation in the body. High tissue concentrations of aluminium (bone, liver, brain) have been found in patients suffering from renal insufficiency.

1.1 ALUMINIUM IN BLOOD AND URINE

Because of the very low level of the element in biological fluids and its ubiquitousness, sample contamination is a major analytical problem and the normal concentrations reported in the literature vary greatly.[54] Mean serum and urinary concentrations in the general population are generally below 1 µg/100 mL and 20 µg/l, respectively.

In order to protect dialysis patients, the CEC has recommended that a level of 20 µg/100 mL plasma should never be exceeded, a level of >10 µg/100 mL should lead to an increased monitoring frequency and health surveillance and a concentration of >6 µg/100 mL should be considered as an excessive build-up of the aluminium body burden.[19]

In subjects with normal renal function, aluminium concentration in urine is a better indicator of aluminium exposure than its serum concentration.[4,62,87,99] The *Deutsche Forschungsgemeinshaft* has adopted 200 µg/l urine (end-of-shift) as Biological Tolerance Value (BTV) for workers exposed to aluminium.[24]

2 ARSENIC

Arsenic and its compounds may enter the human organism through inhalation or ingestion, the rate of absorption being highly dependent on the solubility of the compound and probably also on the valence state of arsenic.[34,54] Some compounds can be absorbed through the skin.[33]

Absorbed arsenic is rapidly excreted via the kidney either unchanged or after detoxication by methylation into monomethylarsonic acid (MMA) and dimethylarsinic or cacodylic acid (DMA). A significant fraction is also excreted via the bile.[59]

In the general population mean serum and urine levels of total arsenic depend on the level of the seafood content in the diet as well as on the arsenic content in drinking water. Urinary levels varying from less than 10 µg/l to several hundred µg/l have been reported. A single crustaceans meal may lead to a urinary excretion of more than 1000 µg/l urine.[66] Various marine organisms may effectively contain very high concentrations of organoarsenicals (e.g., arsenobetaine) of negligible toxicity which are also rapidly excreted via the kidney.[14]

2.1 ARSENIC IN URINE

Specific determination of inorganic arsenic (Asi), monomethylarsonic acid (MMA) and cacodylic acid (DMA) in urine is the method of choice to assess exposure to inorganic arsenic compounds. The sum of these three metabolites reflects the amount of inorganic arsenic recently absorbed. However, since seafood might still influence the excretion rate of DMA, the workers should preferably refrain from eating seafood during 48 hours prior to urine collection.[15] In persons nonoccupationally exposed to inorganic arsenic and who have not recently eaten seafood, the sum of Asi, MMA and DMA generally is less than 10 µg/g creatinine. Values around 50 µg/l or even higher have been reported in Japan.[98,115] It has been estimated that in the absence of seafood consumption a time-weighted-average exposure to 10 and 50 µg/m³ inorganic arsenic leads to mean urinary excretions of the sum of As metabolites (Asi, MMA, DMA) in postshift urine sample of 30 and 50 µg/g creatinine, respectively.[68]

The ACGIH recommends a Biological Exposure Index (BEI) for arsenic and soluble compounds including arsine of 50 µg/g creatinine (inorganic arsenic and its methylated metabolites, end of workweek).[3]

2.2 ARSENIC IN HAIR AND NAILS

Arsenic concentration in hair and nails is a good indicator of the amount of inorganic arsenic absorbed during their growth period. Organic marine arsenic does not appear to be taken up in hair and nails to the same degree as inorganic arsenic. Since there is no method currently available to distinguish between endogenous arsenic and arsenic externally adsorbed on the hair, this analysis is not recommended when air is contaminated by arsenic. Arsenic level in hair is usually below 1 mg/kg. Nails are also subjected to external contamination, however toenails may be less affected than fingernails.[94]

3 LEAD

Lead is absorbed via the lung and the gastrointestinal tract. About 35 to 50% of lead deposited in the lower respiratory tract is absorbed, whereas usually less than 10%

of ingested lead passes into the systemic circulation. Many physical (particle size, solubility, concentration, etc.) and biological (age, sex, iron status, fasting state) factors influence the rate of absorption. It has been observed that some lead compounds such as lead naphthenate, lead nitrate solution or even finely powdered lead metal can be slightly absorbed through the skin.[39,57,106] In blood, lead is mainly bound to erythrocytes (about 99%); less than 1% is in plasma either bound to proteins or in diffusible form. The lead body burden consists essentially of three compartments: (1) the blood mass (2%) and some rapidly exchanging tissues (half-life of about 35 days), (2) soft tissues (half-life of about 40 days), and (3) the bones (90% of the body burden) including two different pools (half-life of more than 10 years). Absorbed lead is eliminated mainly via the kidney (75%) but also via the bile, gastrointestinal secretions, hair, nails, and sweat. Lead crosses the blood/brain and the placental barriers and is excreted in milk in breastfeeding mothers.

3.1   Biological Tests Reflecting the Exposure and/or the Amount Stored in Tissue

Lead in blood   After the beginning of exposure, blood lead level rises progressively. In a steady state situation and after at least one month of exposure, lead in blood is considered to be the best indicator of recent exposure although it may also be influenced by the amount of lead mobilized from soft tissues. Lead in blood does not necessarily correlate with the total body burden.

Blood lead level in the general population is generally below 15 µg/100 mL and in some countries is even lower than 10 µg/100 mL due to the progressive reduction of lead in gasoline. In adults, slower peripheral nerve conduction velocity and perturbations of the central nervous system (neuropsychological effects) have been reported to occur at mean blood lead levels of as low as 40–50 µg/100 mL or even lower. In some studies, however, higher blood lead levels were not associated with detectable disturbances of the nervous system. Acute encephalopathy is observed at levels above 100 µg/100 mL.

Children are more sensitive to the toxic action of lead than adults and blood levels exceeding 20 µg/100 mL can be associated with adverse effects on the central nervous system. The U.S. Public Health Service and the Centers for Disease Control designate 10 µg/100 mL as the maximum permissible concentration of blood lead to protect the health of children, and 20 µg/100 mL as the level for medical intervention.

According to OSHA, workers should be removed from exposure when blood lead reaches 50 µg/100 mL or when three consecutive monthly measurements show values exceeding 40 µg/100 mL.

Lead in urine   After the beginning of lead exposure, urinary lead level increases gradually and relatively faster than blood lead level. Lead in urine reflects the amount of lead recently absorbed. It has the advantage that the sampling is not invasive. However, there is considerable interindividual variation in the urinary lead excretion at a certain blood lead level. Contamination of urine during sampling is also more frequent than blood. Urinary lead level in the general population is usually lower than 50 µg/g creatinine.

Lead in urine after administration of a chelating agent   The amount of lead excreted in urine after administration of a chelating agent (e.g., CaEDTA, dimercaptosuccinic acid) reflects the mobilizable pool of lead (i.e., mainly lead stored in soft tissues and trabecular bone). In the general population, the amount of lead excreted in urine after administration of intravenous CaEDTA (1 g) or dimercaptosuccinic acid (20–30 mg/kg) by oral route does not exceed 600 µg/24 hours.

Lead in bone   Measurement of lead in bone (finger, tibia), performed with an X-ray fluorescent technique, seems to be a promising method to assess the lead body burden. This method is still too sophisticated for the routine surveillance of workers exposed to lead.[55]

3.2   MAIN BIOLOGICAL TESTS REFLECTING THE EARLY BIOLOGICAL EFFECTS OF LEAD RELATED TO THE INTERNAL DOSE

δ-aminolevulinic acid in urine (ALAU)   The earliest biochemical effect of lead on heme synthesis is a decrease of the δ-aminolevulinic dehydratase activity which occurs at blood lead level <15 µg/100 mL. Because of this inhibition, delta aminolevulinic acid accumulates and is excreted in greater amounts. ALAU starts to increase at blood lead level of about 40 µg/100 mL. This increase occurs within two weeks after the start of the exposure and declines rapidly after withdrawal of exposure (15 days).

Zinc protoporphyrin in blood (ZPP)   Another effect of lead on heme synthesis is the inhibition of the ferrochelatase enzyme leading to an accumulation of protoporphyrin in erythrocytes. It has been demonstrated that the protoporphyrin that accumulates in erythrocytes is not free but mainly exists as zinc protoporphyrin. In adults, ZPP starts to increase at blood lead levels of 25–35 µg/100 mL. Since the accumulation of ZPP in erythrocytes results from the action of lead in the bone marrow, and since the average life span of erythrocytes is 120 days, there is a lag time between the increase of the erythrocyte ZPP and the rise of lead in blood. In case of high lead body burden, ZPP can remain elevated for years after the end of exposure due to inhibition of heme synthesis by lead released from the bones. The level of ZPP in the general population is generally less than 40 µg/100 mL blood. Increased erythrocyte ZPP can also occur in iron deficiency or sickle cell anemia.

Other biomarkers of exposure to lead   Other bioindicators of lead exposure include erithrocyte δ-aminolevulinic acid dehydratase, pyrimidine-5′-nucleotidase and urinary coproporphyrins.

δ-Aminolevulinic acid dehydratase in red blood cells is too sensitive to the inhibitory action of lead to be used as routine biomarker of occupational exposure. δ-aminolevulinic dehydratase activity may also be decreased in porphyria, liver cirrhosis, alcoholism.

TABLE 1
Recommendations formulated by Agencies
for Workers Occupationally Exposed to Lead

	Parameter	Biological Limit Value	Sampling Time
	ACGI H[3]		
TLV 0.15 mg/m³	lead in blood	50 µg/100 mL	not critical
→ 0.05 mg/m³*		→ 30 µg/100 mL*	
	lead in urine	150 µg/g creat	not critical
	zinc protoporphyrin in blood	250 µg/100 mL GR or 100 µg/100 mL blood	after 1 month exposure
	DFG[24]		
MAK 0.1 mg/m³	lead in blood	700 µg/l 300 µg/l (♀<45 y)	not fixed
	δ-aminolevulinic acid in urine	15 mg/l 6 mg/l (♀<45 y)	not fixed
	WHO[109]		
	lead in blood	40 µg/100 mL ♂ 30 µg/100 mL (♀ child-bearing age)	

* Notice of intended changes

Several studies have shown that erythrocyte pyrimidine-5′-nucleotidase is as sensitive to the inhibitory action of lead as δ-aminolevulinic acid dehydratase.

The urinary excretion of coproporphyrins starts to increase when blood lead concentration reaches a value of about 40 µg/100 mL. Coproporphyrinuria may also occur in hepatitis, cirrhosis, hemolytic anemia, malign hemopathies, infectious diseases, various intoxications, and even after alcohol consumption. In non-occupationally exposed subjects, urinary levels of copropophyrins usually are below 100 µg/g creatinine.

Table 1 summarizes the recommendations formulated by several agencies. It is important to underline that these critical levels might not be sufficiently protective for the fetus.

4   MANGANESE

In industry, manganese is mainly absorbed by the pulmonary route. The gastrointestinal absorption is rather low and is probably regulated by a homeostatic mechanism. Excretion mainly occurs through the bile, only less than 1% of the absorbed dose being eliminated via urine.

Normal concentrations of manganese in urine, blood and serum or plasma are usually less than 2 µg/g creatinine, 1.5 µg/100 mL and 0.1 µg/100 mL, respectively.

The possibility for monitoring exposure to manganese by a biological method is still limited. Most studies have concluded that there is no direct relation between manganese concentration in biological material and the intensity of exposure or the severity of chronic manganese poisoning.

5 MERCURY

5.1 INORGANIC MERCURY

Approximately 80% of inhaled metallic mercury vapor is absorbed. Gastrointestinal absorption of metallic mercury is negligible but can range from 2 to >30% for some mercury salts. In blood, inorganic mercury is equally distributed between plasma and red blood cells. The principal site of deposition of inorganic metallic salts is the kidney. In case of exposure to mercury vapor, mercury not only accumulates in the kidneys but also in the brain and crosses the placenta. Inorganic mercury is eliminated via the urine and feces. Small quantities are also excreted through saliva and perspiration. The biological half-life of mercury is about two months in the kidney but is much longer in the central nervous system.[112]

The urinary and blood mercury levels observed in the general population are generally below 5 µg/g creatinine and 1 µg/100 mL, respectively. These levels are influenced by fish consumption and the presence of amalgam fillings.

Mercury in urine   Urine is the most commonly used media for assessing occupational exposure to metallic mercury. The rate of urinary excretion probably reflects the amount of mercury in the kidneys. The high variability observed in the mercury concentration of urine samples taken at different times from the same subject can be reduced by correcting the concentration for specific gravity or creatinine.

Many studies have examined the relationship between urine mercury levels and neurologic effects. The critical effect levels of mercury in urine and blood are generally considered to be about 50 µg/g creatinine and 2 µg/100 mL, respectively. However, recent studies seem to indicate that some effects on the central nervous system and the kidneys might already occur at urinary levels below 50 µg/g creatinine, but their health significance remains to be established. Clinical signs of poisoning usually do not occur when urinary mercury concentration is kept below 300 µg/l. According to WHO,[112] when exposure is above 80 µg/m³ corresponding to a urinary mercury level of approximately 100 µg/g creatinine, the probability of developing neurological signs and proteinuria is high. Exposure in the range of 25 to 80 µg/m³ corresponding to a urinary level of 30 to 100 µg/g creatinine increases the risk of mild toxic effects such as psychomotor disturbances, tremor, fatigue, irritability, in the absence of overt clinical impairment.

Mercury in blood   Blood concentration of mercury is mainly influenced by exposure to mercury during recent days. Suspicion of temporary high exposure to mercury vapor may be confirmed by immediate blood sampling, whereas urinary level seems less informative.[7]

Table 2 summarizes the recommendations formulated by several agencies for workers exposed to inorganic mercury.

TABLE 2
Recommendations Formulated by Agencies
for Workers Occupationally Exposed to Inorganic Mercury

	Parameter	Biological Limit Value	Sampling Time
	ACGIH[3]		
TLV 0.025 mg/m³	mercury in urine	35 µg/g creat	ES
	mercury in blood	15 µg/l	ES, EW
	DFG[24]		
MAK 0.1 mg/m³	mercury in urine	200 µg/l	not fixed
	mercury in blood	50 µg/l	not fixed

ES = end of shift
EW = end of week

5.2 ORGANIC MERCURY

Organic mercury compounds which are liposoluble are easily absorbed by all routes. Methylmercury crosses the blood-brain and the placental barriers and accumulates in the brain.

Short chain alkylcompounds such as methylmercury are stable whereas aryl or alkoxyalkyl derivatives release inorganic mercury *in vivo*. For the latter compounds, the concentration of mercury in blood or urine is probably indicative of the exposure intensity. In steady-state situations, mercury in whole blood correlates with methylmercury body burden and with the risk of signs of methylmercury poisoning.

It has been estimated that in persons chronically exposed to alkylmercury, the earliest signs of intoxications (paresthesia, sensory disturbances) may occur when the level of mercury in blood exceeds 20 µg/100 mL.[113] Blood values between 20 and 50 µg/100 mL are associated with appreciable risk of slight intoxication while values between 100 and 200 µg/100 mL are usually associated with severe brain damage.[9] The Deutsche Forschungsgemeinshaft (DFG) recommends 100 µg mercury/l blood as the Biological Tolerance Value (BTV) (sampling time not fixed) for subjects exposed to methylmercury (MAK 0.01 mg/m³).[24]

6 THALLIUM

Thallium is easily absorbed by the respiratory and gastrointestinal tracts. Limited evidence supports the notion that significant absorption may occur through the skin; however due to the cumulative toxicity of thallium, even a slow percutaneous uptake may be of concern. Thallium and its soluble salts should therefore be regarded as a skin hazard.[37] Soluble thallium salts are widely distributed in the body, the highest concentration being found initially in the kidneys. Thallium is excreted in both urine and feces and in small part via hair and into milk.[51] Excretion in both feces and urine may persist for many weeks despite low blood levels in poisoned subjects.[51] With time the concentration increases in hair, which may contain the major part of the body burden, thus providing an important additional route for slow excretion.[51]

The determination of thallium in urine is probably a more reliable indicator of exposure than its determination in blood.[88] The concentration of thallium in the urine of the general population is usually less than 1.5 µg/l or 1 µg/g creatinine. Data concerning the relationship between the concentration of thallium in biological media and the occurrence of symptoms are lacking. A urine level of 300 µg/l has been suggested as threshold above which a risk of thallium poisoning exists.[35] However, an alerting level of 50 µg/l requiring a preventive action has been proposed.[60]

Thallium concentration in hair may provide a useful indicator of cumulative absorption of thallium provided external contamination can be excluded.[51]

7 CARBON DISULFIDE

Inhalation represents the main route of absorption of carbon disulfide (CS_2) in case of occupational exposure. CS_2 may also penetrate the skin, mainly when evaporation is prevented or when extensive immersion takes place.[37]

A fraction of 70 to 90% of the absorbed amount is metabolized and excreted in urine; 2-thiothiazolidine-4-carboxylic acid (TTCA), 2-mercapto-2-thiazolin-5-one, thiocarbamide have been identified in urine of subjects exposed to CS_2.[75,76,104,105] Less than 1% is excreted unchanged in urine. The remainder 10–30% is exhaled unchanged. The elimination of CS_2 in breath is rapid and characterized by an initial fast decrease with a half-life of 1 minute and a slower decrease with a half-life of 110 minutes.[82] However, carbon disulfide can still be measured in exhaled air of workers 16 hours after the end of exposure.[17] Carbon disulfide binds to amino acids and proteins in blood and tissues. The implications in biological monitoring are discussed in the chapter by Costa and Manzo in this book.

The iodine azide test is based on the time elapsed between adding the iodine-azide reagent to urine until discoloration of the iodine. It was widely used in the past. A major limitation of the assay is its low sensitivity.

Although TTCA only represents 1.2–6.5% of the carbon disulfide absorbed, it has been shown to be a specific, sensible and reliable indicator of exposure to carbon disulfide.[17,81,84] TTCA appears in urine shortly after the start of exposure; the concentration usually peaks at the end of exposure and then declines.[84] Data suggest that the excretion is at least biphasic with half-times of four–six hours and two–three days.[81,83] In highly exposed workers, TTCA can still be detected in urine even 32 to 64 hours after exposure indicating some accumulation during the workweek.[83,107] Therefore, the measurement of TTCA in the end of shift urine reflects the exposure level of the day, but also overexposure of the previous days. Studies relating urinary levels of TTCA to neurotoxic effects are lacking.

It should be noted that some substances can be metabolized to carbon disulfide. These included the preservative agent captan, the thiuram derivative, disulfiram, which is used as a therapeutic agent to decrease alcohol consumption, some rubber accelerators, and dithiocarbamates (e.g., Thiram).

TABLE 3
Recommendations formulated by Agencies
for Workers Occupationally Exposed to Carbon Disulfide

	Parameter	Biological Limit Value	Sampling Time
ACGIH[3]			
TLV 10 ppm (31 mg/m³)	TTCA in urine	5 mg/g creat	ES
DFG[24]			
MAK 10 ppm (30 mg/m³)	TTCA in urine	8 mg/l	ES
France[100]			
VME 10 ppm (30 mg/m³)	TTCA in urine	5 mg/g creat	ES

ES = end of shift

A TLV of 10 ppm (based on prevention of chronic neurological disturbances) has been proposed by ACGIH.[3] However, some recent data suggest that the occupational exposure limit (OEL) should be lowered at least to 4 ppm to prevent chronic neurologic sequelae.[1,85] The OSHA PEL is 4 ppm. The biological value corresponding to an eight-hour exposure at 4 ppm carbon disulfide has been estimated at 2.8 mg/g creatinine.[21]

Table 3 summarizes the recommendations formulated by several agencies for workers exposed to carbon disulfide.

8   CARBON MONOXIDE

Carbon monoxide (CO) is absorbed and eliminated unchanged via the respiratory tract. It easily crosses the alveolar capillary barrier and binds to hemoglobin to form carboxyhemoglobin (HbCO). The affinity of hemoglobin for CO is 200–250 times that for oxygen.

Normal HbCO concentration in the blood of nonsmokers is usually inferior to 0.5% attributable to an endogenous production of CO due to heme catabolism. CO is an ubiquitous pollutant and a metabolite of methylene chloride.[54]

CO crosses the placenta, and HbCO is cleared from the fetal blood slower than from maternal blood due to a higher affinity of fetal hemoglobin for carbon monoxide. At a steady state, fetal HbCO levels can be 10% higher than maternal levels.[111]

8.1   CARBOXYHEMOGLOBIN

The concentration of HbCO rises from the start of exposure to CO; the increase levels off after three hours of exposure. After the end of exposure, the concentration declines with a half-life of four–five hours. The HbCO level at the end of the shift reflects the exposure level during the shift, provided this exposure has been more or less constant.

Smoking habits greatly affect the HbCO levels. In a sedentary nonsmoker, an exposure to 35 and 50 ppm CO will lead to HbCO levels of about 5 and 8%, respectively, at the end of the shift. Such an HbCO level or even higher can be found in smokers non occupationally exposed to CO. Smoking one pack of cigarettes per day leads to an average HbCO level of 5–6% and two to three packs per day to 7–9%. HbCO levels can reach 20% in the case of cigar smoking.[2] Therefore, the biological limit value recommended for occupational exposure does not apply to tobacco smokers.

CO does not seem to have large and consistent behavioral effects on healthy, young volunteers when HbCO does not exceed 10%. However, exposure to CO resulting in a prolonged HbCO level of 5 to 10% may affect the performance of tasks requiring a high degree of vigilance such as flying an aircraft or attending a control panel.[2] Since cardiovascular effects have been reported at HbCO levels of 5%, the ACGIH has recently recommended 3.5% HbCO as Biological Exposure Index (BEI), this level being most likely reached at the end of an 8-hour exposure to 25 ppm CO (sample collected immediately at the end of exposure).[3]

8.2 CARBON MONOXIDE IN EXHALED AIR

Provided the exposure is more or less constant, CO concentration in alveolar air reaches a plateau in the second half of the eight-hour shift. After exposure, this concentration decreases with a half-life of about five hours (two–seven hours).[77]

Smoking one pack of cigarettes per day leads to an average CO level in alveolar air of 30–35 ppm; two to three packs per day 45–50 ppm. Carbon monoxide alveolar levels can reach up to 130 ppm in the case of cigar smoking.[2,46]

The ACGIH has recently recommended 20 ppm as BEI which should not be applied to samples collected in emergency, during the first three hours of exposure, later than 15 minutes after the end of exposure or in tobacco smokers.[3]

8.3 CARBON MONOXIDE IN BLOOD

Since the amount of CO dissolved in blood represents a negligible fraction (<1%) of the amount bound to hemoglobin, the total amount of CO that can be liberated from blood is directly related to the hemoglobin content of blood and its degree of saturation by CO. Under standard conditions, 1 g of hemoglobin can bind 1.49 mL of CO. Therefore, in a nonsmoker with 15 g of hemoglobin per 100 mL of blood, the volume of CO normaly present in blood is less than 0.15 mL.[54]

Table 4 summarizes the recommendations formulated by several agencies for workers occupationally exposed to carbon monoxide.

9   NITROUS OXIDE

Nitrous oxide (N_2O) is easily absorbed through the lungs. Several observations seem to confirm that for the assessment of current exposure to N_2O, the method of choice

TABLE 4
Recommendations formulated by Agencies
for Workers Occupationally Exposed to Carbon Monoxide

	Parameter	Biological Limit Value	Sampling Time
	ACGIH[3]		
TLV 25 ppm (29 mg/m³)	carboxyhemoglobin	3.5%	ES
	CO in end exhaled air	20 ppm	ES
	DFG[24]		
MAK 30 ppm (33 mg/m³)	carboxyhemoglobin	5%	ES
	CO in blood	1.5 mL/100 mL	
	France[100]		
VME 50 ppm (55 mg/m³)	carboxyhemoglobin	< 8%	ES
	CO in blood	1.5 mL/100 mL	

ES = end of shift

is the measurement of its concentration in urine.[41,91-93] Mean urinary concentrations of 20, 35, and 65 µg/l have been associated with time-weighted-average exposures of 25, 50, and 100 ppm.[41]

10   *n*-HEXANE

Inhalation is the major route of absorption of *n*-hexane in the workplace; however pulmonary retention of inhaled *n*-hexane is relatively low. Values from 5 to 25% have been reported.[10,11,47,63,65,108] Percutaneous absorption of *n*-hexane also occurs. *n*-Hexane is poorly soluble in blood (very short half-life) but highly soluble in fat and repeated exposures lead to accumulation of the solvent in the adipose tissue where its half-life has been estimated at 64 hours.[74] About 10–15% of the amount absorbed is eliminated unchanged through exhaled air. The elimination is biphasic with half lives of 11 and 99 minutes, respectively. Urinary excretion of unchanged *n*-hexane is negligible.[63] The metabolic pathways of *n*-hexane and methyl-n-butyl-ketone are closely related with both agents producing 2,5-hexanedione as the major neurotoxic metabolite. 2-Hexanol, 2,5-dimethylfuran and γ-valerolactone, and 4,5-dihydroxy-2-hexanone have also been detected in urine of workers exposed to *n*-hexane.[30,72,73] Experimental studies have shown that urinary excretion of *n*-hexane metabolites decreases in case of simultaneous exposure to methyl-ethyl-ketone or toluene.[42,67,95,96] The TLV-TWA value of 50 ppm recommended by ACGIH for *n*-hexane is based on preventing occupational polyneuropathy.[3]

TABLE 5
Recommendations formulated by Agencies
for Workers Occupationally Exposed to *n*-hexane

Parameter		Biological Limit Value	Sampling Time
	ACGIH[3]		
TLV 50 ppm (176 mg/m³)	2,5-hexanedione in urine (acid hydrolysis pH < 1)	5 mg/l	ES
	DFG[24]		
MAK 50 ppm (180 mg/m³)	2,5-hexanedione + 4,5-dihydroxy-2-hexanone in urine (acid hydrolysis pH < 0.1)	5 mg/l	ES
	France[100]		
VME 50 ppm (170 mg/m³)	2,5-hexanedione in urine	5 mg/l	ES

ES = end of shift

10.1 2,5-HEXANEDIONE

Determination of 2,5-hexanedione in urine samples collected at the end of the workweek seems to provide the best information on the intensity of exposure and the risk of neurotoxic effects. A correlation was found between the concentration of 2,5-hexanedione in end-of-shift samples and electroneuromyographic scoring.[36]

2,5-hexanedione appears in urine immediately after the start of exposure; its concentration rises slowly during exposure and peaks about three hours after the end of exposure. Its elimination half-life is 13–14 hours.[43,74] It seems that a certain amount of 2,5-hexanedione detected in urine is produced from an intermediate metabolite, 4,5-dihydroxy-2-hexanone, during the analytical procedure involving acid hydrolysis; the release of 2,5-hexanedione depends upon the pH-value of the urine which is critical.[30,31,71] The formation 2-acetylfuran in the hydrolyzed urine, interfering with 2,5-HD in the gas chromatographic analysis and thus increasing the background 2,5 HD level has also been suggested.[50]

Table 5 summarizes the recommendations formulated by several agencies for workers occupationally exposed to *n*-hexane.

11 METHYL-*N*-BUTYL-KETONE (2-HEXANONE)

2-hexanone absorption can occur via all routes of exposure. Its metabolism is similar to that of *n*-hexane. Data concerning biological monitoring of workers exposed to methyl-n-butyl-ketone are scanty. However, since 2,5-hexanedione is probably the main active metabolite, the same biological limit value as proposed for *n*-hexane might be suggested.

In Germany the DFG has proposed the same BTV for subjects exposed to 2-hexanone as for subjects exposed to n-hexane : 5 mg 2,5-hexanedione + dihydroxy-2-hexanone/l urine (end of shift).[24]

12   CYANIDE AND NITRILES

Cyanide salts can enter the organism via inhalation and ingestion. Hydrogen cyanide is easily absorbed by the respiratory tract. Hydrogen cyanide and soluble cyanide salts may also penetrate the skin, although no detailed data on their rates of penetration are available. However, due to the considerable acute toxicity of cyanide salts, they should be regarded as skin exposure hazard.[37] Most of the nitriles are taken up by inhalation, ingestion or percutaneous contact. Acrylonitrile may penetrate the skin in significant quantities and may cause both local effects and systemic toxicity.[37]

Exposure to cyanides and aliphatic nitriles that can release cyanide *in vivo* causes an increased concentration of thiocyanate in plasma and urine and of cyanide in blood.[20] In nonsmoking, not occupationally exposed subjects, blood cyanide and plasma thiocyanate values are usually below 10 μg/100 mL and 0.6 mg/100 mL, respectively. In smokers, values as high as 52 μg/100 mL and 1.5 mg/100 mL for blood cyanide and plasma thiocyanate concentrations have been reported.[54] Nitriles can also be measured in urine. A significant increase of acrylonitrile concentration in urine has also been associated with the number of cigarettes smoked.[40]

Acrylonitrile exposure might also be assessed by determining cyanoethylmercapturic acid concentration in urine.[45]

13   ACRYLAMIDE

Acrylamide is readily absorbed by ingestion and inhalation. It may penetrate the skin in significant quantities, and dermal absorption of this compound has caused or contributed to numerous intoxications.[37] Data on acrylamide metabolism in humans are scanty. Free acrylamide in plasma, S-(2-carboxyethyl)cysteine in urine and hemoglobin adducts (acrylamide adducts to N-terminal valine in hemoglobin) have been used as biomarkers of occupational exposure to acrylamide, as discussed by Costa and Manzo in this book. The determination of hemoglobin adducts was estimated to be the best predictor of acrylamide induced peripheral neuropathy.[8,16]

14   ETHYLENE OXIDE

Exposure to ethylene oxide mainly occurs via inhalation. Its pulmonary retention lies around 75–80%.[12,13] Two groups of subjects characterized by a significant difference in the rate of metabolism have been identified: those with a rapid disappearance of ethylene oxide from blood (the "conjugators"), and the "nonconjugators."[32,56]

An increase in ethylene glycol concentration in blood has been observed in workers exposed to ethylene oxide.[114] It has been estimated that a time-weighted-average exposure to 2 mg/m³ would lead at the end of the shift to mean alveolar and blood concentrations of ethylene oxide of 0.5 μg/l and 0.8 μg/100 mL, respectively.[12,13] This is in agreement with the estimation of the DFG:[24] exposure to mean air concentrations of 0.92, 1.83, 3.66, 5.49, 7.32 and 9.15 mg/m³ leading to mean alveolar and whole blood ethylene oxide concentrations of 0.22, 0.44, 0.88, 1.32, 1.76 and 2.2 mg/m³ and 3, 6.1, 12.1, 18.1, 24.3, and 30.3 µg/l, respectively (sampling time: during an exposure of at least four hour-duration).

The analysis of hemoglobin adducts may be a sensitive and specific integrative parameter for the biological monitoring of exposure to ethylene oxide.[29,38,49,69,79,86,97] It seems that determination of hydroxyethyl mercapturic acid in urine may be suitable for the differentiation between ethylene oxide exposed and nonexposed groups, but not for individual biomonitoring.[78]

15   ORGANOPHOSPHORUS PESTICIDES

Organophosphorus pesticides (OP) may enter the body via all routes of exposure: inhalation, ingestion, cutaneous contact. The absorption rate depends upon the chemical structure and formulation of the compound. Their biotransformation occurs as a complex multistep process.[25] The organophosphorus compounds and their metabolites are generally rapidly excreted mainly through the kidneys and to a lesser extent via the bile.

15.1   CHOLINESTERASE ACTIVITY

OP are cholinesterase inhibitors, plasma cholinesterase (pseudocholinesterase) is usually more susceptible to inhibition by OP than true acetylcholinesterase, but pseudocholinesterase activity may be reduced in situations other than exposure to OP or carbamate pesticides such as early hepatitis, alcoholic cirrhosis, other liver diseases and various drug intakes. The measurement of the red blood cell acetylcholinesterase activity is an indirect measure of the enzyme activity that exists in the nervous tissue which is the critical target organ of OP. The occurrence and severity of toxic effects depend not only on the degree but also on the rate of enzyme inhibition. There is considerable interindividual variability in plasma and erythrocyte cholinesterase activity between unexposed persons.

A reduction of red cell acetylcholinesterase activity to 70% (inhibition 30%) of the individual baseline value has been adopted as BEI by the ACGIH[3], BTV by the DFG[24] and Biological Indicator of Exposure in France.[100]

15.2   ALKYLPHOSPHATES IN URINE

Exposure to some OP may be assessed by measurement of their water soluble metabolites excreted in urine. Depending upon the chemical structure of the OP, various alkylphosphates can be detected. However, no relationship between the levels of these metabolites and the risk of toxic effects has yet been established.[25]

16   CARBAMATE PESTICIDES

Carbamate pesticides can enter the body through inhalation, ingestion and percutaneous absorption. They are active cholinesterase inhibitors but the inhibition is usually rapidly reversible. Hence, if the blood sample has been stored for a certain time before analysis, cholinesterase activity may not necessarily reflect that in the circulating blood at the time of blood withdrawal. Carbamates are eliminated rather rapidly, after biotransformation, mainly in urine. The determination of urinary metabolites may be a more practical method for monitoring exposure to certain carbamates than blood cholinesterase measurement.[54]

Exposure to carbaryl (1-naphthyl-N-methylcarbamate) can be monitored by the determination of 1-naphthol in urine. Sampling should be carried out by collecting the urine voided four–eight hours after the end of exposure.[61] Data are limited, but it seems that when 1-naphthol concentration in urine does not exceed 10 mg/l at the end of the exposure, the risk of occurrence of symptoms or signs of clinical intoxication is low.[110] Acetylcholinesterase inhibition might be expected to occur at oral doses equal to or greater than 2 mg/kg entailing a urinary excretion of 1-naphthol of about 50 mg/l; after repeated oral exposures the no-effect level was demonstrated at 9 mg/day, corresponding to a 1-naphthol concentration of about 2–3 mg in 24 hour urine.[61]

2-Isopropoxyphenol can be detected in urine of persons exposed to isopropoxyphenyl-N-methylcarbamate (propoxur).[23,101] Recovery of 2-isopropoxyphenol in urine is about 40% of the absorbed dose. A time-weighted-average exposure of 0.5 mg/m^3 would lead to a urinary excretion of 1–2 mg of 2-isopropoxyphenol in the first 8 hour urine, and 0.5–1 mg in the subsequent hours. [61] Depression of blood acetylcholinesterase is expected to occur at or above urinary concentrations of 5–10 mg/l of isopropoxyphenol, which correspond to an absorbed dose of about 0.2–0.3 mg/kg b.w. of propoxur. [61] Sampling should be carried out by collecting the urine of the first eight–10 hours after the beginning of exposure.[61]

17   ORGANOCHLORINE PESTICIDES

Most of organochlorine pesticides are lipid-soluble compounds, easily absorbed by all routes of exposure. They are very persistent in the environment and tend to accumulate in adipose tissue. Exposure to these chemicals can be assessed by their measurement in blood or in adipose tissue. For some compounds, urinary metabolites have also been identified. Many of these pesticides have progressively been replaced by other less persistent products.

Exposure to lindane (γ-hexachlorocyclohexane) can be assessed by its determination in the blood or plasma (serum). The main metabolites identified in the urine of workers exposed to lindane are 2,3,5-, 2,4,5-, and 2,4,6-trichlorophenols; products such as other trichlorophenols, monochlorophenols, tetrachlorophenols, dihydroxychlorobenzene and pentachlorophenol have also been detected.[5,26,28] It seems that neurological symptoms and signs (paresthesia, headache, giddiness, tremor, etc.) may already occur at lindane blood level >2 μg/100 mL.[22,48,64] The DFG has adopted 20 μg γ-hexachlorocyclohexane/l in blood and 25 μg γ-hexachlorocyclohexane/l in serum or plasma as BTV (samples collected at the end of the shift).[24]

Hexachlorobenzene is a very stable and cumulative toxin; its concentration in blood is mainly a reflection of the body burden of this pesticide. The main metabolite identified in humans is pentachlorophenol.[52] The BTV recommended by the DFG is 150 μg hexachlorobenzene/l in plasma or serum (time of sampling not fixed).[24]

The use of aldrin, endrin, and dieldrin is currently limited because of their high toxicity. After absorption, aldrin is rapidly converted to dieldrin which accumulates in the adipose tissue and therefore is rarely found in blood or tissues. After biotransformation, dieldrin is primarily excreted in the feces via the bile and to a lesser extent in urine.[54,102] The major metabolite identified in the feces of workers exposed to aldrin and dieldrin is a hydroxy derivative.[80,102] The determination of dieldrin in blood is the most relevant test for the assessment of aldrin and dieldrin absorption and is an indicator of dieldrin body burden. A blood dieldrin concentration of 20 µg/100 mL has been proposed as the value requiring remedial action as there is danger of imminent or overt intoxication.[116] Anti-12-hydroxyendrin (conjugated with glucuronic acid) is a urinary metabolite of endrin.[6,53,70] In view of the rapid metabolism of endrin, blood concentration of endrin is an indicator of short term overexposure, while urinary anti-12-hydroxyendrin is a more valid parameter to assess long term exposure. No cases of intoxication due to short term endrin overexposure have been reported at endrin concentrations in blood below 5–10 µg/100 ml, (samples collected at the end of exposure).[44,102,103] The highest urinary anti-12-hydroxyendrin concentration measured (1.4 mg/g creatinine) in a sample collected 20 hours after the end of an exposure to endrin was not associated with clinical signs or symptoms.[103]

18   ORGANIC SOLVENTS

Many organic solvents have the common property to depress the central nervous system. Some of them may also affect the peripheral nervous system. Table 6 lists the main neurotoxic organic solvents for which a biological limit value for either the compound itself or its metabolites has been recommended by the ACGIH, DFG and in France for occupationally exposed subjects. In the occupational setting, these substances are mainly absorbed via inhalation; however, in some circumstances they can penetrate the skin in significant amounts. They are usually quite rapidly excreted either via the kidneys, generally after biotransformation or unchanged with the expired air. After repeated daily exposure, there is a possible progressive accumulation of certain solvents in the body (e.g., tetrachloroethylene, trichloroethylene) during the workweek.

Exposure generally is assessed by measuring the concentration of the parent compound in blood or in alveolar air or the concentration of the metabolites in blood or in urine. Knowledge of the time of sampling is critical for the interpretation of the results.

TABLE 6
Biological Limit Values Proposed by Agencies
for Subjects Occupationally Exposed to Some Solvents

agent	parameter	ACGIH [3]	DFG [24]	France [100]
methanol	methanol in urine	15 mg/l (ES)	30 mg/l (ES)	15 mg/l (ES)
	formic acid in urine			80 mg/g creat (PLSW)
perchlorethylene	perchlorethylene in exhaled air	10 ppm → 5 ppm* (PLSW)	9,5 mL/m³ (PNS)	
	perchlorethylene in blood	1 mg/l → 0.5 mg/l* (PLSW)	1 mg/l (PNS)	1 mg/l (PLSW)
	trichloroacetic acid in urine	7 mg/l → 3.5 mg/l* (EW)		7 mg/l (EW)
styrene	styrene in blood	0.55 mg/l (ES) 0.02 mg/l (PNS)		0.55 mg/l (ES) 20 µg/l (PNS)
	phenylglyoxylic acid in urine	240 mg/g creat (ES) 100 mg/g creat (PNS)		240 mg/g creat (ES) 100 mg/g creat (PNS)
	mandelic acid in urine	800 mg/g creat (ES) 300 mg/g creat (PNS)	2 g/l (ES)	800 mg/g creat (ES) 300 mg/g creat (PNS)
	phenylglyoxylic + mandelic acid in urine		2.5 g/l (ES)	
toluene	toluene in blood	1 mg/l (ES)**	1.7 mg/l (ES)	1 mg/l (ES)
	hippuric acid in urine	2.5 g/g creat (ES)**		2.5 g/g creat (ES)
1,1,1 trichloroethane	1,1,1 trichloro-ethane in exhaled air		20 mL/m³ (PNS)	
	1,1,1 trichloro-ethane in blood		550 µg/l (PNS)	
	trichloroacetic acid in urine			10 mg/l (EW)
	trichloroethanol in urine			30 mg/l (ES, EW)
	trichloroethanol in blood			1 mg/l (ES, EW)

TABLE 6 (continued)
Biological Limit Values Proposed by Agencies
for Subjects Occupationally Exposed to Some Solvents

agent	parameter	ACGIH [3]	DFG [24]	France [100]
trichloroethylene	trichloroethanol in blood		5 mg/l (ES)	
	free trichloro-ethanol in blood	4 mg/l (ES, EW)		4 mg/l (ES, EW)
	trichloroacetic acid in urine	100 mg/g creat (EW)	100 mg/l (ES)	100 mg/g creat (EW)
	trichloroacetic acid + trichloro-ethanol in urine	300 mg/g creat (ES, EW)		300 mg/g creat (ES, EW)
xylenes	xylene in blood		1.5 mg/l (ES)	
	methylhippuric acid in urine	1.5 g/g creat (ES)	2 g/l (ES)	1.5 g/g creat (ES)

ES = end of shift
EW = end of workweek
PNS = prior to next shift
PLSW = prior to last shift of workweek

*   notice of intended changes
** BEIs withdrawn based on reduction of chemical substances TLV from 100 ppm to 50 ppm
BEI revision under review

19   REFERENCES

1. Aaserund, O., Hommeren, O., Tvedt, B., Nakstad, P., Mowe, G., Efskind, J., Russel, D., Jörgensen, E., Nyberg-Hansen, R., Rootwelt, K., and Gjerstad, L., "Carbon disufide exposure and neurotoxic sequelae among viscose rayon workers," *Am. J. Ind. Med.,* 18, 25, 1990.
2. ACGIH, American Conference of Governmental Industrial Hygienists, *Documentation of the Threshold Limit Values and Biological Exposure Indices,* Cincinnati, OH, 6th edition, 1991.
3. ACGIH, American Conference of Governmental Industrial Hygienists, *Threshold Limit Values and Biological Exposure Indices 1994–1995,* Cincinnati, OH, 1994.
4. Alessio, L., Mussi, I., Calzaferri, G., and Buratti, M., "Serum and urinary aluminium concentrations in occupationally exposed subjects" (in Italian), *Convegno sulla Patologia da Tossici Ambentali ed Occupazionali,* Cagliari, 26-27 maggio 1983, Edigraf, Torino, Italy, 285, 1983.
5. Angerer, J., Maas, R., and Heinrick, R., "Occupational exposure to hexachlorocyclohexane. VI. Metabolism of γ-hexachlorocyclohexane in man," *Int. Arch. Occup. Environ. Hlth.,* 52, 59, 1983.

6. Baldwin, M., and Hutson, D., "Analysis of human urine for a metabolite of endrin by chemical oxidation and gas-liquid chromatography as an indicator of exposure to endrin," *Analyst*, 105, 60, 1980.

7. Barregard, L., *Occupational exposure to inorganic mercury in chloralkali workers. Studies on metabolism and health effects*, Thesis: University of Göteborg, Department of Occupational Medicine, Göteborg, Sweden, 1991.

8. Bergmark, E., Calleman, C., He, F., and Costa, L. G., "Determination of hemoglobin adducts in humans occupationally exposed to acrylamide," *Toxicol. Appl. Pharmacol.*, 120, 45, 1993.

9. Berlin, M., "Mercury Organic Compounds," in *Encyclopaedia of Occupational Hlth. and Safety*, International Labour Office, Parmeggiani, L., Ed., Geneva, Switzerland, 1983.

10. Brugnone, F., Perbellini, F., Grigolini, L., and Apostoli, P. "Solvent exposure in a shoe upper factory. 1. n-Hexane and acetone concentration in alveolar air and environmental air and in blood," *Int. Arch. Occup. Environ. Hlth.*, 42, 51, 1978.

11. Brugnone, F., Perbellini, L., and Apostoli, P., "Alveolar air in monitoring industrial exposure to volatile organic compounds" (in Italian), *Medicina del Lavoro*, 74, 1, 1983.

12. Brugnone, F., Perbellini, L., Faccini, G., and Pasini, F., "Concentration of ethylene oxide in the alveolar air of occupationally exposed workers," *Am. J. Ind. Med.*, 8, 67, 1985.

13. Brugnone, F., Perbellini, L., Faccini, G., Pasini, F., Bartolucci, G., and De Rosa, E., Ethylene oxide exposure. Biological monitoring by analysis of alveolar air and blood," *Int. Arch. Occup. Environ. Hlth.*, 58, 105, 1986.

14. Buchet, J.-P., Lauwerys, R., and Roels, H., "Comparison of several methods for the determination of arsenic compounds in water and in urine," *Int. Arch. Occup. Environ. Hlth.*, 46, 11, 1980.

15. Buchet, J.-P., Pauwels, J., and Lauwerys, R., "Assessment of exposure to inorganic arsenic following ingestion of marine organisms by volunteers," *Environ. Res.*, 66, 44, 1994.

16. Calleman, C., Wu, Y., He, F., Tian, G., Bergmark, E., Zhang, S., Deng, H., Wang, Y., Crofton, K., Fennell, T., and Costa, L. G., "Relationships between biomarkers of exposure and neurological effects in a group of workers exposed to acrylamide," *Toxicol. Appl. Pharmacol.*, 126, 361, 1994.

17. Campbell, L., Jones, A., and Wilson H., "Evaluation of occupational exposure to carbon disulfide by blood, exhaled air and urine analysis," *Am. J. Ind. Med.*, 8, 143, 1985.

18. CEC, Commission of the European Communities, "Alzheimer's disease: has aluminium any causal role? Toxicity of aluminium. Maximal admissible concentration of aluminium in drinking water," *Scientific Advisory Committe to Examine the Toxicity and Ecotoxicity of Chemical Compounds, CSTE/90/22/COM*, Luxembourg, 1991.

19. CEC, Commission of the European Communities, Resolution of the Council and the Representatives of the Member States. "Protection of dialysis patients by minimizing the exposure to aluminium," 86/C 184/09, *Official Journal of the European Communities*, C 184 Vol. 29, 1986.

20. Chandra, H., Gupta, B., Bhargava, S., Clerk, S., and Mahendra, P., "Chronic cyanide exposure, a biochemical and industrial hygiene study," *J. Analyt. Toxicol.*, 4: 1261, 1980.

21. Cox, C., Lowry, L., and Que Hee, S., "Urinary 2-thiothiazolidine-4-carboxylic acid as a biological indicator of exposure to carbon disulfide: Derivation of a biological exposure index," *Appl. Occup. Environ. Hygiene*, 7, 672, 1992.

22. Czegledi-Janko, G., and Avar, P., "Occupational exposure to lindane: clinical and laboratory findings," *Br. J. Ind. Med.*, 27, 283, 1970.
23. Dawson, J., Heath, D., Rose, J., Thain, E., and Ward, J., The excretion by humans of the phenol derived *in vivo* from 2-isopropoxyphenyl-N-methylcarbamate," *Bulletin of WHO*, 30, 127, 1964.
24. DFG, Deutsche Forschungsgemeinshaft, *Maximum Concentrations at the Workplace and Biological Tolerance Values for Working Materials,* Report No. 30, VCH, Weinheim, 1994.
25. Dillon, H., and Ho, M., "Review of the biotransformation of organophosphorus pesticides," in *Biological Monitoring of Exposure to Chemicals, Organic Compounds,* Ho, M. and Dillon, H., Eds., Wiley-Interscience, New York, 1987, 227.
26. Drummond, L., Gillanders, E., and Wilson, H., "Plasma γ-hexachlorocyclohexane concentrations in forestry workers exposed to lindane," *Br. J. Ind. Med.,* 45, 493, 1988.
27. Elinder, C.-G., Ahrengart, L., Lidums, V., Pettersson, E., and Sjögren, B., "Evidence of aluminium accumulation in aluminium welders," *Br. J. Ind. Med.,* 48, 735, 1991.
28. Engst, R., Macholz, R., and Kujawa, M., "Metabolism of lindane in microbial organisms, warmblooded animals and humans," *Gigiena i Sanitariya,* 10, 64, 1979.
29. Farmer, P., Bailey, E., Gorf, S., Törnqvist, M., Osterman-Golkar, S., Kautiainen, A., and Lewis-Enright, D., "Monitoring human exposure to ethylene oxide by the determination of haemoglobin adducts using gas chromatography-mass spectrometry," *Carcinogenesis,* 7, 637, 1986.
30. Fedtke, N., and Bolt, H., "The relevance of 4,5-dihydroxy-2-hexanone in the excretion kinetics of *n*-hexane metabolites in rat and man," *Arch. Toxicol.,* 61, 131, 1987.
31. Fedtke, N., and Bolt, H., "Methodological investigations on the determination of *n*-hexane metabolites in urine," *Int. Arch. Occup. Environ. Hlth.,* 57, 149, 1986.
32. Föst, U., Hallier, E., Ottenwalder, H., Bolt, H., and Peter, H., "Distribution of ethylene oxide in human blood and its implication for biomonitoring," *Hum. Exp. Toxicol.,* 10, 25, 1991.
33. Garb, L., and Hine, C., "Arsenical neuropathy residual effects following acute industrial exposure," *J. Occup. Med.,* 19, 567, 1977.
34. Georis, B., Cardenas, A., Buchet, J.-P., and Lauwerys, R., "Inorganic arsenic methylation by rat tissues slices," *Toxicology,* 63, 73, 1990.
35. Glomme, J., "Thallium and compounds," in *Encyclopaedia of Occupational Hlth. and Safety,* International Labour Organization, Geneva, Switzerland, 1983, 2170.
36. Governa, M., Calisti, R., Coppa, G., Tagliaventa, G., Colombi, A., and Troni, W., "Urinary excretion of 2,5-hexanedione and peripheral neuropathies in workers exposed to *n*-hexane," *J. Toxicol. Environ. Hlth.,* 20, 219, 1987.
37. Grandjean, Ph., *Skin Penetration: Hazardous Chemicals at Work,* Taylor & Francis, London, 1990, 173.
38. Hagmar, L., Welinder, H., Linden, K., Attewell, R., Osterman-Golkar, S., and Törnqvist, M., "An epidemiological study of cancer risk among workers exposed to ethylene oxide using haemoglobin adducts to validate environmental exposure assessments," *Int. Arch. Occup. Environ. Hlth.,* 63, 271, 1991.
39. Hine, C., Cavalli, R., and Beltran, S., "Percutaneous absorption of lead from industrial lubricants," *J. Occup. Med.,* 11, 568, 1969.
40. Houthuijs, D., Remijn, B., Willems, H., Boleij, J., and Biersteker, K., "Biological monitoring of acrylonitrile exposure," *Am. J. Ind. Med.,* 3: 313, 1982.
41. Imbriani, M., Ghittori, S., Pezzagno, G., and Capodaglio, E., "Nitrous oxide in urine as biological index of exposure in operating room personnel," *App. Ind. Hygiene,* 3: 223, 1988.

42. Iwata, M., Takeuchi, Y., Hisanaga, N., and Ono, Y., "Changes of *n*-hexane neurotoxicity and its urinary metabolites by long term co-exposure with methyl ethyl ketone or toluene," *Int. Arch. Occup. Environ. Hlth.*, 54, 273, 1984.

43. Iwata, M., Takeuchi, Y., Hisanaga, N., and Ono, Y., "A study on biological monitoring of *n*-hexane exposure," *Int. Arch. Occup. Environ. Hlth.*, 51, 253, 1983.

44. Jager, K., "Aldrin, Dieldrin, Endrin and Telodrin: an epidemiological and toxicological study of long term occupational exposure," Elsevier Publishing Co., New York, 1970.

45. Jakubowski, M., Linhart, I., Pielas, G., and Kopecky, J., "2-Cyanoethylmercapturic acid (CEMA) in the urine as a possible indicator of exposure to acrylonitrile," *Br. J. Ind. Med.*, 44, 834, 1987.

46. Jones, R., Coppin, C., and Guz, A., "Carbon monoxide in alveolar air as an index of exposure to cigarette smoke," *Clin. Sci. Mol. Med.*, 51, 495, 1976.

47. Jorgensen, N., and Cohr K. "*n*-Hexane and its toxicologic effects," *Scan. J. Work Environ. Hlth.*, 7, 157, 1981.

48. Kashyap, S., "Health surveillance and biological monitoring of pesticide formulators in India," *Toxicol. Lett.*, 33, 107, 1986.

49. Kautiainen, A., and Törnqvist, M., "Monitoring exposure to simple epoxides and alkenes through gas chromatographic determination of hemoglobin adducts," *Int. Arch. Occup. Environ. Hlth.*, 63, 27, 1991.

50. Kawai, T., Yasugi, T., Mizunuma, K., Horiguchi, S., Uchida, Y., Iwami, O., Iguchi, H., and Ikeda, M., "2-Acetylfuran, a confounder of 2,5-hexanedione in the urine of workers exposed to *n*-hexane," *Int. Arch. Occup. Environ. Hlth.*, 63, 213, 1991.

51. Kazantzis, G., "Thallium and Tin," in *Biological Indicators for the Assessment of Human Exposure to Industrial Chemicals,* Alessio, L., Berlin, A., Roi, R., and van der Venne, M.-Th., Eds., European Commission, Luxembourg, EUR 14815 En, 1994, 79.

52. Koss, G., Reuter, A., and Koransky, W., "Excretion of metabolites of hexachlorobenzene in the rat and in man," in *Hexachlorobenzene: Proceedings of an International Symposium,* Morris, C., and Cabral, J., Eds., IARC Scientific Publication No. 77, Lyon, 1986, 261.

53. Kummer, R., and van Sittert, N., "Field studies on health effects from the application of two organophosphorus insecticide formulations by hand-held ULV to cotton," *Toxicol. Lett.*, 33, 7, 1986.

54. Lauwerys, R., and Hoet, P., *Industrial Chemical Exposure, Guidelines for Biological Monitoring,* Lewis Publishers, Boca Raton, FL, 1993.

55. *Lead in bone.* International "Workshop on Lead in Bone: Implications for dosimetry and toxicology," *Environ. Hlth. Perspect.*, 91, 1991.

56. Leutbecher, M., Langhof, Th., Peter, H., and Föst, U., "Ethylene oxide: metabolism in human blood and its implication to biological monitoring," *The Eurotox Congress, Book of abstracts,* Maastricht, 1991.

57. Lilley, S., Florence, T., and Stauber, J., "The use of sweat to monitor lead absorption through the skin," *Sci. Total Environ.*, 76, 267, 1988.

58. Ljünggren, K., Lidums, V., and Sjögren, B., "Blood and urine concentrations of aluminium among workers exposed to aluminium flake powders," *Br. J. Ind. Med.*, 42, 106, 1991.

59. Mahieu, P., Buchet, J.-P., Lauwerys, R., "Evolution clinique et biologique d'une intoxication orale aiguë par l'anhydride arsénieux et considérations sur l'attitude thérapeutique," *J. Toxicol. Clin. Expérim.*, 7, 273, 1987.

60. Marcus, R., "Investigation of a working population exposed to thallium," *J. Soc. Occup. Med.*, 35, 4, 1985.

61. Maroni, M., "Carbamate pesticides," in *Biological Indicators for the Assessment of Human Exposure to Industrial Chemicals,* Alessio, L., Berlin, A., Boni, M., and Roi, R., Eds., Commission of the European Communities, Luxembourg, EUR 11478 En, 1988, 27.

62. Mussi, I., Calzaferri, G., Buratti, M., and Alessio, L., "Behaviour of plasma and urinary aluminium levels in occupationally exposed subjects," *Int. Arch. Occup. Environ. Hlth.*, 54, 155, 1984.

63. Mutti, A., Falzoi, M., Lucertini, S., Arfini, G., Zignani, M., Lombardi, S., and Franchini, I., "*n*-Hexane metabolism in occupationally exposed workers," *Br. J. Ind. Med.*, 41, 533, 1984.

64. Nigam, S., Karnik, A., Majunder, S., Visweswariah, K., Suryanarayana Raju, G., Muktha Bai, K., Lakkad, B., Thakore, K., and Chatterjee, B., "Serum hexachlorocyclohexane residues in workers engaged at a hexachlorocyclohexane manufacturing plant," *Int. Arch. Occup. Environ. Hlth.*, 57, 315, 1986.

65. Nomiyama, K., and Nomiyama, H., "Respiratory retention, uptake and excretion of organic solvents in man. Benzene, toluene, *n*-hexane, trichloroethylene, acetone, ethylacetate, and ethylalcohol," *Int. Arch. Arbeit (Int. Arch. Occup. Environ. Hlth.)*, 32, 75, 1974.

66. Norin, H., and Vather, M., "A rapid method for the selective analysis of total urinary metabolites of inorganic arsenic," *Scan. J. Work Environ. Hlth.*, 7, 38, 1981.

67. Nylen, P., Ebendal, T., Eriksdotter-Nilsson, M., Hansson, T., Henschen, A., Johnson, A., Kronevi, T., Kvist, U., Sjöstrand, N., Höglund, G., and Olson, L., "Testicular atrophy and loss of nerve growth factor-immunoreactive germ cell line in rats exposed to *n*-hexane and a protective effect of simultaneous exposure to toluene or xylene," *Arch. Toxicol.*, 63, 296, 1989.

68. Offergelt, J., Roels, H., Buchet, J.-P., Boeckx, M., and Lauwerys, R., "Relationship between airborne arsenic trioxide and urinary excretion of inorganic arsenic and its methylated metabolites," *Br. J. Ind. Med.*, 49, 387, 1992.

69. Osterman-Golkar, S., and Bergmark, E., "Occupational exposure to ethylene oxide: relation between *in vivo* dose and exposure dose," *Scand. J. Work Environ. Hlth.*, 14, 372, 1988.

70. Ottevanger, C., and van Sittert, N., "Relation between anti-12-hydroxyendrin excretion and enzyme induction in workers involved in the manufacture of endrin," in *Chemical Porphyria in Man,* Strik, J., and Koeman, J., Eds, Elsevier, Amsterdam, 1979, 23.

71. Perbellini, L., Pezzoli, G., Brugnone, F., and Canesi, M., "Biochemical and physiological aspects of 2,5-hexanedione: endogenous or exogenous product?" *Int. Arch. Occup. Environ. Hlth.*, 65, 49, 1993.

72. Perbellini, L., Brugnone, F., and Faggionato, G., "Urinary excretion of the metabolites of *n*-hexane and its isomers during occupational exposure," *Br. J. Ind. Med.*, 38, 20, 1981.

73. Perbellini, F., Brugnone, F., and Pavan, I., "Identification of the metabolism of *n*-hexane," cyclohexane and their isomers in mens' urine," *Toxicol. Appl. Pharmacol.*, 53, 220, 1980.

74. Perbellini, L., Mozzo, P., Brugnone, F., and Zedde, A., "Physiologico-mathematical model for studying human exposure to organic solvents: kinetics of blood/tissue*n*-hexane concentrations and of 2,5-hexanedione in urine," *Br. J. Ind. Med.*, 43, 760, 1986.

75. Pergal, M., Vukojevic, N., and Djuric, D., "Carbon disulfide metabolites excreted in the urine of exposed workers. II. Isolation and identification of thiocarbamide," *Arch. Environ. Hlth.*, 25, 42, 1972.

76. Pergal, M., Vukojevic, N., Girin-Popov, N., Djuric, D., and Bojovic, T., "Carbon disulfide metabolites excreted in the urine of exposed workers. I. Isolation and identification of 2-mercapto-2-thiazolinone-5," *Arch. Environ. Hlth.*, 25, 38, 1972.

77. Peterson, J., and Stewart, R., "Predicting the carboxyhemoglobin levels resulting from carbon monoxide exposures," *J. Appl. Physiol.*, 39, 633, 1975.

78. Popp, W., Vahrenholz, C., Przygoda, H., Brauksiepe, A., Goch, S., Müller, G., Schell, C., and Norpoth, K., "DNA-protein cross-links and sister chromatid exchange frequencies in lymphocytes and hydroxyethyl mercapturic acid in urine of ethylene oxide-exposed hospital workers," *Int. Arch. Occup. Environ. Hlth.*, 66, 325, 1994.

79. Ribeiro, L., Salvadori, D., Rios, A., Costa, S., Tates, A., Törnqvist, M., and Natarajan, A., "Biological monitoring of workers exposed to ethylene oxide," *Mutation Res.*, 313, 81, 1994.

80. Richardson, A., and Robinson, J., "The identification of a major metabolite of HEOD (dieldrin) in human feces," *Xenobiotica*, 1, 213, 1971

81. Riihimäki, V., Kivistö, H., Peltonen, K., Hepliö, E., and Aitio, A., "Assessment of exposure to carbon disulfide in viscose production workers from urinary 2-thiothiazolidine-4-carboxylic acid determinations," *Am. J. Ind. Med.*, 22, 85, 1992.

82. Rosier, J., Veulemans, H., Masschelein, R., Van Hoorne, M., and Van Peteghem, C., "Experimental human exposure to carbon disulfide: I. Respiratory uptake and elimination of carbon disulfide under rest and physical exercise," *Int. Arch. Occup. Environ. Hlth.*, 59, 233, 1987.

83. Rosier, J., Van Hoorne, M., Grosjean, R., Van de Walle, E., Billemont, G., and Van Peteghem, C., "Preliminary evaluation of urinary 2-thiothiazolidine-4-carboxylic acid levels as a test for exposure to carbon disulphide," *Int. Arch. Occup. Environ. Hlth.*, 51, 159, 1982.

84. Rosier, J., Veulemans, H., Masschelein, R., Van Hoorne, M., and Van Peteghem, C., "Experimental human exposure to carbon disulfide: II. Urinary excretion of 2 thiothiazolidine-4-carboxylic acid TTCA during and after exposure," *Int. Arch. Occup. Environ. Hlth.*, 59, 243, 1987.

85. Ruijten, M., Salle, H., Verbeck, M., and Muijser, H., "Special nerve functions and colour discrimination in workers with long term low level exposure to carbon disulfide," *Br. J. Ind. Med.*, 47: 589, 1990.

86. Sarto, F., Törnqvist, M., Tomanin, R., Bartolucci, G., Osterman-Golkar, S., and Ehrenberg, L., "Studies of biological and chemical monitoring of low level exposure to ethylene oxide," *Scand. J. Work Environ. Hlth.*, 17, 60, 1991.

87. Savory, J., and Wills, M., "Biological monitoring of aluminium," in *Biological Monitoring of Toxic Metals, Rochester Series on Environmental Toxicity*, Clarkson, T., Friberg, L., Nordberg, G. F., and Sager, P., Eds., Plenum Press, New York, 1988.

88. Schaller, K.-H., Manke, G., Raithel, H., Bühlmeyer, G., Schmidt, M., and Valentin, H., "Investigations of thallium exposed workers in cement factories," *Int. Arch. Occup. Environ. Hlth.*, 47, 223, 1980.

89. Sjögren, B., Lidums, V., Hakansson, M., and Hedström, L., "Exposure and urinary excretion of aluminium during welding," *Scand. J. Work Environ. Hlth.*, 11, 39, 1985.

90. Sjögren, B., Elinder, C.-G., Lidums, V., and Chang, G., "Uptake and urinary excretion of aluminium among welders," *Int. Arch. Occup. Environ. Hlth.*, 60, 77, 1988.

91. Sonander, H., Stenqvist, O., and Nilsson, K., "Urinary nitrous oxide as a measure of biologic exposure to nitrous oxide anesthetic contamination," *Ann. Occup. Hygiene,* 27, 73, 1983.

92. Sonander, H., Stenqvist, O., and Nilsson, K., "Exposure to trace amounts of nitrous oxide. Evaluation of urinary gas content monitoring in anesthetic practice," *Br. J. Anesthesiol.,* 55, 1225, 1983.

93. Stevens, M.-P., Walrand, J., Buchet, J.-P., and Lauwerys, R., "Evaluation de l'exposition à l'halothane et au protoxyde d'azote en salle d'opération par des mesures d'ambiance et des mesures biologiques," *Cahiers de Médecine du Travail,* 34, 41, 1987.

94. Suzuki, T., "Hair and nails: advantages and pitfalls when used in biological monitoring," in *Biological Monitoring of Toxic Metals, Rochester Series on Environmental Toxicity,* Clarkson, T., Friberg, L., Nordberg, G. F., and Sager, P., Eds., Plenum Press, New York, 1988, 623.

95. Takeuchi, Y., Ono, Y., Hisanaga, N., Iwata, M., Aoyama, M., Kitoh, J., and Sugiura, Y., "An experimental study of the combined effects of *n*-hexane and methyl ethyl ketone," *Br. J. Ind. Med.,* 40, 199, 1983.

96. Takeuchi, Y., Hisanaga, N., Oni, Y., Shibata, E., Saito, I., and Iwata, M., "Modification of metabolism and neurotoxicity of hexane by co-exposure of toluene," *Int. Arch. Occup. Environ. Hlth.,* 65, S227, 1993.

97. Tates, A., Grummt, T., Törnqvist, M., Farmer, P., van Dam, F., van Mossel, H., Schoemaker, H., Osterman-Golkar, S., Uebel, C., Tang, Y., Zwinderman A., Natarajan, A., and Ehrenberg, L., "Biological and chemical monitoring of occupational exposure to ethylene oxide," *Mutation Res.,* 250, 483, 1991.

98. Vahter, M., "Arsenic," in *Biological Monitoring of Toxic Metals, Rochester Series on Environmental Toxicity,* Clarkson, Th., Friberg, L., Nordberg, G., and Sager, P., Eds., Plenum Press, New York, 1988.

99. Valentin, H., Preusser, P., and Schaller, K., "Die analyse von Aluminium im Serum and Urin zur Überwachung exponierter Personen," *Int. Arch. Occup. Environ. Hlth.,* 38, 1, 1976.

100. "Indicateurs Biologiques d'Exposition. Valeurs-guides utilisables en France," *Cahiers de Notes Documentaires,* 151, 237, 1993.

101. Vandekar, M., Hedayat, S., Plestina, R., and Ahmady, G., "A study of the safety of *O*-isopropoxyphenyl-N-methylcarbamate in an operational field-trial in Iran," *Bulletin of World Hlth. Organization,* 38, 609, 1968.

102. van Sittert, N., and Tordoir, W., "Aldrin and dieldrin," in *Biological Indicators for the Assessment of Human Exposure to Industrial Chemicals,* Alessio, L., Berlin, A., Boni, M., and Roi, R., Eds., Commission of the European Communities, Luxembourg, EUR 11135 En, 1987, 3.

103. van Sittert, N., and Tordoir, W., "Endrin," in *Biological Indicators for the Assessment of Human Exposure to Industrial Chemicals,* Alessio, L., Berlin, A., Boni, M., and Roi, R., Eds., Commission of the European Communities, Luxembourg, EUR 11135 En, 1987, 63.

104. Van Doorn, R., Delbressine, L., Leijdekkers, C., Vertin, P., and Henderson, P., "Identification and determination of 2-thiothiazolidine-4-carboxylic acid in urine of workers exposed to carbon disulfide," *Arch. Toxicol.,* 47, 51, 1981.

105. Van Doorn, R., Leijdekkers, C., Henderson, P., Van Hoorne, M., and Vertin, P., "Determination of thiocompounds in urine of workers exposed to carbon disulfide," *Arch. Environ. Hlth.,* 36, 289, 1981.

106. Van Peteghem, Th., and De Vos, H., "Toxicity study of lead naphtenate," *Br. J. Ind. Med.,* 31, 233, 1974.

107. Van Poucke, L., Van Peteghem, C., and Van Hoorne, M., "Accumulation of carbon disulphide metabolites," *Int. Arch. Occup. Environ. Hlth.,* 62, 479, 1990.

108. Veulemans, H., Van Vlem, E., Janssens, H., Masschelein, R., and Leplat, A., "Experimental human exposure to *n*-hexane. Study of the respiratory uptake and elimination of *n*-hexane concentrations in peripheral venous blood," *Int. Arch. Occup. Environ. Hlth.,* 49, 251, 1982.

109. WHO, World Health Organization, *Report of a study group: Recommended Hlth.-based Limits in Occupational Exposure to Heavy Metals,* Technical Report Series 647, Geneva, Switzerland, 1980.

110. WHO, World Health Organization, *Carbaryl,* Data Sheets on Pesticides, Geneva, Switzerland, 1975.

111. WHO, World Health Organization, Carbon Monoxide, *Environmental Hlth. Criteria 13,* Geneva, Switzerland, 1979.

112. WHO, World Health Organization, Inorganic mercury, *Environmental Hlth. Criteria 118,* Geneva, Switzerland, 1991.

113. WHO, World Health Organization, Mercury, *Environmental Hlth. Criteria 1,* Geneva, Switzerland, 1976.

114. Wolfs, P., Dutrieux, M., Scailteur, V., Haxhe, J.-J., Zumofen, M., and Lauwerys, R., "Surveillance des travailleurs exposés à l'oxyde d'éthylène dans une entreprise de distribution de gaz stérilisants et dans des unités de stérilisation de matériel médical," *Arch. Maladies Professionnelles,* 44, 321, 1983.

115. Yamato, N., "Concentrations and chemical species of arsenic in human urine and hair," *Bullet. Environ. Contam. Toxicol.,* 40, 633, 1988.

116. Zielhuis, R., "Epidemiological toxicology of pesticides exposure — Report of an international workshop," *Arch. Environ. Hlth.,* 25, 399, 1972.

4 Biomarkers in Occupational Neurotoxicology

Lucio G. Costa and Luigi Manzo

CONTENTS

1 BIOMARKERS: GENERAL CONCEPTS

There has been an increasing interest in the field of biomarkers in the past several years; specific research funding programs have been developed both in the U.S.A. and in Europe, and at least three new journals devoted to this area have been published. This area of research has also led to a large number of research publications, books and monographs.[1-10] A discussion of biomarkers should start from an exact definition of this term. In animal research, a number of biochemical and/or morphological alterations in tissues of animals exposed to toxicants are often considered biomarkers of target organ toxicities. Most, if not all, of these biomarkers are of great importance to identify specific toxic effects during acute or chronic animal studies, but do not have direct applications to occupational toxicology. Several end-points of toxicity in humans are sometimes referred to as biomarkers. For example, the presence of chloracne may be considered a biomarker of exposure to, and health effect of, dioxin-like compounds; similarly, behavioral alterations in

TABLE 1
A Tiered Approach to the Development, Validation and Application
of Biomarkers to Occupational Epidemiology

1. Toxicology Experiments	• Biomarker development
	• Dose–effect studies
	• Reproducibility
2. Controlled Human Experiments	• Confirmation of biomarker identification in
(where ethical) and/or	humans
Clinical Studies on Heavily Exposed Humans	• Dose–effect estimation
3. Epidemiologic Application	• Improved dosimetry
	• Characterization of biological response markers
	• Identification of susceptibility markers

children exposed to lead may be considered a biomarker of its subtle neurotoxic effects. In this chapter, the term biomarker is used to mean biological/biochemical/molecular markers, which can be measured by chemical, biochemical or molecular biological techniques. As the discussion will focus on biomarkers in occupational neurotoxicology, such markers must be present in tissues, such as blood or urine, that can be ethically obtained in humans by minimally invasive approaches.

A biomarker is a change in a biological system that can be related to an exposure to, or effects from, a specific xenobiotic or a class of toxic compounds.[5] Furthermore, other biomarkers may be indicators of a particular sensitivity to the effects of toxic chemicals. Biomarkers are thus usually divided in three categories: biomarkers of exposure, of effects and of susceptibility.[1] These subdivisions should not be seen as rigid, as overlaps between different types of biomarkers exist and should be considered. For example, DNA adducts can be considered biomarkes of exposure, but may also be considered biomarkers of effects related to carcinogenesis.

General criteria for the selection and use of biomarkers have been discussed in several recent reviews.[1-10] These involve issues related to the development, validation and application of biomarkers to epidemiological studies, the issues of sensitivity and specificity, and their relationship with the time of exposure and/or the progression of the health effect (Table 1). These considerations would also apply to bio-markers for neurotoxicity and some aspects will be discussed in the following sections.

2   BIOMARKERS IN NEUROTOXICOLOGY

Concern for the acute and long-term effects of chemicals on the nervous system has been growing in the past several years.[11,12] The discipline of neurotoxicology, which bridges neurosciences and toxicology, plays a very important role in the increasing efforts aimed at understanding how the brain and the nervous system work, how to intervene to prevent damage and restore function, and how environmental factors may play a role in central nervous system disorders. Since the nervous system controls movement, vision, hearing, speech, thought, emotions, heart function, respiration and many other physiological functions, it is particularly vulnerable to toxic substances, and even minor changes in its structure and function may have profound neurobiological and behavioral consequences.

Neurotoxicity is commonly defined as any permanent or reversible adverse effect on the structure or function of the central and/or peripheral nervous system by a biological, chemical or physical agent. With regard to biomarkers, it is generally acknowledged that the area of neurotoxicity has been progressing more slowly than other fields. Indeed, this topic has been addressed by only a limited number of reviews in the last decade,[13-18] as compared to the large number of publications devoted, for example, to biomarkers related to chemical carcinogenesis. An analysis of the literature, however, suggests that such apparent lack of progress pertains primarily to the area of biomarkers of effects, rather than of biomarkers of exposure, where biological indicators for a large number of neurotoxicants exist, and have been extensively used in several occupational epidemiology studies. The complexity of the nervous system and its distinctive peculiarities, together with problems associated with the multiplicity of manifestations of neurotoxic effects and the determination of the precise targets for neurotoxicants, are certainly responsible for the slow advancement in the development of biomarkers of effects. With regard to biomarkers of susceptibility, the paucity of examples available may be due, for the most part, to the limited attention that these have received within neurotoxicology. However, as most of such markers relate to enzymes involved in xenobiotic metabolism, it is plausible that they may play a role in susceptibility to several neurotoxicants. In this chapter, biomarkers of exposure, effect, and susceptibility as they may apply to neurotoxic compounds will be initially discussed. This will be followed by a more detailed analysis of a selected number of compounds for which research has been active in the past few years, as examples of possible research approaches in this area. As another chapter of this volume is devoted to "traditional" biomonitoring approaches, the present chapter will emphasize some novel aspects of biomarkers and their role as potential new tools in occupational neurotoxicology investigations.

2.1 BIOMARKERS OF EXPOSURE

An ideal biomarker of exposure has been defined as one that is "chemical–specific, detectable in trace quantities, available by non-invasive techniques, inexpensive to assay and quantitatively relatable to prior exposures."[5] This general concept certainly applies also to biomarkers of exposure for neurotoxicants. Traditional biomonitoring for exposure to neurotoxic chemicals has relied on chemical measurements of the compound of interest and/or its metabolites in biological fluids, such as blood or urine, or in other accessible tissues, such as hair or dentine pulp. These measurements are still of great value and, in many instances, still the best or only valid and reliable exposure markers. Examples of these type of biomarkers abound, and can be found in the area of metals, pesticides and solvents, which represent three major classes of chemicals that include several neurotoxic compounds. For metals, blood levels or levels in urine are commonly used. In some cases, measurements of metal concentration in dentine pulp (e.g., lead) or hair (e.g., mercury or arsenic) have proven to be useful, as they would reflect prior and/or cumulative exposure rather than recent exposures.[19] Exposure to neurotoxic solvents is usually monitored by measuring their concentration in blood or in breath, or levels of metabolites in urine, and urine metabolites are also useful for assessing exposure to pesticides. For compounds that are lipophilic and tend to accumulate in fat tissue (e.g., solvents and organochlorine pesticides), fat biopsy is a way to assess the body burden due to prior and/or prolonged exposure, though because of its invasiveness it is rarely used in occupational neurotoxicology.

Binding to macromolecules has proven useful in monitoring exposure to toxic, and particularly genotoxic, compounds, as it would reflect the dose of a certain agent and/or its metabolites that escapes detoxification and reaches its target protein or DNA.[20,21] Recent approaches of this strategy to neurotoxic compounds have occurred with n-hexane and acrylamide, and an application of the latter to occupationally exposed workers is discussed in a following section. As red blood cells are long-lived (approximately four months in humans), binding to hemoglobin is considered a good biomarker to measure cumulative internal dose due to repeated exposures. Adducts to albumin can also be measured; as albumin has a shorter lifetime in blood (20 to 25 days), these measurements will reflect more recent exposure than hemo-globin adducts. One advantage of albumin adducts is that potential active metabolites can interact with this protein directly upon their release in the blood stream, without having to penetrate a cell membrane.[5] Thus, albumin adducts may offer a better, more sensitive marker for detecting reactive metabolite in blood. On the other hand, if adducts are meant to reflect the levels of neurotoxicant at the target site, then hemoglobin adducts would be a more precise biomarker of target tissue dose. Ideally, both measurements could be carried out and compared, however, there are no examples of such strategy for neurotoxic compounds. The limitation of macromol-ecule adducts as biomarkers of exposure lies in the fact that their measurement is often difficult and time-consuming, and that they are limited to compounds and/or their metabolites capable of forming covalent bonds with proteins. On the other hand, an additional positive factor of macromolecule adducts measurements is that they may be also considered biomarkers of effect, when a similar chemical step is involved in the pathogenesis of neurotoxicity, as shown, in animal studies, for the case of n-hexane.[22]

Various biochemical measurements that relate to biological effects of neurotoxic chemicals may be considered biomarkers of exposure rather than biomarkers of effect, as they do not directly relate to neurotoxicity. For example, in case of lead, the activity of erythrocyte δ-aminolevulinic acid dehydratase, which is inhibited by this metal, is often utilized as a biomarker; however, because of wide interindividual variability, it may not be well suited for lead exposure at or below lead levels of 10 μg/dl, which have been associated with behavioral dysfunction.[23,24] Elevated zinc protoporphyrin (ZZP) or the ZZP/hemoglobin ratio are also well correlated with blood lead level, but only at concentrations higher than 40 μg/dl.[25] In addition, these biomarkers should be utilized cautiously, as their alterations may also be associated with the nutritional state of the organism, e.g., iron deficiency.[26] The interaction of neurotoxic metals, particularly lead and mercury, with the heme biosynthetic path-way has also been exploited to measure concentrations of porphyrins in urine.[25] Since different metals inhibit this pathway at different steps, it is possible to "fin-gerprint" exposure to a specific metal based on the urinary porphyrin profile.[25,27]

These measurements reflect biological effects of metals, but are not related to their known mechanisms of neurotoxicity. Nevertheless, they represent additional means to assess exposure to neurotoxic metals. For example, urinary porphyrin changes in dentists exposed to low-level mercury vapors have been found to be a useful measure of cumulative effects of mercury on specific tests of neurobehavioral function.[28,29]

2.2 BIOMARKERS OF SUSCEPTIBILITY

The relevance of genetic factors in the response to toxic chemicals has been often recognized, however, it has not received much attention in neurotoxicological studies.[30] Indeed, animal studies utilize mostly only a single strain of mice or rats, so that genetic variations in response, if present, would not be noted. Some studies have shown that strain differences exist among laboratory animals in response to several xenobiotics, including neurotoxic agents. For example, strain differences in MPTP neurotoxicity in mice have been associated with differences in melanin binding or monoamine oxidase activity,[31] and genetic influences in the developmental neurotoxicity of alcohol are also apparent.[32] In humans, interindividual variations in response to drugs and other xenobiotics are widely observed, and the study of such genetic variability encompasses the fields of pharmacogenetics and ecogenetics, respectively.[33,34]

A number of enzymes involved in xenobiotic metabolism, for example, several members of the cytochrome P450 (CYP), glutathione transferase (GST) and N-acetyltransferase (NAT) families, display genetic polymorphisms.[35,36] Many studies have revealed associations between a certain genotype and increased risk for smoking- or other xenobiotic-related cancers. For example, mutants of CYP1A1 and GSTM1 have been associated with an increased incidence of lung cancer, while slow acetylators (NAT2 mutants) have an increased risk for bladder cancer. Presumably, these individuals have an altered ability to metabolize polycyclic aromatic hydrocarbons and arylamines, respectively. Genetic polymorphisms have also been identified for other enzymes involved in xenobiotic metabolism, such as epoxide hydrolase, alcohol and aldehyde dehydrogenases, various esterases, and methyltransferases.[35,36]

Organic neurotoxic compounds can be bioactivated or detoxified by these same enzymes, and genetic polymorphisms could certainly play a role in differential sensitivity to their effects on the central and/or peripheral nervous system. With very few exceptions, biomarkers of susceptibility have, however, not been investigated with regard to neurotoxicants. Yet, these may be of great importance where low exposures leading to subtle behavioral effects are investigated. Furthermore, in addition to their major hepatic localization, several of these enzymes are also expressed in the nervous system,[37] where they may contribute to variations in the *in situ* activation or detoxication of neurotoxicants.

Some research on genetic polymorphism has been carried out in the context of studies on the role of environmental factors in the etiology of neurodegenerative diseases, in particular Parkinson's disease. The hypothesis underlying these studies is that genetically determined metabolic differences may contribute to an increased risk for Parkinson's disease, as a result of exposure to still unidentified environmental neurotoxic agents. Several studies have investigated whether impaired debrisoquine hydroxylation, which is due to a cytochrome P450 (CYP2D6), or polymorphism of monoamine oxidase B, may be genetic susceptibility factors for Parkinson's disease; however, the results of such studies have often yielded contrasting results.[38-42]

Genetic polymorphisms in enzymes not associated with xenobiotic metabolism but involved in the metabolism of endogenous neurotransmitters (e.g., catechol-o-methyltransferase) also may affect the response to certain neurotoxicants. Similarly, genetic variations in receptors or other enzymes involved in cellular functions, including repair mechanisms, may be relevant to fully predict and assess neurotoxic outcome. Information in these areas are still scant, but will certainly increase as the knowledge of the human genome progresses, and will offer new tools for investigation.

2.3 BIOMARKERS OF EFFECTS

Biomarkers of effects should reflect early biochemical modifications that precede structural or functional damage. As such, knowledge of the mechanism(s) that lead to ultimate toxicity is necessary, or, at least extremely important, to develop specific and useful biomarkers. These markers should identify early and reversible biochemical events which may also be predictive of later responses.[17] Unfortunately, the exact mechanism of action for most neurotoxic chemicals is still unknown, which slows progress in the area of biomarker research. Furthermore, finding sensitive and specific surrogate markers for the central and peripheral nervous system in readily accessible tissues can be problematic. The complexity of the nervous system and the diversity of manifestations of neurotoxicity, together with the existence of multiple cellular and biochemical targets for many chemicals, make it unlikely that "generic" markers for neurotoxicity will be developed. The problem is, thus, different from that encountered with genotoxic compounds, where, for example, changes in sister chromatide exchange in circulating lymphocytes may reflect a significant biological effect, common to many chemicals. Despite these obvious limitations, different strategies have been employed over the years in the attempt to develop peripheral biochemical indicators of neurotoxic effects.

An area that has received some attention in the past several years is that of neurotransmission. Several chemicals affect various steps of neurotransmission, including neurotransmitter metabolism, receptor interactions, second messenger systems or other relevant enzymes.[13,43,44] As changes due to neurotoxicant exposure cannot be measured in target tissue, suitable peripheral cell systems must be identified that would mirror identical neurochemical parameters in the nervous system (Table 2). Cell types that have been used include platelets, erythrocytes, lymphocytes and fibroblasts,[4,44] and this strategy has found applications in the field of biological psychiatry and asthma research.[13] The oldest, and probably still the best, example of the application of such strategy to neurotoxic compounds, is represented by the measurement of red blood cell acetylcholinesterase following exposure to organophosphorus insecticides. This and other examples of this approach for some neurotoxic chemicals are discussed in following sections. The development of such markers should follow the usual steps of characterization and validation. It is necessary to determine whether the particular enzyme or receptor studied in blood cells is

TABLE 2
Some Parameters of Neurotransmission in Blood Cells

Platelets	Monoamine oxidase B
	Alpha$_2$–adrenoceptors
	Serotonin uptake
Red Blood Cells	Acetylcholinesterase
	Na$^+$/K$^+$–ATPase
Lymphocytes	Beta–adrenoceptors
	Muscarinic cholinergic receptors
Plasma	Cholinesterase
	Dopamine beta hydroxylase

indeed the same entity present in the nervous system, and whether it is similarly affected *in vitro* by a neurotoxicant. Animal studies should then be carried out to investigate whether a good correlation exists between changes in the peripheral markers and those observed in the nervous system; dose — response and time — course studies should be included to assess sensitivity of measurements, and the duration of the observed changes. Pilot studies in humans should determine inter-individual variations, as well as sensitivity upon exposure to occupationally relevant concentrations. In addition to this "background" work, one should also consider limitations and pitfalls of this approach. For example, it may be difficult at times to determine whether alterations of a neurotransmitter metabolite in blood reflect a primary effect in the central and/or peripheral nervous system. Furthermore, one should carefully consider whether other endogenous or exogenous agents (hormones, drugs), pathological conditions or genetic make-up can affect the measured parameters. Additionally, there is the possibility that falsely positive results may be obtained with compounds that do not cross the blood-brain barrier.

A few additional biomarkers for neurotoxic effects have been recently proposed. For example, animal studies have shown that levels of the astrocyte-specific Glial Fibrillary Acidic Protein (GFAP) increase following neuronal damage, and this biochemical measurement has acquired notable importance as an indicator of neurotoxicity.[45] Blood GFAP antibody levels have been found to be significantly elevated in a cohort of workers exposed to lead;[46] however, it is unclear whether this change is related to a damage of the blood-brain barrier, rather than a biomarker of neuronal damage. Indeed, levels of another protein, creatine kinase isoenzyme BB (CK-BB) were found to be increased in blood of boxers and head trauma patients.[47,48] Since levels of CK-BB are very low in control individuals, this increase may indicate disruption of the blood-brain barrier. The cerebrospinal fluid (CSF) is another compartment that may be accessible in humans, though its use in occupational neuroepidemiology studies is, at least, premature. A very recent study has reported the appearance of myelin basic protein (MBP) in CSF of rats following intracerebellar injection of lysolecithin.[49] As this treatment causes extensive demyelination, the increase of MBP in CSF, which lasted for about four days, may represent a useful marker for this effect.

3 APPLICATIONS OF BIOMARKERS
TO OCCUPATIONAL NEUROTOXICOLOGY

3.1 STYRENE

Styrene is a very important solvent used in the manufacture of numerous polymers and copolymers, including polystyrene, styrene — acrylonitrile or styrene — butadiene rubber.[50] Like other organic solvents, acute exposure to high levels of styrene causes irritation (of both skin and respiratory tract) and central nervous system depression, while upon chronic exposure, styrene may have carcinogenic and reproductive toxicity.[51,52] Styrene exposure has also been reported to cause neurotoxic effects. Indeed, a large number of studies in occupationally exposed workers have reported signs and symptoms of central nervous system toxicity,[53-58] though some of these findings have been recently criticized.[59]

The biological monitoring of styrene has been extensively studied.[60] Styrene is metabolized to styrene oxide by cytochrome P450s, in particular CYP2E1; styrene oxide is detoxified via glutathione transferases and/or epoxide hydrolase, the latter pathway being more significant in humans than rodents.[61] Mandelic acid and phenylglyoxylic acid are the most prominent urinary metabolites, and, together with levels of styrene in blood or exhaled air, have been commonly used for biological monitoring.[60,62,63] Three-fold differences in the relative urinary excretion of optical enantiomers of mandelic acid have been reported, which may be related to polymorphisms of cytochrome P450s, leading to R or S styrene 7,8-oxide, or of epoxide hydrolase.[64] This observation may be relevant with regard to genotoxicity, as R-styrene 7,8-oxide has a stronger mutagenic effect than its S-enantiomer.[65] Styrene oxide can form adducts to the N-valine residue on hemoglobin,[66-68] and measurement of these adducts has been utilized to monitor exposure to styrene in animals and humans.[69-71] However, styrene oxide is not considered an effective alkylator of hemoglobin compared to other toxicants such as ethylene oxide.[72] Furthermore, the capacity of humans to form styrene oxide is much lower than rats or mice,[61] which results in very low levels of styrene oxide in blood,[73] as well as of styrene oxide–hemoglobin adducts.[69,70]

While these measurements can be used as biomarkers of exposure to styrene, less progress has been made with regard to biomarkers of health effects. The specific neurotoxic targets for styrene, its mechanism(s) of neurotoxicity, and the role played by styrene oxide are, indeed, for the most part, still unknown. Exposure of rats to styrene has been shown to decrease levels of brain glutathione,[74] and this effect has been attributed, for the most part, to detoxication of styrene oxide by glutathione transferases.[75] Following chronic exposure to styrene, and increase in GFAP was found, possibly reflecting reactive astrogliosis to styrene — or styrene oxide — induced neuronal damage.[76] An *in vitro* study in PC12 cells has indeed shown that styrene oxide causes depletion of intracellular glutathione and ATP, followed by elevation of free calcium levels and induction of DNA single strand breaks.[77] A recent study in primary cultures of murine spinal cord–dorsal root ganglia indicated cytotoxicity of styrene and styrene oxide (the latter being about 10 fold more potent), and oxidation of multiple cellular macromolecules was suggested as a mechanism

TABLE 3
Platelet Monoamine Oxidase B Activity
in Workers Exposed to Styrene

Air styrene concentration (ppm)	MAO B activity (pmole/10^8 cells/h)
1.54	5.36
14.84	4.11
41.21	3.75
89.69	3.63

Adapted from Checkoway et al.[86]

of toxicity.[78] None of this limited mechanistic study has, however, led to the development of biomarkers for use in human investigations. Additional animal studies have suggested that styrene can affect the metabolism of dopamine; an increase in dopamine D2 receptors has been reported,[79] that may occur as compensatory reaction to the observed decrease in dopamine levels,[80] possibly due to condensation of dopamine with the styrene metabolite phenylglyoxylic acid.[81] These findings were mechanistically correlated with an increase in prolactin levels found in blood of styrene-exposed workers, as prolactin release from the anterior pituitary gland is chronically inhibited by dopamine.[82-84] Although of interest, this hypothesis should still be seen with caution, and further evidence should be provided in its support. A decrease in monoamine oxidase B (MAO B) was found in brain of rats exposed to styrene.[85] A dose-related decrease of MAO B activity has also been found, in three separate studies, in platelets from workers exposed to styrene.[58,86] The effect appeared to be specific for styrene, as it was not observed in platelets of workers exposed to the solvents perchloroethylene and toluene (Table 3).[86,87] On the other hand, a decrease in serum dopamine beta hydroxylase (DBH) activity was found in workers exposed to both styrene and toluene,[84,87] possibly because of inhibition of its activity by phenol/cresol metabolites of these solvents, but not following exposure to trichloroethylene.[88] Overall, these observations are of interest, because they may lead to effect-related biomarkers for styrene. However, a better understanding of the mechanism of styrene (and/or styrene oxide) neurotoxicity is necessary, before these and other biomarkers of effect can be utilized with confidence in the occupational setting.

3.2 ACRYLAMIDE

Acrylamide is used in the synthesis of polyacrylamides, which have a variety of industrial applications. A number of toxic effects of acrylamide, such as reproductive toxicity and, possibly, carcinogenicity, have been suggested by animal studies;[89] however, the most relevant effect identified in humans is neurotoxicity, most notably a distal axonopathy.[90-92] Earlier studies in rats had indicated that a high level of acrylamide is associated with blood erythrocytes, and that acrylamide is covalently bound to cysteine residues in brain proteins and hemoglobin.[93,94] Based on this information, Bailey *et al.*[95] developed a gas chromatography–mass spectrometry method to measure S-(2-carboxyethyl) cysteine in hydrolyzed globin from acrylamide-treated rats as a potential biomarker to assess exposure to this compound. In the past few years, a series of studies have extended these results to both animals and exposed human populations. In rats injected with increasing doses of acrylamide, hemoglobin adducts increased linearly;[96] furthermore, hemoglobin adducts of an epoxide metabolite of acrylamide, glycidamide, were also identified.[96,97] Glycidamide may be involved in the reproductive toxicity of acrylamide[98] and in its potential carcinogenicity, as this metabolite, but not the parent compound, is mutagenic in the Ames test.[99] Uncertainties still exist, however, on the role of glycidamide in the pathogenesis of acrylamide-induced peripheral neuropathy, since contrasting results have been reported.[100,101]

A method was also developed to determine acrylamide adducts to N-terminal valine in human hemoglobin, which was used to assess exposure to acrylamide in a group of factory workers.[102] Adducts were detected in all exposed workers (0.3–34 nmol/g Hb) and in only one out of 10 controls (0.01 nmol/g Hb). Based on their roles and locations in the production process, workers were divided into four groups, and the highest adduct levels (14.7 nmol/g Hb) were found in workers involved in the synthesis of acrylamide from acrylonitrile and in the transfer of 35% acrylamide solutions into barrels (Table 4). Based on extrapolations from air concentrations of acrylamide, it was calculated that only a fraction of adducts (0.44 nmol/g Hb) would derive from inhalation of acrylamide, indicating dermal absorption as a primary route of exposure.[103] Among the exposed workers, signs and symptoms indicating peripheral neuropathy were found with statistically significant increased frequencies compared to controls. Based on the result of a questionnaire, of a neurological examination and of vibration thresholds and electroneuromyographic measurements, a neurotoxicity index for acrylamide-induced peripheral neuropathy was developed.[103] The neurotoxicity index, which adequately predicted the clinical diagnosis of peripheral neuropathy, was significantly correlated with hemoglobin adducts of acrylamide (Table 4), but not with the concentration of acrylamide in the air or in plasma of exposed workers.[103] A good correlation was also found between the neurotoxicity index and mercapturic acid in the urine (which could, however, result from both acrylamide and acrylonitrile exposure).[103] Overall, these studies indicate that measurement of hemoglobin adducts of acrylamide is a good biomarker of exposure and may be useful as an indicator of acrylamide-induced peripheral neuropathy. Though the mechanism of the peripheral neurotoxicity of acrylamide has not been fully elucidated, these studies suggest that a covalent bond to axonal proteins may present a relevant step in the pathogenesis of its axonopathy. Further studies may also lead to a better definition of the dose–response relationship of hemoglobin adducts formation and their use as predictors of peripheral neurotoxicity.

3.3 ORGANOPHOSPHATES

Organophosphorus compounds are among the most widely used insecticides; their mechanism of action involves inhibition of acetylcholinesterase (AChE), which leads to accumulation of acetylcholine at cholinergic synapses with an ensuing

TABLE 4
Acrylamide–Hemoglobin Adducts in Exposed Workers

Work category	Acrylamide–valine adducts (nmol/g Hb)	Neurotoxicity Index
Synthesis	13.4	19.2
Ambulatory	9.5	11.3
Polymeryzation	7.7	10.0
Packaging	3.9	8.9
Controls	0	0

Adapted from Calleman et al.[103]

cholinergic crisis.[104] AChE is widely distributed throughout the body and is also present in blood cells, such as erythrocytes and lymphocytes, though its physiological role in these cells, which are devoid of synaptic contacts, has not been elucidated. Inhibition of AChE, as well as of pseudocholinesterase in plasma, has been extensively used as a biomarker of exposure to, and effect of, organophosphates.[105,106] When used as a marker of exposure in populations studies, one should consider the issue of interpersonal variability in the activity of these enzymes. In the best circumstances, baseline values should be obtained for each individual, and variations below these activity levels, rather than absolute levels, should be utilized to assess exposure. In the absence of pre-exposure measurements, repeated post-exposure measurements at different intervals should be obtained, rather than comparison with population ranges. If acute exposure to an organophosphate has occurred, AChE activity would increase over time to reach a level considered normal for that individual.[107]

A much debated issue is whether red blood cell AChE or plasma cholinesterase are a better indicator of exposure to organophosphates. Many organophosphates appear to be better inhibitors of cholinesterase, suggesting that this enzyme may be a more "sensitive" indicator of exposure; however, this is not true for all organophosphorous compounds. Furthermore, plasma cholinesterase activity displays a higher variability, as it can be affected by other exogenous agents (e.g., drugs) or physiological and pathological conditions (e.g., pregnancy or liver damage).[108] Genetic variants of human serum cholinesterase also exist; individuals with atypical cholinesterase, which occurs in homozygous form in one out of 3,500 Caucasians, have an a abnormal response to the muscle relaxant succinylcholine;[109] however, it has not been determined whether genetic variants of cholinesterase may affect sensitivity to organophosphates. If blood cholinesterase activity is considered as a biomarker of effect of organophosphates, it is important to determine whether this peripheral measurement reflects similar changes occurring in target tissues, i.e., the central nervous system and the muscles, particularly the diaphragm. Surprisingly, very few studies have directly attempted to address this important question, though those which did concluded that blood cholinesterase is a good indicator of target organ enzyme activity (Table 5).[110-112] Erythrocyte AChE, in particular, was found to be better correlated with brain or diaphragm activity than plasma cholinesterase

TABLE 5
Correlation between Brain and Blood Cholinesterase
Activity after Acute Exposure of Rats to Chlorpyrifos

Days after dosing	Red blood cells	Plasma	Whole blood
1	0.65	0.58	0.68
4	0.81	0.69	0.85
7	0.89	0.89	0.89
21	0.84	0.75	0.91
35	0.67	0.28	0.76

Adapted from Padilla *et al.*[112]

following acute exposure to the insecticides chlorpyrifos and paraoxon.[112] During repeated exposures to the insecticide disulfoton, the strongest correlation was seen between brain and lymphocyte AChE activity.[111] However, in the recovery period, following termination of exposure, red blood cell AChE better reflected brain AChE activity.[111] It would be useful to conduct additional studies comparing blood and target tissue AChE under various exposure conditions; several different organophosphates should be tested, possibly at low doses for extended periods of time, to mimic occupational exposure. Nevertheless, measurement of blood AChE activity remains an excellent biomarker for exposure and effect of organophosphate exposure, under both acute and chronic conditions.

Upon repeated exposure to organophosphates, tolerance to their toxicity has been shown to develop; this phenomenon is mediated, at least in part, by a homeostatic decrease in the density of cholinergic receptors, to compensate for the prolonged increase in acetylcholine levels.[113] These changes in cholinergic receptors, particularly the muscarinic type, may be seen as a protective mechanism by which the organism normalizes function despite challenge from the external environment. On the other hand, balance of neuronal connections may have been altered, and higher brain functions might be compromised by such receptor alterations. Indeed, in animals repeatedly exposed to the organophosphates, memory deficits have been reported.[114,115] As such cognitive impairment was also observed in some, though not all, studies of occupationally exposed workers,[116,117] experiments were carried out to investigate whether a peripheral biomarker of muscarinic receptors could be found, that would reflect changes in central nervous system muscarinic receptors. Lymphocytes were considered as a possible surrogate tissue, as in these cells from rats and humans, muscarinic receptors have been identified at the protein and mRNA levels.[118,119] Following a two week exposure to the organophosphate disulfoton, the density of muscarinic receptors was decreased to a similar degree in hippocampus and cerebral cortex as well as in circulating lymphocytes.[120] However, while a strong correlation between the levels of muscarinic receptors in brain areas and lymphocytes was found during the period of exposure, in the recovery period, following termination of exposure, such correlation was not present, possibly because of the high turnover of lymphocytes.[111] Thus, measurement of lymphocytic muscarinic receptors should be seen as a useful indicator of central nervous system changes only during prolonged exposure, but not afterwards.

A number of organophosphates also cause another type of neurotoxicity, characterized as a central–peripheral distal axonopathy.[121] This syndrome, commonly known as OPIDP (Organophosphate Induced Delayed Polyneuropathy), is totally independent of inhibition of AChE, and is delayed, as symptoms appear about two–three weeks after the initial poisoning, when acute cholinergic signs have subsided. The mechanism of initiation of OPIDP involves the phosphorylation of a protein in the nervous system called NTE (Neuropathy Target Esterase) and the aging of the phosphoryl-enzyme complex.[122] Compounds that age cause OPIDP if the threshold of inhibition of NTE (70–80%) is reached, whereas compounds that do not age (e.g., phosphinates or carbamates) do not cause OPIDP and, when given before a neuropathic compound, actually protect against its delayed neurotoxicity. NTE activity has been found in lymphocytes and platelets[123,124] and within 24 hours after acute exposure, there is a good correlation between lymphocyte and brain NTE in the hen, which is the species of choice for OPIDP studies.[125] Measurement of lymphocyte NTE has been suggested as a potential biomarker to monitor for organophosphate-induced polyneuropathy.[126] The best example of its application in humans is in a case of attempted suicide with the insecticide chlorpyrifos, where, based on 60% inhibition of lymphocyte NTE, it was correctly predicted that a delayed neuropathy would develop, well after recovery from acute cholinergic symptoms.[127] In a study of workers exposed to the defoliant DEF, inhibition of lymphocyte NTE was observed, but was considered a false positive, as no clinical or electrophysiological signs of OPIDP were detected in exposed workers.[128] Recently, compounds that given before a neuropathic organophosphate offer protection, have been found to act as "promoters" (i.e., to potentiate OPIDP), when given afterwards.[129] These findings have challenged the understanding of the mechanisms of OPIDP, and, unfortunately, complicated the use of lymphocyte NTE as a biomarker, as the clinical outcome of a combined exposure to protective/promoter and neuropathic pesticides would be unpredictable.[121]

As organophosphates undergo metabolic activation and detoxication, these pathways have been considered for biological monitoring and for studies on genetic susceptibility. Most organophosphates are activated to their correspective oxygen analog by an oxidative desulfuration reaction, catalyzed by cytochrome P450. Upon phosphorylation of AChE, a portion of the molecule, the so-called leaving group, is released and excreted. Both the parent compound and the oxon can undergo a series of detoxication reactions, mediated by various A esterases (paraoxonase, carboxylesterase), by P450s, and by glutathione transferases. The leaving group, for example p-nitrophenol in case of parathion, which is also generated by hydrolytic cleavage, as well as alkylphosphates, are excreted in the urine, and can be quantified as an index of organophosphate exposure.[130-132] In case of alkylphosphates, studies in humans have indicated that they may represent a sensitive marker of organophosphate exposure, as they may be detected even if no significant changes in blood cholinesterase activity can be measured.[133] Because of the importance of the activation step in the toxicity of organophosphates, genetically determined differences in cytochrome P450 may play a significant role. Though the need of the conversion to the oxon has been known for almost 40 years, the specific P450 isozyme(s) responsible for the activation of thioates have not been identified. Clearly, genetic variations in these enzymes may be important, in particular to explain the exacerbated reactions to organophosphate exposure observed in certain individuals. This is certainly an area where some additional research efforts should be addressed.

A metabolic pathway that has been investigated as a possible source of genetic susceptibility, is the hydrolysis of several organophosphates by paraoxonase. This enzyme, which takes its name from its most studied substrate, paraoxon, is capable of hydrolyzing the oxygen analogs of a number of commonly used organophosphorus insecticides, such as chlorpyrifos oxon and diazinon oxon. It long has been known that human paraoxonase exhibits a substrate-dependent polymorphism.[134] One form hydrolyzes paraoxon with a high turnover number, and the other with a low turnover number.[135] In addition to the observed polymorphism, there is a large variation in enzyme levels (over 10 fold) observed within a genetic class, with a full scale variation of about 60 fold.[135] Several lines of evidence suggest that high levels of serum paraoxonase are protective against poisoning by organophosphorus insecticides whose active metabolites are substrates of this enzyme. For example, rabbits, that have a very high level of serum paraoxonase, are less sensitive than rats to the acute toxicity of paraoxon.[136] Furthermore, when rats or mice are injected with paraoxonase purified from rabbit serum to increase the blood hydrolyzing activity, their sensitivity toward the toxicity of paraoxon and chlorpyrifos oxon as well as of the parent compound chlorpyrifos, is significantly decreased (Table 6).[137,138] These animal experiments provide convincing evidence that serum paraoxonase is an important determinant of susceptibility to organophosphate poisoning. The polymorphism of human paraoxonase has been recently elucidated. One isoform, with arginine at position 192, hydrolyzes paraoxon with a high rate, whereas the other isoform with glutamine at position 192, hydrolyzes paraoxon at a slower rate.[139] Three genotypes have been observed in human populations: individuals homozygous for the low activity allele (48%), individuals homozygous for the high activity allele (9%) and heterozygotes (43%).[135] Certain considerations should be made about the use of paraoxonase status as a genetic marker for susceptibility to poisoning by organophosphates. First, not all substrates exhibit polymorphism (e.g., chlorpyrifos oxon) does not. Second, some substrates (e.g., diazinon oxon) exhibit a "reversed" polymorphism in that high metabolizers of paraoxon are poor metabolizers of diazinon-oxon (Furlong, Costa et al., unpublished observation). Thus, genetic susceptibility cannot be generalized and needs to be characterized with regard to specific organophosphorus compounds.

3.4 OTHER COMPOUNDS

Two additional neurotoxic compounds for which promising biomarkers are being developed are n-hexane and carbon disulfide. The neurotoxicity of n-hexane was first identified in humans and then confirmed by animal studies. Symptoms of n-hexane neurotoxicity are numbness of the extremities, followed by weakness of the intrinsic muscles of the hands and feet and, with continuous exposure, by

TABLE 6
Protective Effect of Paraoxonase
on Organophosphate Toxicity

Treatment	AChE (% of untreated control)	
	Control	+ Paraoxonase
Chorpyrifos oxon (14 mg/kg)		
Brain	12 ± 1	89 ± 3*
Diaphragm	23 ± 2	89 ± 7*
Chlorpyrifos (100 mg/kg)		
Brain	42 ± 4	83 ± 4*
Diaphragm	22 ± 3	57 ± 3*

Paraoxonase was purified from rabbit serum and injected i.v. in mice 30 min. before dermal application of the organophosphates. Serum chlorpyrifos-oxonase activity was increased 35-fold by this treatment. *p<0.05.

Adapted from Li *et al.*[138]

progressive loss of sensory and motor functions. Observations in humans and investigations in animals have characterized this as a distal sensorimotor neuropathy, specifically a distal axonopathy of the so called "dying-back" type.[140,141] Another solvent, methyl n-butyl ketone, also causes a sensorimotor neuropathy, and the formation of the same toxic metabolite, 2,5-hexanedione, from both *n*-hexane and methyl n-butyl ketone, has been firmly demonstrated. The mechanism of *n*-hexane neurotoxicity has been elucidated and has been recently reviewed.[141] Briefly, the reaction of 2,5-hexanedione (and other γ-diketones, but not diketones with other than γ-spacing) with the lysyl groups of proteins leads, via a series of intermediate steps, to the formation of chemical adducts characterized as pyrroles; pyrrole formation is a critical step in the sequence of events that results in the axonopathy.[142]

Exposure to *n*-hexane in occupationally-exposed workers can be assessed by determination of urinary levels of the toxic metabolite 2,5-hexanedione.[143,144] However, this measurement reflects only recent exposure, in the order of a few hours to a few days. On the other hand, the understanding of the molecular mechanisms of neurotoxicity offers the possibility of utilizing measurements of pyrrole adducts to macromolecules as a biomarker of cumulative exposure to *n*-hexane. Experiments in rats exposed to γ-diketones have shown that formation of hemoglobin adducts is proportional to both time and dose.[22] As the formation of adducts has been shown to be causally related to the development of neuropathy, hemoglobin may be seen as a surrogate for neurofilaments during *in vivo* exposure.[141] As such, measurements of hemoglobin pyrrole adducts in case of *n*-hexane are more than a biomarker of exposure, and can be seen as a biomarker of effect. In this regard, more detailed dose-response studies in animals, as well as studies in humans, may be useful to arrive at a better quantitative assessment of neurotoxic risk linked to exposure to *n*-hexane or other γ-diketone precursors. As the cross -linking of neurofilaments is essential for neurofilament-filled axonal swellings,[141] the possibility to measure protein cross-linking in an accessible tissue would provide an additional means of assessing exposure to, and health effects of, n-hexane. A protein present in erythrocytes, spectrin, has proved to be useful in this regard. The α and β subunits of spectrin are closely associated on the cytoplasmic side of the red blood cell membrane. Cross-linking compounds yield α, β-heterodimers that can be identified in SDS polyacrylamide gels.[141] Neurotoxic γ-diketones have been indeed shown to cause spectrin dimerization, while administration of nontoxic diketones does not cause spectrin-cross linking.[141,142] This biochemical change may serve, even more than pyrrole adducts, as a biomarker directly related to pathological changes in the axon.

Exposure to another widely used solvent, carbon disulfide (CS_2), also results in neurotoxicity. The most common effect is a distal sensorimotor neuropathy, characterized as a neurofilamentous axonopathy, and identical to that caused by n-hexane.[141,145] Chronic exposure to CS_2 has also been associated with the development of an encephalopathy,[146] whose mechanism has not been elucidated. Though CS_2 does not form pyrrole adducts, through a series of interactions with amino groups of proteins, it ultimately causes neurofilament cross-linking,[147] which has led to the suggestion that erythrocyte spectrin, which is also cross-linked by CS_2, may be used as a potential biomarker to assess the peripheral neurotoxicity of CS_2.[147] Spectrin dimers could be measured prior to clinical or morphological evidence for neurotoxicity,[148] suggesting that this biomarker may be useful to detect pre-morbid conditions. Carbonyl sulfide, which derives from oxidative metabolism of CS_2 by yet unidentified cytochrome P450(s), can also participate in cross -linking reactions via a protein-bound isocyanate intermediate.[141] The contribution of carbonyl sulfide in the peripheral neurotoxicity of CS_2 is not known, but, in light of possible polymorphisms of the P450(s) involved in the oxidation of CS_2, it would be of interest to define the relative roles of the parent compound and this metabolite.

In the occupational setting, concentration of CS_2 in expired air has been used as a marker of exposure;[149] however, this measurement only reflects very recent exposure. The reaction of CS_2 with amines to yield dithiocarbamates has led to the monitoring of the latter in blood as a method to assess occupational exposure.[150] Furthermore, 2-thiothiazolidine-4-carboxylic acid, a metabolite derived either from dithiocarbamate and/or from trithiocarbonate, has been identified in urine from workers exposed to CS_2, and has been utilized to monitor exposure;[151,152] indeed, the current Biological Exposure Index is based on such measurements.[153] As mentioned earlier, encephalopathy is another manifestation of the neurotoxicity of CS_2 neurotoxicity. Animal studies have shown that in rats, brain dopamine levels are increased following exposure to CS_2,[154] suggesting that, among various possibilities, CS_2 may inhibit dopamine beta hydroxylase (DBH). This biochemical effect has indeed been observed *in vitro*,[155] and *in vivo* in rat adrenals following acute, but not after chronic, exposure.[156,157] On the other hand, in a group of female viscose rayon workers exposed to low levels of CS_2, a decrease in serum DBH was observed only after prolonged exposure (more than one year).[158] There are no studies that have pursued the involvement of alterations of dopamine metabolism in the central neurotoxicity of CS_2; thus, the significance of these observations with regard to possible utilization as biomarkers is unknown. Furthermore, the effects on peripheral dopamine metabolism are probably impossible to discern from those on the central nervous system.

4 CONCLUSIONS

This chapter has provided a discussion of some aspects of biomarkers related to the assessment of exposure and susceptibility to, and effects of, neurotoxic chemicals. By far, the best available tools for use in occupational neurotoxicology studies are in the area of biomarkers of exposure; traditional measurements of neurotoxic chemicals and/or their metabolites in biological fluids provide useful and reliable indicators of exposure. Novel approaches, such as measurements of hemoglobin adducts, may be useful for assessing prior and/or repeated exposures to electrophilic compounds. For these chemicals, further mechanistic studies should be carried out, as the formation of adducts to proteins in the target tissue may represent an essential step in the pathogenesis of neurotoxicity. If this were the case (as discussed for *n*-hexane), measurements of adducts in blood may be also utilized as biomarkers of effect. As several neurotoxicants undergo metabolic activation and/or detoxification by enzymes that are known to exhibit polymorphisms in humans, it is surprising that little attention has been devoted to this topic within neurotoxicology. The examples and suggestions provided may serve to stimulate more research in this area, which may be relevant in light of the lower levels of exposure normally encountered in developed countries. As techniques for genotyping are becoming more accessible, epidemiological studies on occupationally exposed workers will certainly include such measurements. With regard to biomarkers of effects, it is clear that new opportunities will derive mostly, if not solely, from a better understanding of the cellular and molecular targets for neurotoxicants. In the few cases where this has occurred (e.g., organophosphates), useful biomarkers have indeed been developed. Clearly, neurotoxicity can be manifested in a wide variety of ways, each one characterized by different cellular and biochemical substrates. Thus, the development of one or more markers for "neurotoxic effects" does not appear as a feasible option. Rather, biomarkers would be specific for a class of chemicals or certain target cellular process. Still, by understanding the chain of events that ultimately lead to neurotoxicity, it would be possible to develop biomarkers that are indicative of early biochemical alterations preceding irreversible damage. As neurotoxicology, possibly more than other areas of toxicology, requires a multidisciplinary approach, the best use of biomarkers would ultimately be in conjunction with electrophysiological and/or behavioral assessments of exposed populations.

5 ACKNOWLEDGMENTS

Research by the authors was supported by grants from NIEHS, EPA, NIOSH, the European Community and the National Research Council (CNR), Italy. Ms. Chris Sievanen provided secretarial assistance.

6   REFERENCES

1. National Research Council, "Biological markers in environmental health research," *Env. Hlth. Persp.* 74,3, 1987.
2. National Research Council, *Biologic Markers in Immunotoxicology,* National Academy Press, Washington, DC, 1992.
3. National Research Council, *Biologic Markers in Reproductive Toxicology,* National Academy Press, Washington, DC, 1989.
4. Lucier, G. W. and Thompson, C. L., "Issues in biochemical applications to risk assessment: when can lymphocytes be used as surrogate markers?" *Env. Hlth. Persp.* 76,187, 1987.
5. Henderson, R. F., Bechtold, W. E., Bond, J. A., and Sun, J. D., "The use of biological markers in toxicology," *Crit. Rev. Toxicol.* 20, 65, 1989.
6. Hulka, B. S., Wilcosky, T. C., and Griffith J. D., Eds., *Biological Markers in Epidemiology,* Oxford University Press, Oxford, UK, 1990.
7. Schulte, P. A. and Perera, F. P., *Molecular Epidemiology, Principles and Practices,* Academic Press, San Diego, 1993.
8. Schulte, P. A., "Contribution of biological markers to occupational health," *Am. J. Ind. Med.*, 20,435, 1991.
9. Timbrell, J. A., Draper, R., and Waterfield, C. J., "Biomarkers in toxicology: new uses for some old molecules?" *Biomarkers* 1,1, 1996.
10. Wilcosky, T. C., "Criteria for selecting and evaluating markers," in *Biological Markers in Epidemiology,* Hulka, B. S., Wilcosky, T. C., and Griffith, J. D., Eds, Oxford University Press, Oxford, UK, 1990, 28.
11. Office of Technology Assessment, *Neurotoxicity: Identifying and Controlling Poisons of the Nervous System.*, U.S. Congress, 1990.
12. National Research Council, *Environmental Neurotoxicology,* National Academy Press, Washington, DC, 1992.
13. Costa, L. G., "Peripheral models for the study of neurotransmitter receptors: their potential application to occupational health." In *Occupational and Environmental Chemical Hazards,* Foa V., Emmett E. A., Maroni M., Colombi A., Eds, Ellis Horwood, Chichester, UK, 1987, 524.
14. Maroni, M.and Barbieri, F., "Biological indicators of neurotoxicity in central and peripheral toxic neuropathies," *Neurotoxicol. Teratol.* 10, 479, 1988.
15. Slikker, W., "Biomarkers of neurotoxicity: an overview," *Biomed. Environ. Sci.,* 4, 192, 1991.
16. Costa, L. G. and Manzo, L., "Biochemical markers of neurotoxicity: research strategies and epidemiological applications," *Toxicol. Lett.*, 77, 137, 1995.
17. Silbergeld, E. K., "Neurochemical approaches to developing biochemical markers of neurotoxicity: review of current status and evaluation of future prospects," *Environ. Res.* 63, 247, 1993.
18. Costa, L. G., "Biomarker research in neurotoxicology: the role of mechanistic studies to bridge the gap between the laboratory and epidemiological investigations," *Env. Hlth. Persp.* 104(S1), 55, 1996.
19. Clarkson, T. W., "The role of biomarkers in reproductive and developmental toxicology," *Env. Hlth. Perspect.* 74," 103, 1987.
20. Farmer, P. B., Neumann, H. G. and Henschler, D., "Estimation of exposure of man to substances reacting covalently with macromolecules," *Arch. Toxicol.* 60, 25, 1987.

21. Goldring, J. M. and Lucier, G. W., "Protein and DNA adducts." In *Biological Markers in Epidemiology,* Hulka, B. S., Wilcosky, T. C., and Griffith, J. D., Eds., Oxford University Press, Oxford, UK, 1990, 78.

22. Anthony, D. C., Amarnath, V., Simons, G. R., St. Clair, M. B. G., Moody, M. A., and Graham, D. G., "Accumulation of pyrrole residues as the molecular basis of cumulative neurotoxic dose of 2,5-hexanedione," *J. Neuropathol. Exp. Neurol.,* 47, 325, 1988.

23. Goyer, R. A., "Lead toxicity: Current concerns," *Environ. Hlth. Persp,* 100, 177, 1993.

24. Needleman, H. L. and Gatsonis, C. A., "Low-level lead exposure and the IQ of children," *J. AMA,* 263, 673, 1990.

25. Woods, J. S., "Porphyrin metabolism as indicator of metal exposure and toxicity," In *Handbook of Experimental Pharmacology,* Goyer, R. A., Cherian, M. G., eds., Berlin, Springer Verlag 1995, Vol. 115, 19.

26. Hashmi, N. S., Kachru, D. N., and Tandon, S. K., "Interrelationship between iron deficiency and lead intoxication," *Biol. Trace Elem. Res.,* 22, 298, 1989.

27. Bowers, M. A., Aicher, L. D., Davis, H. A., and Woods, J. S., "Quantitative determination of porphyrins in rat and human urine and evaluation of urinary porphyrin profiles during mercury and lead exposures," *J. Lab. Clin. Med.* 120, 272, 1992.

28. Woods, J. S., Martin, M. D., Naleway, C. A., and Echeverria, D., "Urinary porphyrin profiles as a biomarker of mercury exposure: studies in dentists with occupational exposure to mercury vapor," *J. Toxicol. Environ. Hlth.,* 40,239, 1993.

29. Echeverria, D., Heyer, N., Martin, M. D., Naleway, C. A., Woods. J. S., and Bittner, A. C., "Behavioral effects of low-level exposure to Hg° among dentists," *Neurotoxicol. Teratol.,* 17, 161, 1995.

30. Festing, M. F. W., "Genetic factors in neurotoxicology and neuropharmacology: a critical evaluation of the use of genetics as a research tool," *Experientia,* 47, 990, 1991.

31. Sonsalla, P. K. and Heikkila, R. E., Neurotoxic effects of 1-methyl-4-phenyl-1,2,3,6-tetrahydro-pyridine (MPTP) and methamphetamine in several strains of mice," *Neuropsychopharmacol. Behav. Psych.,* 12, 345, 1988.

32. Riley, E. P. and Lockney, E. A., "Genetic influences in the etiology of fetal alcohol syndrome," In *Fetal Alcohol Syndrome,* Abel, E. L., ed., CRC Press, Boca Raton, FL, 1982, Vol. III, 113.

33. Omenn, G. S., "Susceptibility to occupational and environmental exposures to chemicals," *Prog. Clin. Biol. Res.,* 214, 527, 1986.

34. Hirvonen, A., "Genetic factors in individual responses to environmental exposures," *J.O.E. M.,* 37, 37, 1995.

35. Daly, A. K., Cholerton, S., Gregory, W., and Idle, J. R., "Metabolic polymorphisms," *Pharmacol. Ther.,* 57, 129, 1993.

36. Smith, C. A. D., Smith, G. and, Wolf, C. R., "Genetic polymorphisms in xenobiotic metabolism," *Eur. J. Cancer,* 30A, 1921, 1994.

37. Farin, F. M. and Omiecinski, C. J., "Regiospecific expression of cytochrome P450s and microsomal epoxide hydrolase in human brain tissue," *J. Toxicol. Env. Hlth.,* 40, 317, 1993.

38. Barbeau, A., Cloutier, T., and Roy, M., "Ecogenetics of Parkinson's disease: 4-hydroxylation of debrisoquine," *Lancet,* ii, 1213, 1985.

39. Smith, C. A. D., Gough, A. C. and Leigh, P. N., "Debrisoquine hydroxylase gene polymorphism and susceptibility to Parkinson's disease," *Lancet,* 339, 1375, 1992.

40. Kallio, J., Marttila, R. J., Rinne, U. K., Sonninen, V. and Syvalathi, E., "Debrisoquine oxidation in Parkinson's disease," *Acta Neurol. Scand.,* 83,194 1991.

41. Kurth, J. H., Kurth, M. C., Poduslo, S. E. and Schwankhaus, J. D., "Association of monoamine oxidase B allele with Parkinson's disease," *Ann. Neurol.*, 33, 358, 1993.

42. Ho, S. L., Kapadi, A. L., Ramsden, D. B., and Williams, A. C., "An allelic association study of monoamine oxidase B in Parkinson's disease," *Ann. Neurol.*, 37, 403, 1995.

43. Castoldi, A. F., Coccini, T., Rossi, A., Nicotera, P., Costa, L. G., Tan, X. X., and Manzo, L., "Biomarkers in environmental medicine: alterations of cell signaling as early indicators of neurotoxicity," *Funct. Neurol.*, 9, 101, 1994.

44. Manzo, L., Artigas, F., Martinez, M., Mutti, A., Bergamaschi, E., Nicotera, P., Tonini, M., Candura, S. M., Ray, D. E., and Costa, L. G. "Biochemical markers of neurotoxicity. A review of mechanistic studies and applications," *Hum. Exp. Toxicol,* 15(S1) 20, 1996.

45. O'Callaghan, J. P., "Assessment of neurotoxicity: use of glial fibrillary acidic protein as a biomarker," *Biomed. Env. Sci.*, 4, 197, 1991.

46. Abdel Moneim, I., Shamy, M. Y., El-Gazzar, R. M. and El-Fawan H. A. N., "Autoantibodies to neurofilaments, glial fibrillary acidic protein and myelin basic protein in workers exposed to lead," *Toxicologist*, 14, 291, 1994.

47. Phillips, J. P., Jones, H. M., Hitchcock, R., Adams, N., and Thompson, R. J., "Radioimmunoassay of serum creatine kinase BB as index of brain damage after head injury," *Br. Med. J.*, 281, 777, 1980.

48. Brayne, C. E. G., Calloway, S. P., Dow, L., and Thompson, R. J., "Blood creatine kinase isoenzyme BB in boxers," *Lancet*, ii, 1308, 1982.

49. Liu, X., Glynn, P., and Ray, D. E., "Myelin basic protein in cerebrospinal fluid as a monitor of active demylinating lesions in the rat CNS," *Neurotoxicology*, 16, 542, 1995.

50. Miller, R. R., Newhook, R., and Poole, A., "Styrene production, use and human exposure," *Crit. Rev. Toxicol.*, 24, S1, 1994.

51. Bond, J. A., "Review of the toxicology of styrene," *Crit. Rev. Toxicol.*, 19, 227, 1989.

52. Brown, N. A., "Reproductive and developmental toxicity of styrene," *Reprod. Toxicol.*, 5, 3, 1991.

53. Cherry, N. and Gautrin, D., "Neurotoxic effects of styrene: further evidence," *Br. J. Ind. Med.*, 47, 29 1990.

54. Jegaden, D., Amann, D., Simon, J. F., Habault, M., Legoux, B., and Galopin, P., "Study of the neurobehavioral toxicity of styrene at low levels of exposure," *Int. Arch. Occup. Envir. Hlth.*, 64, 527 1993.

55. Matinaiken, E., Forsman-Gronholm, L., Pfaffli, P. and Juntunen, J., "Nervous system effects of occupational exposure to styrene: a clinical and neurophysiological study," *Env. Res.*, 61, 84, 1993.

56. Edling, C., Anundi, H., Johanson, G., and Nilsson, K., "Increase in neuropsychiatric symptoms after occupational exposure to low levels of styrene," *Br. J. Ind. Med.*, 50, 843, 1993.

57. Pahwa, R. and Kalra, J., "A critical review of the neurotoxicity of styrene in humans," *Vet. Human Toxicol.*, 35, 516, 1993.

58. Checkoway, H., Costa, L. G., Camp, J., Coccini, T., Daniell, W. E. and Dills, R. L., "Peripheral markers of neurochemical functions among styrene-exposed workers," *Br. J. Ind. Med*, 49, 560, 1992.

59. Rebert, C. S. and Hall, T. A., "The neuroepidemiology of styrene: a critical review of representative literature," *Crit. Rev. Toxicol.*, 24, S57, 1994.

60. Guillemin, M. P. and Berode, M., "Biological monitoring of styrene: a review," *Am. Ind. Hyg. Assoc. J.*, 49, 497, 1988.

61. Sumner, S. J. and Fennell, T. R., "Review of the metabolic rate of styrene," *Crit. Rev. Toxicol.*, 24, S1, 1994.

62. Gobba, F., Galassi, C., Ghittori, S., Imbriani, M., Pugliese, F., and Cavalleri, A., "Urinary styrene in the biological monitoring of styrene exposure," *Scand. J. Work. Environ. Hlth.*, 19, 175, 1993.

63. Ong, C. N., Shi, C. Y., Chia, S. E., Chua, S. C., Ong, H. Y., Lee, B. L., Ng, T.P., and Teramoto, K., "Biological monitoring of exposure to low concentrations of styrene," *Am. J. Ind. Med.*, 25, 719, 1994.

64. Hallier, E., Goergens, H. W., Karels, H., and Golka, K., "A note on individual differences in the urinary excretion of optical enantiomers of styrene metabolites and of styrene-derived mercapturic acids in humans," *Arch. Toxicol.*, 69, 300, 1995.

65. Seiler, J. P., "Chirality-dependent DNA reactivity as the possible cause of the differential mutagenicity of the two components in an enantiomeric pair of epoxides," *Mutat. Res.*, 245, 165, 1990.

66. Nordqvist, M. B., Lof, A., Osterman-Golkar, S., and Walles, S. A. S., "Covalent binding of styrene and styrene 7,8-oxide to plasma proteins, hemoglobin and DNA in the mouse," *Chem. Biol. Inter.*, 55, 63, 1985.

67. Hemminki, K., "Covalent binding of styrene oxide to amino acids, human serum proteins and hemoglobin," *Prog. Clin. Biol. Res.*, 207, 159, 1986.

68. Ting, D., Smith, M. T., Doane-Setzer, and Rappaport, S. M., "Analyses of styrene oxide globin adducts based upon reactions with Raney nickel," *Carcinogenesis*, 11, 755, 1990.

69. Christakopoulos, A., Bergmark, E., Zorcec, V., Norppa, H., Maki-Paakkanen, J., and Osterman-Golkar, S., "Monitoring occupational exposure to styrene from hemoglobin adducts and metabolites in blood," *Scand. J. Work Environ. Hlth.*, 19, 255, 1993.

70. Severi, M., Pauwels, W., Van Hummelen, P., Roosels, D., Kirsh-Volders, M., and Veulemans, H., "Urinary mandelic acid and hemoglobin adducts in fiberglass reinforced plastics workers exposed to styrene," *Scand. J. Work Env. Hlth.*, 20, 451, 1994.

71. Osterman-Golkar, S., Christakopoulos, A., Zorcec, V., and Svennson, K. "Dosimetry of styrene 7,8-oxide in styrene — and styrene oxide — exposed mice and rats by quantification of hemoglobin adducts," *Chem. Biol. Inter.*, 95, 79, 1995.

72. Phillips, D. H. and Farmer, P. B., "Evidence for DNA and protein binding by styrene and styrene oxide," *Crit. Rev. Toxicol.*, 24, 535, 1994.

73. Korn, M., Gforer, W., Filser, J. G., and Kessler, W., "Styrene-7,8-oxide in blood of workers exposed to styrene," *Arch. Toxicol.*, 68, 524, 1994.

74. Coccini, T., Di Nucci, A., Tonini, M., Maestri, L., Costa, L. G., Liuzzi, M., and Manzo, L, "Effects of ethanol administration on cerebral non-protein sulfhydryl content in rats exposed to styrene vapour," *Toxicology*, 106, 115, 1996.

75. Trenga, C. A., Kunkel, D. D., Eaton, D. L., and Costa, L. G., "Effect of styrene oxide on rat brain glutathione," *NeuroToxicology*, 12, 165, 1991.

76. Rosengren, L. E. and Haglid, K. G., "Long term neurotoxicity of styrene, A quantitative study of glial fibrillary acidic protein (GFA) and S-100," *Br. J. Ind. Med.*, 46, 316, 1989.

77. Dypbukt, J. M., Costa, L. G., Manzo, L., Orrenius, S., and Nicotera, P., "Cytotoxic and genotoxic effects of styrene 7,8-oxide in neuroadrenergic PC12 cells," *Carcinogenesis*, 13, 417, 1992.

78. Kohn, J., Minotti, S., and Durham, H., "Assessment of the neurotoxicity of styrene, styrene oxide and styrene glycol in primary cultures of motor and sensory neurons," *Toxicol. Lett.*, 75, 29, 1995.

79. Agrawal, A. K., Srivastava, S. P., and Seth, P. K., "Effect of styrene on dopamine receptors," *Bull. Environ. Contam. Toxicol.*, 29, 400, 1982.
80. Mutti, A., Falzoi, M., Romanelli, A., and Franchini, I., "Regional alterations of brain catecholamines by styrene exposure in rabbits," *Arch. Toxicol.*, 55, 173, 1984.
81. Mutti, A., Falzoi, M., Romanelli, A., Bocchi, M. C., Ferroni, C., and Franchini, I., "Brain dopamine as a target for solvent toxicity: effects of some monocyclic aromatic hydrocarbons," *Toxicology*, 49,77, 1988.
82. Mutti, A., Vescovi, P. P., Falzoi, M., Arfini, G., Valenti, G., and Franchini, I., "Neuroendocrine effects of styrene on occupationally exposed workers," *Scand. J. Work Environ. Hlth.*, 10, 225, 1984.
83. Arfini, G., Mutti, A., Vescovi, P., Ferroni, C., Ferrari, M., Giaroli, C., Passeri, M., and Franchini, I., "Impaired dopaminergic modulation of pituitary secretion in workers occupationally exposed to styrene: further evidence from PRL response to TRH stimulation," *J. Occup. Med.*, 29, 826, 1987.
84. Bergamaschi, E., Mutti, A., Cavazzini, S., Vettori, M. V., Renzulli F. S., and Franchini I., "Peripheral markers of neurochemical effects among styrene-exposed workers," *Neurotoxicology*, 17, 753, 1996.
85. Husain, R., Srivastava, S. P., Mushtag, M., and Seth, P. K., "Effect of styrene on levels of serotonin, noradrenaline, dopamine and activity of acetylcholinesterase and monoamine oxidase in rat brain," *Toxicol. Lett.*, 7, 47, 1980.
86. Checkoway, H., Echeverria, D., Moon, J. D., Heyer, N., and Costa, L. G., "Platelet monoamine oxidase B activity in workers exposed to styrene," *Int. Arch. Occup. Env. Hlth.*, 66, 359, 1994.
87. Smargiassi, A., Mutti, A., Bergamaschi, E., Belanger, S., Truchon, G., and Mergler, D., "Pilot study of peripheral markers of catecholaminergic systems among workers occupationally exposed to toluene," *Neurotoxicology*, 17, 769, 1996.
88. Nagaya, T., Ishikawa, N., and Hada, H., "No change in serum dopamine-β-hydroxylase activity in workers exposed to trichloroethylene," *Toxicol. Lett.*, 54, 221, 1990.
89. Dearfield, K. L., Abernathy, C. O., Ottley, M. S., Brantner, J. H., and Hayes. P. F., "Acrylamide: its metabolism, developmental and reproductive effects, genotoxicity and carcinogenesis," *Mutat. Res.*, 195, 45, 1988.
90. Tilson, H. A., "The neurotoxicity of acrylamide: an overview," *Neurobehav. Toxicol. Teratol.*, 3, 445, 1981.
91. Miller, M. S. and Spencer, P. S., "The mechanism of acrylamide axonopathy," *Ann. Rev. Pharmacol. Toxicol.*, 25, 643, 1985.
92. He, F., Zhang, S., Wang, H., Li, G., Zhang, Z., Li, F., Dong, X., and Hu, F., "Neurological and electroneuromyographic assessment of the adverse effects of acrylamide on occupationally exposed workers," *Scand. J. Work Environ. Hlth.*, 15, 125, 1989.
93. Hashimoto, K. and Aldridge, W. N., "Biochemical studies on acrylamide, a neurotoxic agent," *Biochem. Pharmacol.*, 19, 2591, 1970.
94. Miller, M. J., Carter, D. E., and Sipes, I. G., "Pharmacokinetics of acrylamide in Fischer–344 rats," *Toxicol. Appl. Pharmacol.*, 63, 36, 1982.
95. Bailey, E., Farmer, P. B., Bird, I., Lamb, J. H., and Peal, J. A., "Monitoring exposure to acrylamide by the determination of S-(2-carboxyethyl) cysteine in hydrolyzed hemoglobin by gas chromatography–mass spectrometry," *Anal. Biochem.*, 157, 241, 1986.
96. Bergmark, E., Calleman, C. J. and Costa, L. G., "Formation of hemoglobin adducts of acrylamide and its epoxide metabolite glycidamide in the rat," *Toxicol. Appl. Pharmacol.*, 111, 352, 1991.

97. Calleman, C. J., Bergmark, E., and Costa, L. G., "Acrylamide is metabolized to glycidamide in the rat: evidence from hemoglobin adduct formation," *Chem. Res. Toxicol.*, 3, 406, 1990.

98. Costa, L. G., Deng, H., Gregotti, C., Manzo, L., Faustman, E. M., Bergmark, E., and Calleman, C. J., "Comparative studies on the neuro- and reproductive toxicity of acrylamide and its epoxide metabolite glycidamide in the rat," *Neurotoxicology*, 13, 219, 1992.

99. Hashimoto, K. and Tanii, H., "Mutagenicity of acrylamide and its analogues in *Salmonella typhimurium.*," *Mutat. Res.*, 158, 129, 1985.

100. Abou Donia, M. B., Ibrahim, S. M., Corcoran, J. J., Lack, L., Friedman, M. A., and Lapadula, D. M., "Neurotoxicity of glycidamide, an acrylamide metabolite, following intraperitoneal injections into rats," *J. Toxicol. Env. Hlth.*, 39, 447, 1993.

101. Costa, L. G., Deng, H., Calleman, C. J., and Bergmark, E., "Evaluation of the neurotoxicity of glycidamide, an epoxide metabolite of acrylamide: behavioral, neurochemical and morphological studies," *Toxicology*, 98, 151, 1995.

102. Bergmark, E., Calleman, C. J., He, F., and Costa, L. G., "Determination of hemoglobin adducts in humans occupationally exposed to acrylamide," *Toxicol. Appl. Pharmacol.*, 120, 45, 1993.

103. Calleman, C. J., Wu, Y., He, F., Tian, G., Bergmark, E., Zhang, S., Deng, H., Wang, Y., Crofton, K.M., Fennell, T., and Costa, L. G., "Relationship between biomarkers of exposure and neurological effects in a group of workers exposed to acrylamide," *Toxicol. Appl. Pharmacol.*, 126, 361, 1994.

104. Gallo, M. A. and Lawryk, N. J., "Organic Phosphorus Pesticides," In *Handbook of Pesticide Toxicology*, Hayes, W. J. and Laws, E. R., Eds. Academic Press, San Diego, 1991, 917.

105. Coye, M. J., Lowe, J. A., and Maddy, K. T., "Biological monitoring of agricultural workers exposed to pesticides: I. Cholinesterase activity determinations," *J. Occup. Med.*, 28, 619, 1986.

106. Ames, R. G., Brown, S. K., Mengle, D. C., Kahn, E., Stratton, J. W., and Jackson, R. J., "Cholinesterase activity depression among California agricultural pesticide applicators," *Am. J. Ind. Med.*, 15, 143, 1989.

107. Coye, M. J., Barnett, P. G., Midtling, J. E., Velasco, A. R., Romero, P., Clements, C. L., and Rose, T. G., "Clinical confirmation of organophosphate poisoning by serial cholinesterase analyses," *Arch. Int. Med.*, 147, 438, 1987.

108. Chatonnet, A. and Lockridge, O., "Comparison of butyrylcholinesterase and acetylcholinesterase," *Biochem. J.*, 260, 625, 1989.

109. Lockridge, O., "Genetic variants of human serum cholinesterase influence metabolism of the muscle relaxant succinylcholine," *Pharmacol. Ther.*, 47, 35, 1990.

110. Pope, C. N. and Chakraborti, T. K., "Dose-related inhibition of brain and plasma cholinesterase in neonatal and adult rats following sublethal organophosphate exposure," *Toxicology*, 73, 35, 1992.

111. Fitzgerald, B. B. and Costa, L. G., "Modulation of muscarinic receptors and acetylcholinesterase activity in lymphocytes and in brain areas following repeated organophosphate exposure in rats," *Fund. Appl. Toxicol.*, 20, 210, 1993.

112. Padilla, S., Wilson, V. Z., Bushnell, P. J., "Studies on the correlation between blood cholinesterase inhibition and "target tissue" inhibition in pesticide-treated rats," *Toxicology*, 92, 11, 1994.

113. Costa, L. G., Schwab, B. W., and Murphy, S. D., "Tolerance to anticholinesterase compounds in mammals," *Toxicology*, 25, 79, 1982.

114. McDonald, B. E., Costa, L. G., and Murphy, S. D., "Spatial memory impairment and central muscarinic receptor loss following prolonged treatment with organophosphates," *Toxicol. Lett.*, 40, 47, 1988.

115. Bushnell, P. J., Padilla, S. S., Ward, T., Pope, C. N., and Orszyk, V. P. "Behavioral and neurochemical changes in rats dosed repeatedly with diisopropylfluorophosphate," *J. Pharmacol. Exp. Ther.*, 256, 741, 1991.

116. Gershon, S. and Shaw, F. H., "Psychiatric sequelae of chronic exposure to organophosphorus insecticides," *Lancet*, i, 1371, 1961.

117. Metcalf, D. R., and Holmes, J. H., "EEG, psychological and neurological alterations in humans with organophosphorus exposure," *Ann. NY Acad Sci.*,160, 357, 1969.

118. Costa, L. G., Kaylor, G., and Murphy, S. D., "Muscarinic cholinergic binding sites on rat lymphocytes," *Immunopharmacology*, 16, 139, 1988.

119. Costa, P., Auger, C. B., Traver, D. J., and Costa, L. G., "Identification of m3, m4 and m5 subtypes of muscarinic receptor mRNA in human blood mononuclear cells," *J. Neuroimmunol.*, 60, 45, 1995.

120. Costa, L. G., Kaylor, G., and Murphy, S. D., "*In vitro* and *in vivo* modulation of cholinergic muscarinic receptors in rat lymphocytes and brain by cholinergic agents," *Int. J. Immunopharmacol.*, 12, 67, 1990.

121. Lotti, M., "The pathogenesis of organophosphate polyneuropathy," *Crit. Rev. Toxicol.*, 21, 465, 1992.

122. Johnson, M. K., "The target for initiation of delayed neurotoxicity by organophosphorus ester: biochemical studies and toxicological applications," *Rev. Biochem. Toxicol.*, 4, 141, 1982.

123. Bertoncin, D., Russolo, A., Caroldi, S., and Lotti, M., "Neuropathy target esterase in human lymphocytes," *Arch. Env. Hlth.*, 40, 139, 1985.

124. Maroni, M. and Bleecker. M. L., "Neuropathy target esterase in human lymphocytes and platelets," *J. Appl. Toxicol.*, 6, 1, 1986.

125. Schwab, B. W. and Richardson, R. J., "Lymphocyte and brain neurotoxic esterase: dose and time dependence of inhibition in the hen examined with three organophosphorous esters," *Toxicol. Appl. Pharmacol.*, 83, 1, 1986.

126. Lotti, M., "Organophosphate-induced delayed polyneuropathy in humans: perspectives for biomonitoring," *Trends Phamacol. Sci.*, 8, 175, 1987.

127. Lotti, M., Moretto, A., Zoppellari, R., Dainese, R., Rizzuto, N., and Barusco, G., "Inhibition of lymphocytic Neuropathy Target Esterase predicts the development of organophosphate induced delayed neuropathy," *Arch. Toxicol.*, 59, 176, 1986.

128. Lotti, M., Becker, C. E., Aminoff, M. J., Woodrow, J. E., Seiber, J. N., Talcott, R. E., and Richardson, R. J. "Occupational exposure to the cotton defoliants DEF and Merphos: a rational approach to monitoring organophosphorus-induced delayed neurotoxicity," *J. Occup. Med.*, 25, 517, 1983.

129. Lotti, M., Moretto, A., Capodicasa, E., Bertolazzi, M., Peraica, M., and Scapellato, M. L. "Interactions between Neuropathy Target Esterase and its inhibitors and the development of polyneuropathy," *Toxicol. Appl. Pharmacol.*, 122, 165, 1993.

130. Brokopp, C. D., Wyatt, J. L., and Gabica, J. "Dialkyl phosphates in urine samples from pesticide formulators exposed to disulfoton and phorate," *Bull. Environ. Contam. Toxicol.*, 26, 524, 1981.

131. Coye, M. J., Lowe, J. A., and Maddy, K. J., "Biological monitoring of agricultural workers exposed to pesticides: II. Monitoring of intact pesticides and their metabolites," *J. Occup. Med.*, 28, 628, 1986.

132. Vasilic, Z., Drevenkar, V., Frobe, Z., Stengl, B., and Tkalcevic, B., "The metabolites of organophosphorus pesticides in urine as an indicator of occupational exposure," *Toxicol. Environ. Chem.*, 14, 111, 1987.

133. Levy, Y., Graner, F., Levy, S., Chuners, P., Gruener, N., Marzuk, J., and Richter, E. D., "Organophosphate exposures and symptoms in farm workers and residents with "normal" cholinesterases." In *Environmental Quality and Ecosystem Stability: Vol IV-A, Environmental Quality.*, Luria, M., Steinberger, Y., and Spanier, E., Eds., ISEEQS Publ., Jerusalem, 1989, 299.

134. Geldmacher-von Mallinckrodt, M. and Diebgen, T. L., "The human serum paraoxo-nase-polymorphism and specificity," *Toxicol. Environ. Chem.*, 18, 79, 1988.

135. Furlong, C. E., Richter, R. J., Seidel, S. O., Costa, L. G., and Motulsky, A. G., "Spec-trophotometric assay for the enzymatic hydrolysis of the active metabolites of chlo-rpyrifos and parathion by plasma paraoxonase/arylesterase," *Anal. Biochem.*, 180, 242, 1989.

136. Costa, L. G., Richter, R. J., Murphy, S. D., Omenn, G. S., Motulsky, A. G., and Fur-long, C. E., "Species differences in serum paraoxonase correlate with sensitivity to paraoxon toxicity," In *Toxicology of Pesticides: Experimental, Clinical and Regula-tory Perspectives*, Costa, L. G., Galli, C. L., and Murphy, S. D., Eds., Springer Verlag, Heidelberg, 1987, 263.

137. Costa, L. G., McDonald, B. E., Murphy, S. D., Omenn, G. S., Richter, R. J., Motulsky, A. G., and Furlong, C. E., "Serum paraoxonase and its influence on paraoxon and chlorpyrifos-oxon toxicity in rats," *Toxicol. Appl. Pharmacol.*, 103, 66, 1990.

138. Li, W. F., Furlong, C. E., and Costa, L. G., "Paraoxon protects against chlorpyrifos toxicity in mice," *Toxicol. Lett.*, 76, 219, 1995.

139. Humbert, R., Adler, D. A., Distecha, C. M., Hassett, C., Omiecinski, C. J., and Fur-long, C. E., "The molecular basis of the human serum paraoxonase activity polymor-phism," *Nature Genet.*, 3, 73, 1993.

140. Spencer, P. S., Schaumburg, H. H., Sabri, M. L. and Veronesi, B. V., "The enlarging view of hexacarbon neurotoxicity," *Crit. Rev. Toxicol.*, 7, 279, 1980.

141. Graham, D. G., Amarnath, V., Valentine, W. M., Pyle, S. J., and Anthony, D. C., "Pathogenetic studies of hexane and carbon disulfide neurotoxicity," *Crit. Rev. Toxi-col.*, 25, 91, 1995.

142. Genter, M. B., Szakal–Quin, G., Anderson, C. W., Anthony, D. C., and Graham, D. G., "Evidence that pyrrole formation is a pathogenetic step in γ-diketone neuropathy," *Toxicol. Appl. Pharmacol.*, 87, 351, 1987.

143. Kawai, T., Yasugi, T., Mizunuma, K., Horiguchi, S., Uchida, Y., Iwassi, O., Oguchi, H., and Ideka, M., Dose-dependent increase in 2,5-hexanedione in the urine of workers exposed to *n*-hexane, *Int. Arch. Occup. Env. Hlth.*, 63, 285, 1991.

144. Saito, I., Shibata, E., Huang, J., Hisanaga, N., Ono, Y., and Takeuchi, Y., "Determi-nation of urinary 2,5-hexanedione concentration by an improved analytical method as an index of exposure to *n*-hexane," *Br. J. Ind. Med.*, 48, 568, 1991.

145. Beauchamp, R. O., Bus, J. S., Popp, J. A., Boreiko, C. J., and Goldberg, L., "A critical review of the literature on carbon disulfide toxicity," *Crit. Rev. Toxicol.*, 11, 169, 1983.

146. Aaserud, O., Hommeren, O. J., and Tuedt, B., "Carbon disulfide exposures and neu-rotoxic sequelae among viscose rayon workers," *Am. J. Ind. Med.*, 18, 25, 1990.

147. Valentine, W. M., Amarnath, V., Graham, D. G., and Anthony, D. C., "Covalent cross-linking of proteins by carbon disulfide," *Chem. Res. Toxicol.*, 5, 254, 1992.

148. Valentine, W. M., Graham, D. G., and Anthony, D. C., "Covalent cross-linking of eryth-rocyte spectrin by carbon disulfide *in vivo*," *Toxicol. Appl. Pharmacol.*, 121, 71, 1993.

149. Phillips, M., "Detection of carbon disulfide in breath and air: a possible new risk factor for coronary artery disease," *Int. Arch. Occup. Environ. Hlth.*, 64, 119, 1992.

150. Lam, C. W. and DiStefano, V., "Blood-bound carbon disulfide: an indicator of carbon disulfide exposure, and its accumulation in repeatedly exposed rats," *Toxicol. Appl. Pharmacol.*, 70, 402, 1983.

151. Riihimaki, V., Kivisto, H., Peltonen, K., Helpio, E., and Aritio, A., "Assessment of exposure to carbon disulfide in viscose production workers from urinary 2-thiothiazolidine-4-carboxylic acid determinations," *Am. J. Industr. Med.*, 22, 85, 1992.

152. Kitamura, S., Ferrari, F., Vides, G., and Filho, D. C. M., "Biological monitoring of workers occupationally exposed to carbon disulfide in a rayon plant in Brazil: validity of 2-thiothiazolidine-4-carboxylic acid (TTCA) in urine samples taken at different times, during and after the real exposure period," *Int. Arch. Occup. Environ. Hlth.*, (Suppl.) 65, S177, 1993.

153. Cox, C., Lowry, L. K., and Que Hee, S. S., "Urinary 2-thiothiazolidine-4-carboxylic acid as a biological indicator of exposure to carbon disulfide: derivation of a biological exposure index," *Appl. Occup. Environ. Hyg.*, 7, 672, 1992.

154. Magos, L. and Jarvis, J. A. E., "The effects of carbon disulfide exposure on brain catecholamine in rats," *Br. J. Pharmacol.*, 39, 26, 1970.

155. McKenna, M. J. and Distefano, V., "Carbon disulfide II. A proposed mechanism for the action of carbon disulfide on dopamine beta hydroxylase," *J. Pharmacol. Exp. Ther.*, 202, 253, 1977.

156. Caroldi, S., Jarvis, J. A. E., and Magos, L., "*In vivo* inhibition of dopamine-β-hydroxylase in rat adrenals during exposure to carbon disulphide," *Arch. Toxicol.*, 55, 265, 1984.

157. Caroldi, S., Jarvis, J. A. E., and Magos, L., "Stimulation of dopamine β-hydroxylase in rat adrenals by repeated exposures to carbon disulphide," *Biochem. Pharmacol.*, 33, 1933, 1984.

158. Wasilewska, E., Stanosz, S., and Bargiel, Z., "Serum dopamine-beta-hydroxylase activity in women occupationally exposed to carbon disulfide," *Ind. Hlth.*, 27, 89, 1989.

5   Epidemiological Methods in Occupational Neurotoxicology

Harvey Checkoway and Mark R. Cullen

CONTENTS

1   INTRODUCTION

The recognition of neurotoxicity in the workplace historically has been prompted by case reports of affected workers, generally manifesting profound functional impairment. Clinical identification of hazards by the case series method remains an important aspect of occupational medicine practice, although inferences drawn from these occurrences may have limited scientific generalizability. In particular, case series are essentially anecdotal reports that may at times give a distorted impression of the true magnitude of risk in an exposed population. Increasingly, formal epidemiologic methods are being exploited to determine disease occurrence rates and to characterize dose-response relations for known hazards, with emphasis on risks

0-8493-9231-4/98/$0.00+$.50
© 1998 by CRC Press LLC

related to low-level exposures. Epidemiologic methods also are applied to monitor for emerging neurotoxic hazards.

In this chapter, we will outline the research techniques that epidemiologists use to investigate occupational neurotoxicity. The various study designs will be presented in order of frequency of use, which in large measure parallels their historical development in occupational epidemiology. We will also highlight the relative advantages and limitations of the commonly used designs, with particular attention given to the suitability of each approach for investigating occupational associations for various health end-points.

2   EPIDEMIOLOGIC STUDY DESIGNS

The most direct method for studying occupational neurotoxicity is to perform an epidemiologic investigation among a defined group from a particular workplace(s) where neurotoxic exposures are known to occur. This study selection strategy, as opposed to studies conducted in the general population, enables efficient determination of dose-response relations among what are, ordinarily, the most heavily exposed persons (exceptions, of course, are situations where environmental contamination poses widespread, severe hazards to entire communities). Studies conducted in the workplace frequently focus on disease signs and symptoms or functional impairment that require direct contact and measurement of study subjects. An alternative approach is to examine associations between neurotoxicant exposures and health in the general population. Community-based studies typically are conducted to explore possible relations of occupational exposures with chronic neurodegenerative disorders.

2.1   Workplace Studies

2.1.1 Cross-Sectional Studies

By far, the most frequently applied epidemiologic design in workplace investigations is the cross-sectional study. This approach involves sampling of study subjects on the basis of exposure level and comparisons of health status among groups. The assessment of neurotoxicity is often guided by past literature, but may be influenced by case reports suggesting specific effects in a particular workforce. Often in cross-sectional studies, a battery comprised of neurobehavioral tests, symptom reports, and neurophysiologic function tests are administered to evaluate a range of adverse neurologic effects.

An example of a typical occupational cross-sectional study is a comparison of neurobehavioral test performance and color vision discrimination between styrene-exposed boat builders and non-exposed carpentry workers from the same plant.[1] Noteworthy findings of the study were relative performance decrements for some tests, including color vision discrimination (Table 1). Comparisons against a non-exposed reference group, as in the preceding example, represent a gross assessment of neurotoxicity. More convincing evidence of adverse effects is obtained when dose-response relations are detected. Another study of styrene-related neurotoxicity by

TABLE 1
Neurobehavioral Test Performance Among Styrene-Exposed Workers and Nonexposed Carpentry Workers*

Test	Exposed (N = 21)	Referents (N = 21)
Benton Visual Retention	6.0 (0.08)[†]	7.7 (0.07)
Digit span	11.7 (0.07)	15.6 (0.06)
Digit symbol	26.3 (0.08)	38.0 (0.07)
Total color difference score	164.0 (0.04)	131.8 (0.04)
Finger tapping		
dominant hand	41.8 (0.05)	44.3 (0.05)
non-dominant hand	39.5 (0.05)	42.5 (0.04)

* Adapted from: Chia S. E., *et al.*, 1994, *Am. J. Ind. Med.* 26: 481–488.
† Mean (standard error)

TABLE 2
Sensory Nerve Conduction Velocity among Styrene-Exposed Workers*

Exposure level (ppm)	Sensory nerve (m/sec)		
	Median	Ulnar	Sural
≤50 (N = 30)	45.6 (4.7)[†]	45.8 (4.5)	42.7 (3.1)
51–100 (N = 15)	43.5 (4.3)	44.6 (4.7)	39.9 (3.2)
≥100 (N = 14)	42.5 (5.2)	41.4 (5.5)	39.0 (3.0)

* Adapted from: Cherry, N. and Gautrin, D., 1990, *Brit. J. Ind. Med.* 47: 29–37.
† Mean (standard deviation)

Cherry and Gautrin[2] demonstrated apparent dose-response relations for sensory nerve conduction velocities in relation to air concentrations of styrene (Table 2).

2.1.2 Longitudinal Studies

Cross-sectional studies, by design, are limited to an examination of exposure and health outcome at a single point in time. A more comprehensive determination of association requires a longitudinal design involving repeated samplings of workers' exposures and health status. There are several significant advantages to longitudinal studies relative to cross-sectional studies, the most important of which is the possibility for determining whether observed neurotoxic effects are likely to be transient (reversible) or persistent. Another, less obvious, advantage of the longitudinal design is the opportunity to estimate the reliability of test procedures. Uncertain reliability is especially problematic in studies relying on neurobehavioral test batteries that

TABLE 3
Longitudinal Study of Peripheral Nerve Conduction in Lead-Exposed Workers*

		Examination					
	Baseline	1-yr exam		2-yr exam		4-yr exam	
		Blood Pb		Blood Pb		Blood Pb	
Nerve		<30	≥30	<30	≥30	<29	≥29
parameter	(N = 23)	(N = 12)	(N = 11)	(N = 8)	(N = 7)	(N = 5)	(N = 5)
Median							
MCV[†]	58.8**	61.5	55.7	60.4	54.8	64.5	59.1
SCV[‡]	63.7	64.2	60.5	60.7	58.5	61.1	59.9
Ulnar							
MCV[†]	60.7	61.5	57.0	62.2	59.3	62.9	62.8
SCV[‡]	62.1	64.6	59.2	60.8	59.9	66.1	63.2

* Adapted from: Seppalainen AM, *et al.*, 1983, *NeuroToxicol 4*: 181-192.
† Maximal motor conduction velocity (m/sec)
‡ Sensory conduction velocity (m/sec)
** Mean

are vulnerable to variability introduced by changes in test administration conditions, subject learning, and the inherent variations in human performance.[3] Repeated measurement of the same subjects is also advantageous for minimizing confounding effects of extraneous factors (e.g., alcohol consumption), as each person can serve as his or her own referent.

In a longitudinal study, the outcomes of greatest interest are changes in neurologic function or symptoms over time in relation to exposure. It is best, although not always practical, to include a non-exposed reference group that is evaluated in the same manner as the exposed index population. The reference group provides a basis for determining expected temporal changes in health outcome that may be due to test learning phenomena, aging in the case of very long-term studies, or other unmeasured confounders. Seppalainen et al.[4] four-year longitudinal study of peripheral nerve conduction velocity in lead-exposed battery workers reveals some of the strengths and limitations of this method. There was clear evidence of exposure-related effects on median and ulnar nerve conduction (Table 3). However, temporal patterns of change in neurophysiological function, in relation to blood lead level, were difficult to discern because of out-migration from the study over time. Only five subjects from each of the high and low lead exposure groups were available for the year four evaluation. Study group attrition over time can arise for numerous reasons, including impaired health related to workplace exposures, and will limit the ability to identify adverse effects.

2.1.3   Cohort Studies

Cohort studies require enumeration and long-term follow-up of defined occupational groups. Cohorts may be composed of workers from a single plant, workers

TABLE 4
Mortality from Neurodegenerative Disorders
Among Styrene-Exposed Workers*

| | Exposure Group | | | | | |
| | Non-exposed | | Low | | High | |
	(Obs)	RR[†]	(Obs)	RR[†]	(Obs)	RR[†]
Duration of employment (yr)						
<1[‡]	(6)	1.0	(8)	0.9	(8)	1.9
≥1	(9)	1.3	(21)	1.7	(8)	2.5
Time since first employment (yr)						
<10[‡]	(1)	1.0	(3)	1.9	(3)	3.8
≥10	(14)	4.6	(26)	5.3	(13)	8.0

* Adapted from: Kolstad HA, *et al.*, 1995, *Occup. Environ. Med. 52:* 320–327.
† Mortality rate ratio
‡ Reference category

from multiple plants within the same industry, or trade or professional associations. A special type of cohort can consist of workers with histories of prior exposure-related effects (e.g., pesticide poisoning) who are followed to determine long-term sequelae. The majority of occupational cohort studies are historical in design (i.e., follow-up is from the past to the present), and focus on fatal diseases because mortality data are routinely available in many locales. Cohort studies are best suited for investigations of chronic disease risks, such as cancer and cardiovascular diseases, for which diagnoses inferred from death certificates are reasonably accurate. Data on neurologic disorders and other causes of death that may be of interest for studies of neurotoxic effects (e.g., suicides) can also be obtained in such studies.

An illustrative example is a comparative study of mortality among 36,610 workers exposed to styrene in reinforced plastics manufacturing and 14,293 workers from other industries without styrene exposure.[5] During a 20-year follow-up, there were 3,031 deaths from all causes in the exposed cohort and 1453 total deaths in the reference cohort; however, the numbers of deaths which were identified as "degenerative disorders of the nervous system," including Alzheimer's disease, multiple sclerosis, Parkinsonism, and motor neuron disease, were only 45 and 15, respectively in the two groups. Relative mortality excesses for neurodegenerative disorders were observed in the styrene-exposed cohort (Table 4), yet the interpretation of this finding is complicated by the absence of quantitative exposure data and the acknowledged heterogeneity of the pathogenetic mechanisms of the various neurologic diseases. This study suffers from two fairly common problems that beset efforts to identify causal relations with neurologic diseases in occupational cohort studies: uncertain diagnoses from death certificate information, and low statistical power to detect associations with specific disorders due to the failure of mortality data to provide an adequate accounting of neurologic disease incidence.

TABLE 5
Nested Case-Control Study of Multiple Sclerosis
Among Automotive Carburetor Plant Workers*

Exposure	Relative risk[†]	(95% Confidence interval)
Die casting	4.2	(1.4–10.8)
Mercury	1.0	(0.20–3.6)
Solvents	2.6	(0.75–7.1)
Coolants	2.6	(0.75–7.1)
Metal fumes	2.3	(0.74–6.0)
Organophosphates	4.2	(1.4–10.8)

* Adapted from: Krebs, J. M., *et al.*, 1995, *Arch. Environ. Health 50*: 190–195.

† Relative to the 90th percentile of the controls' cumulative exposure

2.1.4   Nested Case-Control Studies

A principal use of occupational cohort studies is to estimate disease excesses and deficits relative to baseline rates occurring in a non-exposed population. More in-depth examination of exposure/disease associations may require detailed inspection of past exposures for cases of diseases of prior interest or diseases which appear, from the cohort analysis, to have occurred at an excessive rate. The nested case-control design is an efficient strategy for generating and testing etiologic hypotheses. The case group is defined as all identified cases of the index disease from the base occupational cohort (e.g., plant population), and controls are selected from among cohort members free of that condition. Matching controls to cases with respect to age, gender, and ethnicity is commonly done to eliminate confounding bias from these factors. A limitation of this design is that verification of disease free status among controls may be problematic in studies of the more common neurologic diseases (e.g., Alzheimer's disease). Comparisons of cases' and controls' past exposures is then performed to identify high risk jobs and exposures.

The nested case-control design can be especially useful to examine disease clusters more rigorously than a mere case series report. An example is a nested case-control study of multiple sclerosis among automotive carburetor plant workers[6] that was conducted to identify potential workplace risk factors after an earlier report of an excessive rate in the workforce.[7] Compared to a control group of workers matched on age, gender, and ethnicity, the multiple sclerosis cases had a greater frequency of past employment in die casting operations where there are exposures to fumes of zinc, aluminum, and to hydraulic fluids containing organophosphates (Table 5).[6]

2.2   COMMUNITY-BASED STUDIES

Epidemiologic studies of neurologic disorders and occupational exposures may not be feasible when there is limited access to occupational groups or when the expected numbers of cases of specific disorders (especially of neurodegenerative

TABLE 6
Registry-Based Case-Control Study of Motor Neuron Disease*

Exposure	Relative risk	(95% Confidence interval)
Lead	5.7	(1.6–30)
Chemicals, solvents	3.3	(1.3–10)
Pesticides	1.4	(0.6–3.1)
Minerals, ores	3.5	(0.7–34)

* Adapted from: Chancellor, A. M., *et al.* (1993). *J. Neurol. Neurosurg. Psych.* 56: 1200–1206.

diseases) are small within available occupational cohorts. An alternative approach is to perform community-based case-control studies wherein cases are identified from hospital or other patient registers, and controls are selected from the population-at-large. As in case-control studies nested within defined occupational cohorts, there is the requirement that the controls are free of the index disease. Typically in community-based studies, information on exposures is obtained from interviews of cases or controls or their proxies (e.g., next-of-kin). As such, the specificity of associations with occupational exposures, and the applicability of findings for estimating dose-response relations, are generally considerably lower in community-based than in industry-based cohort or nested case-control studies. An example is a Scottish case-control study of motor neuron disease (MND).[8] Cases (103) were identified from the Scottish Motor Neuron Disease Register, and an age- and gender-matched control group of equal size was selected from the patient rosters of the general practitioners who provided routine care for the cases. Based on subjects' responses to a questionnaire, which elicited data on lifetime occupational history, as well as non-occupational factors (e.g., cigarette smoking), associations were detected with exposures to lead and the nonspecific categories of chemicals and solvents, and pesticides (Table 6).

Notable advantages of community-based case-control studies are the opportunity to assemble much larger case series for rare neurologic diseases than would occur in a particular worker cohort, and much improved diagnostic accuracy than that provided by death certificate data. The principal limitation of this method is that the community-based design will only yield etiologic insights for exposures that are relatively common in the population and can be identified with minimal ambiguity (e.g., lead). Associations, however strong, with specific agents that affect only small segments of the population will thus be difficult, if not impossible, to discern in community-based studies.

3 EXPOSURE ASSESSMENT

In workplace-based studies, especially those involving cross-sectional or longitudinal health assessments, exposure classification frequently is based on measured concentrations of contaminants in environmental media. Typical examples are air concentrations of solvent vapors, metal dusts and fumes, or noise levels. An inherent limitation of reliance on workplace environmental exposures is that, depending on when and how often measurements are taken, they may not be adequately represen- tative of workers' actual exposures. In particular, measurements made at a single point in time may fail to account for within-worker variability that results from day- to-day differences in job demands and work habits. There is empirical evidence that the within-worker exposure variation can greatly exceed between-worker variation for airborne toxicants, and that such factors as breathing rate and the use of protective equipment may be significant determinants of a worker's exposure profile.[9] Repeated exposure measurements for individual workers is a means of improving exposure classification, although this may not be practical in many instances. Field application of biological monitoring of toxicants or their metabolites in bodily tissues, or assays of biomarkers of response (e.g., acetyl cholinesterase depression among workers exposed to organophosphate pesticides) to improve internal dose estimation has gained more prominence in recent years. In the case of lead, there exist several biomonitoring methods that can assess recent intakes (blood levels), and long-term body burden (X-ray fluorescence).[10] The choice of exposure biomarker should be guided by toxicokinetic principles and the presumed mechanism of toxicity.

When environmental measurement or biomarker data are not available for expo- sure assessment, exposure surrogates, such as job title and duration of employment, are necessitated. Occupational cohort studies of chronic disease mortality generally are restricted to these cruder exposure metrics because valid and representative historical exposure measurements, spanning relevant years of employment, are rarely available in most industries. Job title and employment duration designations are useful for identifying gross differences in risk among a workforce, but are poorly suited for quantitation of dose-effect relations.

Exposure assessment in community-based studies is fraught with even greater uncertainty than in industry-based studies, mainly because the principal source of occupational exposure data in community-based studies is self-reports by the study subjects. Even when subjects are capable of reporting lifetime employment histories, information regarding types and amounts of exposures to specific toxicants is fre- quently ambiguous. This problem is compounded further when data are obtained from proxy respondents, as will occur when study subjects are cognitively impaired. One method to balance the quality of exposure information in community-based case-control studies of psychiatric and other neurologic disorders entailing dimin- ished cognition is to obtain exposure data from proxy respondents of both cases and controls.[11] Pointed questions about exposures to known or suspected toxicants may yield reliable data in situations where the agent is recognizable to the subject (e.g., lead in a lead smelter); however, agent-specific exposures and relative levels gener- ally can only be inferred from reported industries and occupations (e.g., welder).

4   SOURCES OF RESEARCH BIAS

Epidemiologic research is vulnerable to a variety of biases that may result in erro- neous scientific conclusions. The effects of research bias can range from failure to detect or underestimation of true effects to generation of exaggerated or in some instances spurious associations. There are numerous manifestations of bias, many of which are subtle. The three major types of bias are: information bias, selection bias, and confounding. These are considered in turn.

4.1   INFORMATION BIAS

Information bias refers to the incorrect assignment of exposure or health status to study subjects. Errors in exposure or disease classification are potentially problematic in all study designs, although the manifestations of information bias will vary according to the nature of the exposure/disease relation under consideration.

4.1.1   Exposure Misclassification

The net effect of incorrect exposure assignment will be distorted observed associations between workplace neurotoxicants and adverse effects. Non-differential exposure misclassification is the situation whereby exposure assessment is imperfect but is unrelated to health status. Often, but not always, non-differential misclassification will cause adverse health outcomes to be under-estimated, or in the extreme, missed.[12] Another scenario is differential misclassification of exposure in which the extent of misclassification varies by health or disease status. Studies that combine biological monitoring with disease screening, and subsequent medical and exposure follow-up, are particularly prone to differential misclassification because exposure assessments will typically be more accurate for cases of neurological disease or impairment than for unaffected workers. Differential misclassification may produce either under- or over-estimated effects attributable to exposure.

4.1.2   Health Outcome Misclassification

Health status misclassification is another especially vexing problem in occupational neuroepidemiology. There can be considerable uncertainty in diagnoses for even the most well recognized neurological disorders. Overlapping clinical signs and symptoms among major neurologic diseases contributes to diagnostic uncertainty. Even more complex is the interpretation of neurological function or performance testing. For example, many of the tests in common use in epidemiologic studies (e.g., neurobehavioral tests) have unsubstantiated reliability, and their relation to clinical impairment is highly uncertain. Substantial test variability may be introduced by unmeasured factors, such as sleep deprivation, medication use, and motivation. Misclassification of health status may be either differential or non-differential with respect to exposure status, and effect estimation may be falsely attenuated or exaggerated, depending on the configuration of misclassification.

In community-based case-control studies, controls are selected from among persons thought to be free of neurologic disease. Confirming the absence of disease will invariably be subject to uncertainty, especially for conditions such as Alzheimer's disease that are diagnosed symptomatically and are highly prevalent at older ages. Correct classification of dementias and other neurodegenerative disorders, such as multiple sclerosis, is a difficult yet evolving clinical challenge that complicates the epidemiological investigator's choice of criteria for inclusion or exclusion of cases. Misclassification bias is less likely to occur in studies of rare diseases with relatively unambiguous diagnoses (e.g., motor neuron disease).

4.2 SELECTION BIAS

The most valid approach for selecting subjects for a cross-sectional or longitudinal epidemiologic study is to sample workers on the basis of their exposure characteristics. The presumed nature of neurotoxic effect, acute, subacute, chronic and so forth, will be the essential determinant of the most relevant exposure index, and hence subject selection strategy. Thus, for example, in a study of chronic neurologic impairment related to long-term pesticide application, cumulative exposure, usually estimated by duration of exposure, will be the metric of interest; thus, subjects should be selected to represent a range of employment duration. Inclusion of a reference group of workers free from the exposure of interest can provide estimates of neurologic health baseline values. A non-exposed reference group may be selected from among workers who are not exposed to the agent(s) under study. However, determination of the absence of exposure (or not exceeding ambient levels) among reference workers requires review of detailed occupational history information which may be obtained from personnel records or by subject interview. For example, in their study of neurotoxicity associated with lead exposure, Baker et al.[13] compared neurobehavioral and electrophysiologic test results between lead-exposed foundry workers and a reference group composed of assembly plant workers not exposed to lead or other known occupational neurotoxicants. The absence of exposure to lead or other neurotoxicants among the assembly subjects was based primarily on a review of detailed job history records. Other selection procedures for reference subjects may be necessitated when it is logistically difficult to locate a non-exposed group. This can occur when exposure permeates an entire workplace such that virtually everyone is exposed to some extent, or when it is not feasible to study workers from another workplace. In some instances, reference subjects may be identified personally by the exposed workers. Rosenstock et al.[14] asked subjects with histories of pesticide poisoning to nominate friends or siblings of the same gender to serve as reference subjects for a study of neuropsychological test performance.

It is especially important in studies purporting to estimate dose-effect relations to include subjects with a relatively wide range of exposure. In reality, however, subject selection seldom proceeds according to best scientific principles. The range of exposures experienced by available subjects may not be optimal for dose-response analysis, or workers' willingness to participate may be motivated by their actual or perceived health and exposure circumstances. Also, workers may reduce their own exposures willingly as a result of adverse neurotoxic effects, such that the observed relation with exposure level may be misleading. In the extreme case, the prevalence of neurologic disorders or impaired function may be inversely associated with current exposure because of self-selection out of heavily exposed jobs. A related selection bias that will cause a spuriously low prevalence of disease or impairment in a workforce can occur when affected workers preferentially leave employment due to adverse exposure-related sequelae. This situation, which has been termed the healthy worker survivor effect,[15] is especially difficult to recognize because many epidemiologic studies are limited to workers actively employed as of a defined point in time; workers who had left employment would not be included in the study. Longitudinal studies that recruit subjects at initial employment and involve repeated testing over time is the best way of minimizing healthy worker selection bias, although unfortunately, this will seldom be feasible. An alternative, albeit less direct, approach to address selection bias due to health-related job transfers is to request from study subjects historical information on the onset of neurologic symptoms, which can then be related to exposures that occurred at relevant times. Deficient subject recall is an obvious potential shortcoming of this strategy.

4.3   CONFOUNDING

Numerous demographic, lifestyle, environmental, and other non-occupational factors may be related to neurologic outcomes and thus need to be considered as possible explanations for observed associations with occupational agents. Confounding is a distortion caused by failure to control for extraneous factors that relate both to disease and the exposure(s) of interest. Important potential confounders in occupational neuroepidemiology studies include age, gender, alcohol use, medical history (e.g., diabetes, head injury), and previous employment.

In studies of effects of mixed neurotoxic exposures, one agent may confound the association observed with another. Solvent use, for example, can vary greatly by type and amount among industrial maintenance workers. In such instances, it will be of interest to examine isolated effects related to specific agents, while controlling for the confounding effects of the others. Nelson et al.[16] were confronted with the difficulty of determining associations with a variety of halogenated and non-halogenated organic solvents in their case-control study of neuropsychiatric and other neurologic disorders among automobile assembly workers. Distinguishing effects related to individual solvent exposures resulting from a variety of jobs, including spray painting, degreasing, and application of adhesives and sealers proved to be an intractable problem. Instead, risks were examined in relation to duration of exposure to an aggregated category of solvents.

Confounding poses a threat to research validity most severely in studies that have anticipated weak or subtle effects.[17] It is important to realize that an association of another extraneous variable with disease is not sufficient to produce confounding. The confounder must also be related to exposure. Thus for example, alcohol consumption would only confound the relations between solvents and neurobehavioral test performance if alcohol use was correlated with solvent exposure level.

Control for confounding in cross-sectional, and longitudinal studies can be achieved by matching exposed and non-exposed subjects on other risk factors. Matching cases and controls with respect to presumed confounders is commonly performed in case-control studies. Restriction of subjects to specific categories of age, alcohol use, or medical history, is a form of matching on a group basis. However, care should be taken to avoid being overly restrictive in subject selection because the neurotoxic effects related to the exposure of interest may only be manifest among workers with exposures to other factors. In other words, some potential confounders may be effect modifiers of neurotoxic exposures. An example might be the case where the neurotoxicity of a given agent may occur among older persons or those with a history of heavy alcohol use, in which case restriction of study subjects to young workers or non-drinkers would fail to identify an association. Other methods to control confounding require statistical adjustment techniques, although data analysis can become complicated when there are many potential confounders.

5 FUTURE RESEARCH NEEDS

In most situations of unambiguous work-related neurotoxicity, conventional epidemiologic research methods, such as cross-sectional and nested case-control designs, have the potential to provide adequate characterization of dose-effect relations. Such scenarios typically pertain to investigations of well-defined, clinically recognized syndromes of nervous system damage in workplace settings where there is ample prior evidence of neurotoxic effects. Increasingly, however, epidemiologic research is being applied to investigations of subtle or preclinical effects, that may or may not presage irreversible damage. Also, community-based epidemiologic studies of occupational risk factors for chronic neurodegenerative disorders are becoming more frequent. Conventional research approaches may not be adequate to address some important scientific and public health questions that emerge from an expanded scope of epidemiologic application. Enhancements of study design, measurement and classification of neurological impairment, and exposure assessment will undoubtedly be needed.

When feasible, longitudinal studies of acute and subacute neurotoxicity should be preferred over cross-sectional studies. Ideally, longitudinal studies provide baseline measurements of exposure and health status to enable determination of effect reversibility in relation to exposure reduction. Important ancillary benefits of longitudinal studies should be much needed data on the reproducibility of many of the neurobehavioral and neurophysiological tests that are in common use, and potentially the development of a data base of normative test values that can be referenced in future studies of working populations. Overcoming some of the forbidding logistical difficulties that have precluded longitudinal studies will require creative planning of very efficient study designs.

Advances in clinical diagnostic techniques for the chronic neurodegenerative disorders will undoubtedly benefit epidemiologic research by reducing misclassification. However, for subclinical neurotoxic sequelae, poorly defined outcomes continue to impede research progress. Continued standardization of symptom report questionnaires and neurophysiological measurement methods must be encouraged, as well as consensus criteria for clinical diagnosis for disorders of interest. Development of disease-specific biomarkers would also be of great benefit to etiologic research.

The scientific value of occupational epidemiology studies is highly dependent on the quality of the underlying exposure assessments. More precise dose estimation will be especially important to quantify dose-response associations for preclinical effects related to low-level exposures. In recent years, biomonitoring has become more widespread in epidemiologic studies conducted in the workplace. Continued efforts should be made to establish the validity and reproducibility of exposure and response biomarkers. Establishing normative values and intra-individual variation in non-exposed populations and characterizing the influences of confounding exposures (e.g., alcohol consumption) on biomarker expression should be high-priority focus areas. Increasing the precision of exposure assessment for community-based studies may prove to be a particularly challenging proposition. Occupational neuroepidemiologists would be well advised to take advantage of continuing efforts to standardize questionnaires and job/exposure data linkage methods that support community-based cancer studies.[18]

6 REFERENCES

1. Chia, S.-E., Jeyaratnam, J., Ong, C.-N., Ng, T.-P., and Lee, H.-S., "Impaired color vision among workers exposed to low concentrations of styrene," *Am. J. Ind. Med.*, 26, 481, 1994.

2. Cherry, N. and Gautrin, D., "Neurotoxic effects of styrene: further evidence," *Br. J. Ind. Med.*, 47, 29, 1990.

3. Vorhees, C. V., "Reliability, sensitivity, and validity of behavioral indices of neurotoxicity," *Neurotoxicology and Teratology*, 9, 445, 1987.

4. Seppalainen, A. M., Hernberg., S., Vesanto, R., and Kock, B., "Early neurotoxic effects of occupational lead exposure: a prospective study," *NeuroToxicology*, 4, 181, 1983.

5. Kolstad, H., A., Juel, K., Olsen, J., and Lynge, E., "Exposure to styrene and chronic health effects: mortality and incidence of solid cancers in the Danish reinforced plastics industry," *Occup. Env. Med.*, 52, 320, 1995.

6. Krebs, J. M., Park, R. M., and Boal W. L., "A neurological disease cluster at a manufacturing plant," *Archi. Env. Hlth.*, 50, 190, 1995.

7. Stein, E. C., Schiffer, R. B., Hall, W. J., and Young, N., "Multiple sclerosis and the workplace: report of an industry-based cluster," *Neurology*, 37, 1672, 1987.

8. Chancellor, A. M., Slattery, J. M., Fraser, H., and Warlow, C. P., "Risk factors for motor neuron disease: a case-control study based on patients from the Scottish Motor Neuron Disease Register," *J. Neurol. Neurosurg. Psych.*, 56, 1200, 1993.

9. Rappaport, S. M., "Assessment of long-term exposures to toxic substances in air," *Ann. Occup. Hyg.*, 35,61, 1991.

10. Chettle, D. R., Scott, M. C., and Somervaille, L., J., "Lead in bone: sampling and quantitation using K X-rays excited by ^{109}Cd," *Env. Hlth. Perspect.*, 91, 49, 1991.

11. Nelson L, Longstreth, W. T., Koepsell, T. D., and van Belle, G., "Proxy respondents in epidemiologic research." *Epidemiologic Reviews*, 12, 71, 1990.

12. Flegal, K. M., Keyl, P. M., and Nieto, F. J., "Differential misclassification arising from nondifferential errors in exposure measurement," *Am. J. Epidem.*, 134, 1233, 1991.

13. Baker, E. L., Feldman, R. G., White, R. A., Harley, J. P., Niles, C. A., Dinse, G. E., and Berkey, C. S., "Occupational lead neurotoxicity: a behavioural and electrophysiological evaluation," *Br. J. Ind. Med.*, 41, 352, 1984.

14. Rosenstock, L., Keifer, M., Daniell, W. E., McConell, R., and Claypoole, K., "Chronic central nervous system effects of acute organophosphate pesticide poisoning," *Lancet*, 338, 223, 1991.

15. Arrighi, H. M. and Hertz-Picciotto, I., "The evolving concept of the healthy worker survivor effect," *Epidemiology*, 5, 189, 1994.

16. Nelson, N. A., Robins, T. G., White, R. A., and Garrison, R. P., "A case-control study of chronic neuropsychiatric disease and organic solvent exposure in automobile assembly plant workers," *Occup. Env. Med.*, 51, 302, 1994.

17. Ahlbom, A., Steineck, G., "Aspects of misclassification of confounding factors," *Am. J. Ind. Med.*, 21, 107, 1992.

18. Stewart, P. A. and Stewart, W. F., "Occupational case-control studies: II. Recommendations for exposure assessment," *Am. J. Ind. Med.*, 26, 313, 1994.

6 Evaluation and Management of Neurotoxicity in Occupational Illnesses

L. Manzo and V. Cosi

CONTENTS

1 INTRODUCTION

The occurrence of neurotoxic disorders caused by occupational exposure to chemicals in industry and agriculture is not uncommon. Nervous system impairment has

0-8493-9231-4/98/$0.00+$.50
© 1998 by CRC Press LLC

often been described in cases of acute poisoning in the workplace (e.g., due to accidental chemical spills, unintentional absorption of large amounts of pesticides, etc.) and in large scale industrial accidents simultaneously involving many workers and/or surrounding communities.[1,2] With modernization, occupational hazards associated with neurotoxic substances typically present in factories, mines and agriculture have come to include apparently "safe" work sites such as office buildings, hospitals and other indoor spaces. Public health implications related to these changes may be relevant given the evidence that some subtle neuropsychological and/or neurophysiological alterations can result from a continuous exposure to low levels of several compounds in the workplace.

The prevalence of work-related neurotoxic illnesses is not known. A labor force survey in the U.K. indicated that about 7% of consultations in general practice were due to work related conditions, with a large representation of neuropsychological disorders.[3] On the other hand, many chronic illnesses actually due to workplace exposures cannot be attributed to any specific industrial process or chemical agent because of their similarity to diseases of different origin.[4] It is well known that occupational diseases often go unnotified to public health agencies and so their frequency is largely underestimated.[5-9]

This chapter summarizes clinical and preventive aspects of neurological disorders associated with occupational chemical exposure. Criteria which may help the clinician in establishing the presence of a work-related neurotoxic syndrome will be considered. For a more details on the topic, the reader is referred to comprehensive reviews and articles from the recent literature.[10-14]

2   CLINICAL SYNDROMES

Solvents, gases, metals and pesticides are the major classes of chemicals involved in the etiology of work-related neurotoxic illness (see Chapter 1). These agents can produce both acute and chronic effects on the central and peripheral nervous systems. Despite the enormous complexity of the nervous system functions, neurological disorders apparently occur only in a relatively low number of clinical syndromes. Notably, signs and symptoms in neurotoxic disorders are virtually indistinguishable from those of different etiology, but in most cases they are reversible if diagnosed early enough.

2.1   CNS SYNDROMES

Central nervous system disorders produced by toxic chemicals can be better described as syndromes involving brain functions and/or structures rather than specific diseases, because the pathophysiological bases of these disorders are usually unknown. Certain symptoms and signs should be considered as characteristic of these syndromes, although a significant individual variability is usually seen in clinical manifestations.

In the following section, diffuse encephalopathies and focal brain lesions will be discussed as distinct clinical expressions of a chemical toxicity.

2.1.1   Diffuse encephalopathies

Chemicals may provoke encephalopaties both by a direct action on the brain tissue (e.g., triethyltin-induced encephalopathy) and by an indirect mechanism such as a decreased availability of factors essential to brain metabolism (e.g., the oxygen, in acute carbon monoxide poisoning).

Manifestations observed in severe cases of acute or subacute intoxications include confusional states, delirium, coma, hallucinations, convulsions, and others. A diffuse cerebral edema with intracranial hypertension (headache, papilledema, etc.) has been described as a usual feature in lead-induced encephalopathy and in poisoning by organic tin compounds.[1]

Behavioral and cognitive abnormalities can be observed in mild intoxications (e.g., exposure to sub-narcotic concentrations of organic solvents) and as transient or residual symptoms during the recovery phase from a severe encephalopathy.[12]

A subtle neurobehavioral toxicity may result from a prolonged exposure to low levels of industrial chemicals. For example, a mild syndrome, characterized by mood disorders and memory and psychomotor function impairment, has been described as an earliest form of intoxication in workers exposed to organic solvents over a long period.[10] A depressive pattern, with complaints of sleep disturbances, increased fatigability and loss of sexual interest, is usually present, constituting an impairment of work performance and a significant interference with the activities of daily life.

Depressive disorders may precede the onset of cognitive abnormalities ("intermediate syndrome" or mild chronic toxic encephalopathy). In this case, in addition to the symptoms of mood impairment, the affected individuals complain of memory troubles and impairment of psychomotor functions, which can be confirmed by formal neuropsychological testing. An impairment of visuospatial abilities and abstract concept formation may also be observed. A physical examination usually does not show any abnormality. An early recognition of this condition is important to allow the discontinuance of the exposure and prevent the progression to the more severe stage of toxic encephalopathy, which implies a higher risk of persistent damages. Moreover, clinical manifestations usually improve over a period of weeks after the cessation of the exposure.[10,26]

2.1.2   Focal Brain Lesions

Damages restricted to a brain area may occur in certain intoxications, due to the specific susceptibility of certain brain areas to the action of the implied agent. Examples include visual system damage in methanol intoxication and the cerebellar syndrome occurring in patients with methyl chloride poisoning.[1,13]

Also the extrapyramidal system may be selectively involved, as we can observe in individuals chronically exposed to high concentrations of manganese[15] or in the delayed syndrome of carbon monoxide poisoning.[16]

2.2   Neuropathy

Many industrial chemicals have been shown to cause peripheral neuropathy[17] (Table 1), usually a polyneuropathy, in a decreasing order a sensory-motor, a pure

---

TABLE 1
Industrial Chemicals Causing Peripheral
Nervous System Disorders in Man

Acrylamide
Allyl alcohol
Arsenic
Carbon disulfide
Dimethylaminopropionitrile
Dithiocarbamates
Ethylene oxide
n-Hexane
Lead
Mercury
Methyl bromide
Methyl n-butyl ketone
Methyl methacrylate
Organophosphorous compounds
Thallium
1,1,1-trichloroethane
Trichloroethylene

References: Baker et al.,[12] Kulig,[10] Spencer and Schaumburg,[1]
Blum and Manzo,[48] Feldman,[11] Rosenberg,[13] Hathaway et al.[46]

sensory, a pure motor polyneuropathy. The clinical picture includes distal weakness and wasting, decrease or loss of tendon reflexes, paresthesias, sensory impairment for all modalities (glove-and-stocking anesthesia), but often with a predominant deficit either of superficial or of vibration and position senses. An autonomic involvement may also occur (mainly postural hypotension, lack of sweating and sexual dysfunctions). These disorders may develop either in acute intoxications, as in thallium, arsenic, and carbon monoxide poisoning, or after prolonged exposures to agents, such as lead and organic solvents. Although the manifestations are fairly consistent from one toxin to another, some specific characteristics are unique to individual agents. For example, lead induces a motor neuropathy primarily involving wrist extensors.[12] A mild peripheral neuropathy associated with concomitant signs of central nervous system involvement often occurs in carbon disulfide intoxication;[51] a similar association was reported in tri-ortho-cresyl phosphate (TOCP) outbreaks. Both motor and sensory disorders occur in neuropathies due to n-hexane, methyl n-butyl ketone and acrylamide, while sensory ataxia, sweating of palms and desquamation (hands and soles) are typical of acrylamide-induced neuropathy. Distal paresthesias and painful limbs are distinctive features of thallium and arsenic neuropathies, while an increased sensitivity to touch of feet and sensory involvement predominate in the relatively rare neuropathy due to alkyl mercury poisoning.[1]

Cranial nerve involvement is uncommon in toxic neuropathies, but trichloroethylene affects trigeminal and facial nerves (facial weakness and numbness).[18]

An unusual manifestation of neurotoxicity was observed several years ago in workers of a polyurethane manufacturing plant, who were exposed to catalyst dimethylaminopropionitrile. It caused an autonomic neuropathy with bladder disorders and sexual dysfunction. These symptoms improved over a period of two years after the removal of the affected individuals from the exposure.[19]

3 ASSESSMENT OF A NEUROTOXIC ETIOLOGY

The assessment of neurotoxic etiology is usually easy and even obvious in most cases of acute intoxication in the workplace, in particular when many workers have been affected at the same time. Apart from these cases, a neurotoxic etiology must be assessed through a careful investigation, which includes as a first step both a general and a neurological examination, completed by routine laboratory investigation and, if necessary, by neuropsychological batteries, neurophysiological tests (EEG, electromyography, electroneuronography, evoked potentials, eye movements, etc.), brain CT or MRI scans. Whenever the diagnosis of a neurological disease cannot be reached, a careful investigation of a possible occupational etiology must be started.

3.1 ACUTE INTOXICATION

The recognition of a neurotoxicity is seldom difficult in acute poisoning. History of the incident must be obtained as well as accurate details about the work environment and the chemical(s) involved in the incident. Besides a routine physical examination, emphasis should be placed on organs and systems most likely to be affected by the specific agent(s).

Chemicals, such as solvents or pesticides, which cause dizziness, incoordination or loss of consciousness in overdose may enhance the risk of an accident resulting in traumatic injury.[49]

Notably, the effects of certain neurotoxic substances (e.g., carbon monoxide, organophosphate insecticides, arsenic, thallium) may be delayed and therefore a prolonged follow-up may become necessary.

In certain cases, any other worker at risk should mandatorily be checked, and the risk of contamination of the environment outside the plant should also be considered.

3.2 CHRONIC EXPOSURE

Occupational neurotoxicity syndromes are more commonly observed in medical practice as the result of a chronic, and often misunderstood, exposure.

In individual cases of suspected neurotoxic disorders casually encountered by the practitioner, a precise diagnosis is absolutely required, since the patient's co-workers may develop similar job-related disorders.[20] Unfortunately, the attribution of a central nervous system impairment to a chronic neurotoxic exposure is complicated by several factors. Symptoms and signs are often insidious and difficult to distinguish from other mild neurobehavioral changes. Moreover, it could be difficult to differentiate symptoms related to an intoxication from those that depend on an idiosyncratic response. In some instances, an occupational exposure could simply produce an exacerbation of a pre-existent chronic disease, for example an angina worsened by the exposure to low levels of carbon monoxide. In addition, a long latency of symptoms from the onset of exposure, which is observed for many neurotoxicants, may lead a physician to rule out neurotoxicity or to not even consider this possibility. A number of factors, including a concomitant abuse of ethanol, tobacco, street drugs, prescription drugs, a family history of neurological diseases, and ultimately normal aging, can offer a plausible explanation for mild syndromes, which are actually due to a neurotoxin.

The problem is most frequently encountered in the controversies about solvent-induced encephalopathy. The term "encephalopathy" is commonly used to describe a potentially reversible state of cerebral dysfunction which may be due to a variety of causes, for example, cirrhosis of the liver, uremia, diabetes and hypertension. If metabolic changes persist or are severe enough, a sustained brain damage may follow. In neurotoxicology, the term "encephalopathy" is often used rather loosely, but it tends to imply the idea of a structural damage. However, a structural deterioration is often undemonstrated in solvent-induced CNS disorders. An association between duration of exposure and outcomes, especially those based on cognitive and behavioral measures, is also equivocal. Moreover, it is problematic to establish the ratio solvent dose/"lesions" burden, because of the short half-life of these substances.

In cases of organic solvent exposures, a diagnosis of toxic encephalopathy relies on a careful medical evaluation, including a detailed occupational history, but mainly on psychiatric findings and cognitive changes assessed by computer-assisted testing techniques.[10,12]

A longitudinal examination, preferably with the support of a psychiatrist and/or of a clinical psychologist, is quite useful in establishing a definite diagnosis in a clinical setting.

4   AN APPROACH TO DIAGNOSIS

A detailed description of the standard methodology used in medical and neurological examination is presented in journal articles[21] and reference textbooks,[22] where the topic has been linked to specific aspects of neurotoxic diseases. The discussion here is addressed to practical aspects and problems to be considered in assessing work-related neurotoxic illnesses in the specific context of clinical toxicology and occupational medicine.

4.1   OCCUPATIONAL HISTORY

Differentiating neurotoxic from non-neurotoxic diseases may be difficult in the presence of common neurological symptoms. Thus, the evaluation of an individual case begins with the collection of patient's clinical history including an accurate evaluation of his exposure history.[8,23] A detailed work- and job-related history may suggest an association between the patient's signs and symptoms and exposure to a toxic agent.

The essentials of an occupational history include the current employment (data about the workplace, employer, type of job hazards, relation of symptoms to job exposure, implementation of control measures) and past job experiences. The latter may be important in disclosing sources of hazardous exposures in view of the fact that many processes in neurotoxic disease have a long latency from the beginning of the exposure to the onset of signs and symptoms.

The patient should be interviewed with respect to possible exposure to substances with known neurotoxic potential as well as jobs in which exposure to these agents is reasonably suspected.[8] The patient can of course be at high risk because of a particular job type (Table 2).

Whenever a physician determines that a patient has a neurological syndrome with no obvious cause, it is imperative that a detailed history of work and leisure-time activities be obtained. The latter are commonly forgotten, but can be extremely important if the patient uses toxic chemicals at home, such as lead glazing in stained glass or ceramic making, and solvents.

Since a differential diagnosis of neurotoxic syndromes involves consideration of hereditary, degenerative, deficiency and metabolic diseases, the patient's family history demands close attention as do the details of his nutritional intake.[22]

Attempts must initially be made to identify the specific agent to which the patient has been exposed. The number of toxic substances is so large that the practitioner cannot be expected to be familiar with every presentation and diagnosis. However, computers have revolutionized access to toxicological data through on-line and CD-ROM services (See Chapter 12). Information concerning major health effects of specific chemicals can be easily obtained from textbooks or through phone consultation with the staff of public health agencies. Moreover, poison centers and environmental toxicology information centers can be utilized as specialized referral sources for toxic identification (type of product, composition, and contained size as a clue to dose), or obtaining data of estimated toxic and fatal doses for a human adult, and specific therapy.[4,6]

According to clinical experience, ingestion is a common route of exposure in cases of acute poisoning whereas inhalation, skin and ocular contact are generally the most significant routes of subacute or chronic entry of hazardous chemicals in an unprotected worker.[6] Substances such as organic solvents and certain agrochemicals are absorbed through the skin in significant amounts, resulting in systemic toxicity. Occupational ingestions are usually accidental or may be associated with smoking, eating or drinking in a contaminated workplace.

4.2   CLINICAL EXAMINATION

Once the exposure history is obtained, clinical examinations are performed to assess the presence and the extent of a neurological dysfunction. One must remember that asymptomatic neurotoxic diseases may be an extremely common phenomenon in the occupational setting.[13] If objective findings are observed, the physician must

TABLE 2
Selected Occupations Associated with Neurotoxic Hazards

Occupation or Industry	Exposures
Battery making and recycling	lead, manganese, mercury
Car painters	solvents
Chloralkali plants	mercury
Construction painters	lead, solvents
Dentists, dental technicians	mercury, waste anesthetic gases, methyl methacrylate
Dry cleaners	perchloroethylene
Electronic workers	solvents
Electroplating	cyanide
Epoxy resins	allyl chloride
Explosives manufacturing	trinitrotoluene
Farm workers	pesticides
Fire fighters	carbon monoxide, cyanide
Foundry workers	lead
Fumigators	methyl bromide, ethylene oxide
Furniture refinishers	n-hexane
Hospital workers	ethylene oxide
Metallurgy	carbon monoxide
Offset printing	solvents
Paint and pigment manufacturing	lead, solvents
Paper manufacturing	acrylamide
Polymer and plastics production	solvents, styrene, acrylamide
Polyurethane foams	dimethylaminopropionitrile
Pottery industry	lead
Rayon industries	carbon disulfide
Rodenticides	thallium
Shoe workers	n-hexane and other solvents
Sterilizer operators	ethylene oxide
Textiles	mercury, carbon disulfide
Welders	metal fumes, lead, manganese
Woodworkers	n-hexane and other solvents

attempt to distinguish between the toxin-induced disorders and those of other etiologies. This process of differential diagnosis is difficult: the physician must ascertain the extent and degree of exposure to a neurotoxic substance to determine its role, if any, in the genesis of the disorder under consideration.

The signs and symptoms must be consistent with the known properties of the suspected noxious agent(s). Aspects to be considered when evaluating a patient with presumed neurotoxic injury include the pattern of preexistent symptoms. It must be determined if the symptoms increase during the working periods vs. the non-working ones; the presence of unusual combinations of neurological and psychiatric symptoms and signs (e.g., manganese poisoning) must be also detected. Noteworthy, not only the nervous system but also other systems and organs may be affected by toxicants. For example, peripheral neuropathy in thallium poisoning is almost invariably associated with hair loss; chlorinated hydrocarbon solvents, such as trichloro-ethylene, are particularly likely to cause hepatocellular damage so the measurement of hepatic enzymes can be useful to detect definite exposure.[10]

Neurotoxicity most commonly manifests itself as a nonfocal syndrome, as neu-rotoxins generally affect the nervous system in a diffuse way. Toxic peripheral neuropathies are generally easy to document on physical examination and by elec-trophysiological studies of peripheral nerve function. For example, *n*-hexane neur-opathy is initially suggested by the insidious onset of numbness at toes and fingers. A sustained exposure results in proximal spread of sensory loss, followed by loss of Achille's tendon and patellar reflexes. Muscle weakness and atrophy develop after the sensory findings have become pronounced.[24]

4.3 SPECIAL DIAGNOSTIC PROCEDURES

The clinician often needs the assistance of laboratory tests and other diagnostic procedures to establish the diagnosis of neurotoxicity and/or to quantify its extent. In most instances, laboratory and instrumental findings are either nondiagnostic or may be equivocal. For example, neurophysiological studies that show the presence of a peripheral neuropathy in a subject with an exposure to *n*-hexane are equivocal in case of concomitant diabetes or excessive alcohol intake, as both these conditions frequently produce a similar nervous involvement.[25]

4.3.1 Neuropsychological tests

It is widely accepted that the earliest detectable neurotoxic manifestations are likely to be behavioral and altered behavior can represent a significant outcome in several work-related disorders.[21]

Neuropsychological test batteries can be a very sensitive tool in identifying subtle nervous system dysfunction and subclinical impairments in subjects with a history of occupational exposure to neurotoxic chemicals[21,26] (See also Chapter 11 in this book). The application of these procedures offers a number of advantages.[50,54] There are standardized rules for administering and scoring these tests. The results can be analyzed objectively, according to published normative data corrected by age and sex. Some of the available tests have been validated in both research and clinical settings, allowing consistent interpretation of the results by independent neuropsychol-ogists. The reliability of the tests (test-retest) is often established. Finally, the tests are used widely and results may be applied in both diagnostic and treatment settings.

The available batteries can be employed for the clinical diagnosis of encephal-opathy in individual patients and/or to screen large groups of exposed subjects. Brain functions assessed by these tests include attention, psychomotor skills, concept formation (reasoning), memory, visuospatial abilities, language, mood, and person-ality. The most commonly used criterion for inclusion of a test in a battery is its confirmed sensitivity to particular neurotoxic agents as previously documented in research literature. In clinical situations, it is generally desirable to choose neuro-psychological tests with the highest sensitivity and specificity, which besides permit the assessment of the broadest spectrum of functions.

4.3.2 Neurophysiological studies

Despite a very large employment in clinical neurology, neurophysiological methods have achieved only limited use in occupational medicine, compared to other testing procedures, such as behavioral methods. Reasons include the relatively high cost of equipment and the need for specially trained personnel to administer tests and interpret data.[27] However, electrophysiological methods can be very useful in the medical evaluation of individual subjects to assess early subclinical alterations caused by toxic chemicals in both central and peripheral nervous systems (See Chapter 9).

Nerve conduction measurements and electromyography have been used to evaluate the extent and severity of peripheral nerve damage in workers exposed to industrial toxins.[28,53]

Non-invasive techniques that measure sensory thresholds for vibration and temperature have recently been developed to monitor diabetic patients for sensory neuropathy. These techniques can also be efficient tools for screening of workers with significant exposure to neurotoxic agents or with early sensory symptoms.

Electroencephalography has been used in the evaluation of workers exposed to neurotoxic agents. Normally these studies have not been so useful as nerve conduction tests. However, they may be of value in the assessment of altered states of consciousness of unknown etiology.

A more promising approach is the measurement of cortical evoked potentials following auditory or visual stimuli. These tests have been proposed as a screening tool for early neurotoxic dysfunction in field studies.[27] Chemicals known to alter visual evoked potentials in man include mercury vapor, lead, n-hexane, toluene, xylene, tetrachloroethylene, and solvent mixtures.[27,29-31]

Chemosensory evoked potentials (a term indicating scalp-recorded electrical activity elicited by chemical stimulation of the olfactory and/or trigeminal nerve) represent another direct method for the objective measurement of functioning of these ways.[31] Recorded in conjunction with psychophysical measures of sensory stimuli, they offer a direct means to evaluate response to chemical exposure. Recent applications include the study of volatile organic compounds in relation to their possible role in "sick building syndrome," a condition which may develop in occupants of new or renovated buildings.[27] Trigeminal somatosensory evoked potentials have been investigated as a possible marker to screen workers for trigeminal lesions resulting from exposure to trichloroethylene.[32]

The assessment of event-related potentials may provide an additional measure of early neurotoxic effects in occupational health studies. Studies of workers meeting criteria for mild solvent encephalopathy (type 2A/2B, as described in Chapter 1), have shown significant delayed latency in the P300 components of the event-related potentials.[33] P300 is known to be generated in response to meaningful stimuli, such as infrequent and unpredictable events. Increased latency of the P300 has typically been associated with a variety of organic disorders and head injury. P300 may represent a valuable marker of central nervous system dysfunction unconfounded by affective or motivational factors.[33] P300 latencies were slightly prolonged in steel smelting workers exposed to manganese.[34]

Visual contrast sensitivity has been investigated in workers with previously diagnosed organic-solvent induced chronic encephalopathy and has proved to be a useful addition in the diagnostic procedure for this disease.[35]

Otovestibular investigations also represent a valuable diagnostic measure especially in cases of suspected CNS involvement due to solvents.

4.3.3   Neuroimaging

Positron emission tomography (PET), single photon emission computed tomography (SPECT) and magnetic resonance imaging (MRI) have been used in the study of subjects occupationally exposed to neurotoxic chemicals including mercury,[36] manganese,[37] carbon monoxide,[38] and organic solvents[39] (See also Chapter 10).

Magnetic resonance imaging provided evidence of cerebral deposits of paramagnetic metals in occupational manganese intoxication.[37] Different brain alterations were found by magnetic resonance imaging studies of solvent abusers presenting abnormalities on neurological examination and brainstem auditory evoked response testing.[40]

By using labelled neurotransmitters and their analogues, the PET scanner can produce images related to a specific metabolic aspect of brain tissue. In this way, brain biochemistry and processes associated with disorders in specific areas can be investigated. Measurements of cerebral blood flow have documented reduced metabolic rate in both subcortical and cortical structures in workers exposed to solvents.[41]

Unfortunately, neuroimaging techniques are quite expensive and have been so far of relatively little benefit in routine evaluation of patients with a neurotoxic illness. Often no changes from normal are seen in patients presenting complaints or dysfunctions. Specificity of abnormalities detected by these techniques is limited, in other words it is difficult and sometimes impossible to ascribe the cause of an abnormality to a toxic exposure.[25]

4.4   ASSESSMENT OF EXPOSURE

Analytical assays can be useful in acute accidents to assess the degree of chemical exposure, because in such cases the substance to which the individual was exposed is normally positively identified. In assessing exposure after accidental or deliberate ingestion of chlorinated hydrocarbons, it must be noted that these agents are densely radiopaque when in the gastrointestinal tract.[42]

A chronic exposure to neurotoxins may occur at such low levels that clinical laboratory tests may fail to establish a body burden for that particular chemical. When exposure data are available, the clinician should consult with an occupational medicine specialist or an industrial hygienist engineer to clarify their significance in the individual case and determine whether additional chemical investigations are required.

It is unusual for a worker to be exposed over a long time to only one toxic agent or class of chemicals; moreover, it may be impossible to identify prior exposures in workers with known or suspected chronic toxicity even when an accurate industrial plant hygiene evaluation scheme is normally implemented.

As illustrated in another chapter (Chapter 4), a number of biochemical and molecular markers have been developed in the recent years as valuable means to assess the patient's exposure and the early effects of hazardous agents including neurotoxic substances. In perspective, biomarkers can represent a significant contribution to early detection and prevention of neurotoxic disorders. Clearly, caution should be exercised in the utilization of nonspecific, poorly reproducible or nonvalidated indicators in the diagnosis of controversial clinical situations.[8] While several markers of exposure are currently used in monitoring of workers exposed to lead and neurotoxic solvents (See Chapter 3), only few markers of effects have undergone sufficient validation to merit practical use in occupational neurotoxicology. Measurement of red blood cell cholinesterase concentrations is recommended in acute intoxication from organophosphate insecticides and is also used in occupational settings for surveillance of exposed workers.

In most cases, establishing a link between exposure and clinical dysfunction still relies on a focused history of exposure and careful medical examination rather than on clinical laboratory.[14]

5   CLINICAL MANAGEMENT

Once the diagnosis of work-related toxic disease is made, the physician should be able to initiate a proper medical treatment and, in addition, successfully grapple with the complex array of social, economic, ethical and legal implications arising from the diagnosis and its appropriate follow-up.[8,43,49]

5.1   ACUTE EXPOSURE

If the patient is acutely ill, symptomatic management of acute effects of poisoning is required: that is, in general, establish a patent airway, assist ventilation if needed, give supplemental oxygen, and monitor the vital signs and electrocardiogram. If the patient has not been decontaminated at the scene, simple removal of clothing and showering or flushing with copious amounts of water are adequate emergency measures for most chemical exposures. Any contaminated clothing must be bagged and saved for possible chemical identification. Data on general toxicity and human health effects of occupational chemicals and treatment of acute poisoning can be found in several reference books[44-46] and can be obtained from easily available electronic sources (See Chapter 12).

5.2   CHRONIC EXPOSURE

Clinical management plays a limited but important role in the control of occupational disease caused by long-term exposure to neurotoxic agents in the workplace. Recognition and early diagnosis of disease are important in view of the fact that other workers in the same workplace have been or will be affected in the same way. The goal of management is prevention of the progression of disease in the patient and prevention of subsequent illness in the co-workers.[43]

Typically, when clinical or subclinical effects of the exposure to a toxic agent are assessed, the clinician should recommend the removal from a further occupational exposure to the agent. Continued exposure, even if better protection is used, is not safe.[10] The patient should be informed that the diagnosed illness may be job-related and educated regarding short-term and long-term health implications.

Work restriction may offer the best hope for cure or control of occupational diseases but it often results in economic hardships or job loss.[8] This is a crucial and often difficult decision which may affect the patient's employment status and should be made with the patient's full knowledge and agreement. An interaction with the patient's employer is always desirable.

Depending on specific local regulations, the physician may be required to report the occurrence of occupational disease to health authorities or a governmental agency. Such regulations are designed not only to benefit the affected individual, but also to prevent similar health problems in co-workers with the same exposure. It has been noted that the physician often neglects to carefully document the patient's history and other aspects which are necessary to enable a subsequent legal judgment for a compensation purpose.[8,47]

All these patients should undergo a regular follow-up, to monitor their progressive improvement until the disappearance of all symptoms and signs, which is expected in most cases.

6 CONCLUSIONS

Neurotoxic disorders are probably only a small percentage of all occupational illnesses but their importance should not be underestimated, because manifestations associated with acute poisoning, such as seizures, disorders of neuromuscular transmission and coma, are life-threatening. In chronic exposure, failure to recognize early neurological changes, with consequent absence of withdrawal, may result in serious and irreversible health damage for the worker. Moreover, a missed diagnosis often equals missed opportunity to protect other workers who are similarly at risk. Thus, the implications of diagnosing an occupational neurotoxic illness go beyond simple treatment.[3,4] This notion is further supported by two general considerations:

(a) Occupational diseases are preventable in most cases by appropriate intervention and control strategies.[8,43] Effective prevention requires adequate knowledge of hazards posed by chemical agents and the prudent use of the same agents in the workplace. Physicians, especially those engaged in primary care, should be familiar with the basic principles of medical toxicology and occupational medicine; they should be able to recognize and diagnose diseases that are caused or exacerbated by occupational toxins[4] and fully aware of the social, ethical, and legal implications of the diagnosis of work-related diseases.[43]

(b) Clinical studies of patients with work-related illness have been of considerable importance in the advancement of neurotoxicology as a scientific discipline and, in many instances, they have represented the initial step to the identification of new disease entities. Following these observations, efforts have been directed by the basic scientist to try to better understand the effects of the offending chemical, which in some cases have led to more effective treatment or prevention.

A great deal of our understanding of the subtle neurobehavioral effects of environmental chemicals has been gained through the examination of workers exposed to low concentrations of solvents, metals and pesticides in the workplace.[55] The importance of future studies in this area is emphasized by preliminary data implicating environmental chemicals in the initiation, progression and/or exacerbation of neurodegenerative disorders (See Chapter 7).

7 REFERENCES

1. Spencer, P. S. and Schaumburg, H. H., Eds., *Experimental and Clinical Neurotoxicology*, Williams & Wilkins, Baltimore, MD, 1980.
2. Jeyaratnam, J., "Occupational health issues in developing countries." *Environ. Res.*, 60, 207, 1993.
3. Seaton, A., "Diagnosing and managing occupational disease." *Br. Med. J.*, 310, 1282, 1995.
4. Landrigan, P. J. and Baker D. B., "The recognition and control of occupational disease." *J. AMA*, 266, 676, 1991.
5. Blanc, P. D., Remple, D., Maizlish, N., Hiatt, P., Olson, K. R., "Occupational illness: case detection by poison control surveillance," *Ann. Int. Med.*, 111, 238, 1989.
6. Litovitz, T., Oderda, G., White, J. D. and Sheridan, M. J., "Occupational and environmental exposures reported to poison centers," *Am. J. Publ. Hlth.*, 83, 739, 1993.
7. Hinnen, U., Hotz, P., Gossweiler, B., Gutzwiller, F., and Meier, P. J., "Surveillance of occupational illness through a national poison control center: an approach to reach small-scale enterprises?" *Int. Arch. Occup. Environ. Hlth.*, 66, 117, 1994.
8. Newman, L. S., "Occupational illness," *N. Engl. J. Med.*, 333, 1128, 1995.
9. Fernando, R., "Pesticide poisoning in the Asia-Pacific region and the role of a regional information network," *J. Toxicol. Clin. Toxicol.*, 33, 677, 1995.
10. Kulig, K., "Clinical neurotoxicology of industrial and agricultural chemicals," In *Neurotoxicology: Approaches and Methods*, "Chang, L. W. and Slikker, W., Jr., Eds., Academic Press, Orlando, FL, 1995, p. 629.
11. Feldman, R. G., "Effects of toxins and physical agents on the nervous system," in *Neurology in Clinical Practice*, Bradley, D. M., Ed., Butterworths, London, 1988.
12. Baker, E. L., Feldman, R. G. and French J. G., Environmentally related disorders of the nervous system. *Med. Clin. North Am.* 74, 325, 1990.
13. Rosenberg, N. L., "Basic principles of clinical neurotoxicology." In *Neurotoxicology: Approaches and Methods*, Chang, L. W. and Slikker W., Jr., Eds., Academic Press, Orlando, FL, 1995, Chapter 39.
14. Schaumburg, H. H. and Spencer, P. S., "Recognizing neurotoxic disease." *Neurology*, 37, 276, 1987.
15. Calne, D. B., Chu N.-S. Huang C.-C., Lu C.-S., and Olanow W., "Manganism and idiopathic Parkinsonism: Similarities and differences." *Neurology*, 44, 1583, 1994.
16. Butera, R., Candura, S. M., Locatelli, C., Varango, C., Li, B., and Manzo, L., "Neurological sequelae of carbon monoxide poisonning. Role of hyperbaric oxygen." *Indoor Environ.*, 4, 134, 1995.
17. Ludolph, A. C. and Spencer, P. S., "Toxic neuropathies and their treatment," in *Clinical Neurology*, Hartung, H. P., Ed., Baillière Tindall, London, 1995, p. 505.
18. Seppalainen, A. M., "Halogenated hydrocarbons," in *Neurotoxicology*, Blum, K. and Manzo, L., Eds., Marcel Dekker, New York, 1985, Chapter 22.

19. Keogh, J. P., Pestronk, A., Wertheimer, D., and Moreland, R., "An epidemic of urinary retention caused by dimethylaminopropionitrile," *JAMA*, 243, 746, 1980.

20. Rosenstock, L. and Cullen, M. R., *Clinical Occupational Medicine*, Saunders, Philadelphia, 1986.

21. Feldman, R. G. and White, R. F., "Role of the neurologist in hazard identification and risk assessment," *Environ. Hlth.. Perspect.*, 104(S2), 227, 1996.

22. Sterman, A. B. and Schaumburg, H. H., "The neurological examination," in *Experimental and Clinical Neurotoxicology*, Spencer P. S. and Schaumburg, H. H., Eds, Williams & Wilkins, Baltimore, MD, 1980, Chapter 46.

23. Goldman, R. H. and Peters, J. M., "The occupational and environmental health history," *J. AMA*, 246, 2831, 1981.

24. Spencer, P. S., Couri, D., and Schaumburg, H. H., "*n*-Hexane and methyl *n*-butyl ketone," in *Experimental and Clinical Neurotoxicology*, Spencer, P. S. and Schaumburg, H. H., Eds, Williams & Wilkins, Baltimore, MD, 1980, Chapter 32.

25. Prockop, L. D., "Neuroimaging in neurotoxicology." In *Neurotoxicology: Approaches and Methods*, Chang, L. W. and Slikker, W., Jr, Eds., Academic Press, Orlando, FL, 1995, Chapter 50.

26. Fiedler, N., Feldman, R. G., Jacobson, J., Rahill, A., and Wetherell, A., "The assessment of neurobehavioral toxicity: SGOMSEC joint report," *Environ. Hlth. Perspect.*, 104 (S2), 179, 1996.

27. Otto, D. A. and Hudnell, H. K, "The use of visual and chemosensory potentials in environmental and occupational health," *Environ. Res.*, 62, 159, 1993.

28. Seppalainen, A. M. and Hernberg, S., "Subclinical lead poisoning. Electrophysiological aspects at different blood lead levels," in *Advances in Neurotoxicology*, Manzo, L., Ed., Pergamon Press, Oxford, UK, 1980, p. 35.

29. Andersen, A., Ellingsen, D. G., Morland, T., and Kjuus, H., "A neurological and neurophysiological study of chloralkali workers previously exposed to mercury vapour," *Acta Neurol. Scand.*, 88, 427, 1993.

30. Chang, Y. C., Yeh, C., and Wang, J. D., "Subclinical neurotoxicity of mercury vapor revealed by a multimodality evoked potential study of chloralkali workers," *Am. J. Ind. Med.*, 27, 271, 1995.

31. Discalzi, G. L., Capellaro, F., Bottaro, L., Fabbro, D., and Mocellini, A., "Auditory brainstem evoked potentials (BAEPs) in lead exposed workers," in *Current Issues in Nervous System Toxicology*, Mutti, A., Costa, L. G., Manzo, L., and Cranmer, J., Eds., Intox Press, Little Rock, AR, 1992, p. 207.

31. Kobal, G. and Hummel, T., "Cerebral chemosensory evoked potentials elicited by chemical stimulation of the human olfactory and respiratory mucosa" *Electroenceph. Clin. Neurophysiol.*, 71, 241, 1988.

32. Barret, L., Garrel, S., Danel, V. and Debru, J., "Chronic trichloroethylene intoxication: A new approach by trigeminal-evoked potentials?" *Arch. Environ. Hlth.*, 42, 297, 1987.

33. Morrow, L., Steinhauer, S. R., and Hogdson, M. J., "Delay in P300 latency in patients with organic solvent exposure," *Arch. Neurol.*, 49, 315, 1992.

34. Wennberg, A., Hagman, M., and Johansson, L., "Preclinical neurophysiological signs of Parkinsonism in occupational manganese exposure," in *Current Issues in Nervous System Toxicology*, Mutti, A., Costa, L. G., Manzo, L., and Cranmer, J., Eds., Intox Press, Little Rock, AR, 1992, p. 271.

35. Donoghue, A. M., Dryson, E. W., and Wynn-Williams, G., "Contrast sensitivity in organic-solvent-induced chronic toxic encephalopathy," *J. Occup. Environ. Med.*, 37, 1357, 1995.

36. White, R. F., Feldman, R. G., Moss, M. B., and Proctor, S. P., "Magnetic resonance imaging (MRI), neurobehavioral testing, and toxic encephalopathy: two cases." *Environ. Res.*, 61, 117, 1993.

37. Nelson, K., Golnick, J., Korn, T., and Angle, C., "Manganese encephalopathy. Utility of early magnetic resonance imaging," *Br. J. Ind. Med.*, 50, 510, 1993.

38. Murata, S., Asaba, H., and Hiraishi, K., "Magnetic resonance imaging findings in carbon monoxide intoxication," *J. Neuroimag.*, 3, 128, 1993.

39. Triebig, G. and Lang, C., "Brain imaging techniques applied to chronically solvent-exposed workers: current results and clinical evaluation," *Environ. Res.*, 61, 239, 1993.

40. Rosenberg, N. L., Spitz, M. C., Filley, C. M., Davis, K. A., and Schaumburg, H. H., "Central nervous system effects of chronic toluene abuse. Clinical, brainstem evoked response and magnetic resonance imaging studies," *Neurotoxicol. Teratol.*, 10, 489, 1988.

41. Morrow, L., Callender, T., Lottenberg, S., Buchsbaugh, M., Hogdson, M. J., and Robin, N., "PET and neurobehavioral evidence of tetrabromoethane encephalopathy," *J. Neuropsychiat. Clin. Neurosci.*, 2, 431, 1990.

42. Sottili, S., Pizzi, A., Locatelli, C., and Abbondati, G. G., "Trichloroethylene is radio-paque." *J. Toxicol. Clin. Toxicol.*, 31, 657, 1993.

43. Rosenstock, L., Rest, K. M., Benson, J. A., Cannella, J. M., Cohen, J., Cullen, M. R., Davidoff, F., Landrigan, P. J., Reynolds, R. C., Clever, L. H., Ellis, G. B., and Goldstein, B. D., "Occupational and Environmental Medicine," *N. Eng. J. Med.*, 325, 924, 1991.

44. Borak, J., Callan, M., and Abbott, W., *Hazardous Materials Exposure: Emergency Response and Patient Care*, Brady, 1991.

45. Sullivan, J. B. and Krieger, G. R., *Hazardous Materials Toxicology. Clinical Principles of Environmental Hlth.*, Williams & Wilkins, Baltimore, MD, 1992.

46. Hathaway, G., Procter, N. H., Hughes, J. P., and Fischman, M., Eds, *Chemical Hazards of the Workplace*, 3rd ed., Van Nostrand Reinhold, New York, 1991.

47. Cornes, P. and Aitken, R. C. B., "Medical reports on persons claiming compensation for personal injury," *J. R. Soc. Med.*, 85, 329, 1992.

48. Blum, K. and Manzo, L., Eds., *Neurotoxicology*, Marcel Dekker, New York, 1985.

49. Becker, C. E. and Olson, K. R., *Initial evaluation of the patient with occupational chemical exposure, in Poisoning and Drug Overdose*, Olson, K. R., Ed., Appleton & Lange, East Norwalk, 1994, p. 387.

50. Iregren, A., "Behavioral methods and organic solvents: questions and consequences," *Environ. Hlth. Perspect.*, 104(S2), 361, 1996.

51. Seppalainen, A. M. and Haltia, M., "Carbon disulfide," in *Experimental and Clinical Neurotoxicology*, Spencer P. S. and Schaumburg, H. H., Eds, Williams & Wilkins, Baltimore, MD, 1980, Chapter 25.

52. Manzo, L. and Sabbioni, E., "Thallium toxicity and the nervous system," in *Metal Neurotoxicity*, Bondy, S. C. and Prasad, K. N., Eds., CRC Press, Boca Raton, FL, 1988, Chapter 3.

53. Le Quesne, P. M., "Clinical expression of neurotoxic injury and diagnostic use of electromyography," *Environ. Hlth. Perspect.*, 26, 89, 1978.

54. Spurgeon, A., "Current approaches to neurobehavioural testing in occupational health," *Occup. Environ. Med.*, 53, 721, 1996.

55. White, R. F. and Proctor, S. P., "Solvents and Neurtoxicity," *Lancet.* 349. 1239. 1997.

7 Role of Occupational Neurotoxicants in Psychiatric and Neurodegenerative Disorders

Stefano M. Candura, Luigi Manzo, and Lucio G. Costa

CONTENTS

0-8493-9231-4/98/$0.00+$.50
© 1998 by CRC Press LLC

1   INTRODUCTION

A large number of chemicals encountered in the workplace or in the general environment can cross the blood-brain barrier and exert deleterious effects on the central nervous system (CNS). Some substances perturb ionic channels or synaptic mechanisms required for the orderly transfer of electrochemical information within the CNS. Others disrupt sites necessary for the maintenance of cellular integrity, resulting in degenerative responses of neurons and glial cells. The delicate neural vasculature is an additional site of vulnerability.[1]

Chemicals which damage the vasculature, myelin or neuronal membranes generally cause encephalopathies that represent a generalized response, while interactions with specialized structures with limited distribution in the CNS or specific cell groups result in selective toxicity. Some of these effects are short-lasting and rapidly reversible; others, especially those that are associated with structural damage, may be permanent, in that restoration of nervous function through cell division or regeneration of cell processes is difficult or impossible. Functional recovery following toxic exposures may, however, occur through partial regrowth of axons and dendritic processes. Additionally, other neuronal pathways may take over the function. Usually, complete recovery can occur at low exposure level, while larger doses result in permanent damage.[1,2]

Chemical-induced CNS damage may result in a wide variety of neurologic and psychiatric effects, ranging from subtle behavioral disturbances to severe coma and death. Since anatomical damage to a certain nervous structure or the impairment of a particular function is reflected by the same signs and symptoms irrespective of the cause, the clinical presentation of both acute and chronic intoxications, as well as of their *sequelae*, may be very similar to that of neuropsychiatric illnesses of unknown origin, or secondary to conditions such as congenital defects, infections, physical trauma, nutritional and metabolic disorders, vascular diseases, and tumors: delicate diagnostic problems may arise.

The ability of the CNS to recover functionally from irreversible structural damage can be accounted for by redundancy of nerve cells involved in certain functions, or the recruitment of other neurons. Thus, some types of CNS damage may not be reflected in any clinical manifestation. However, this type of damage may deplete reserve capacity and, therefore, make the brain or the spinal cord more vulnerable to additional insult. Physiological loss of neurons with aging may be accelerated, and may result in functional changes at a time when the relationship to earlier damage is unrecognized. Thus, environmental or occupational exposures to chemicals may be involved in the causation of human neurodegenerative diseases which are usually labelled as "idiopathic."[2,3]

This chapter presents three typical examples of human CNS disease caused by chemicals contained in food or present in the living environment, and then focuses on the occupational setting. Particularly, it describes the neuropsychiatric effects induced by some metals, solvents and pesticides, and reviews the debated role of occupational exposures in the etiopathogenesis of the most common neurodegenerative disorders (Parkinson's disease, Alzheimer's disease, amyotrophic lateral sclerosis).

2 ENVIRONMENTAL CHEMICALS IN NEUROPSYCHIATRIC/ NEURODEGENERATIVE DISORDERS

2.1 LATHYRISM

Excessive consumption of the seed of *Lathyrus sativus* (chickling or grass pea) and related species is associated with an irreversible neurological disease called lathyrism. Paralysis of the lower limbs occurs in both genders and at all ages, but it appears more prevalent among young males.[4] The pathological features have not been well described, but seem to be dominated by degeneration of the pyramidal pathways. The causative agent is currently believed to be β-N-oxalylamino-L-alanine (BOAA), a potent neuroexcitatory aminoacid found in the plant.[5]

Lathyrism is one of the oldest neurotoxic diseases known to humans and was once prevalent in Europe, northern Africa, the Middle East, and parts of the Far East. One of the best described outbreaks of the disease occurred in a labor camp during World War II, when a large population of political prisoners had subacute spastic limb weakness that was traced to a high dietary intake of *Lathyrus sativus*.[6] After the war, several hundreds of these victims migrated to Israel, where they have been followed for 40 years. After decades of clinical stability, some of the patients began to have an additional lower motor neuron deficit, whose final clinical picture resembled slowly progressive motor neuron disease.[7]

Nowadays, the disease is restricted to India, Bangladesh, and Ethiopia.[4] Lathyrism tends to develop in times of flood or drought when sources of food are scarce and the chickling pea becomes the major dietary staple.

2.2 GUAM'S DISEASE

A clinical variant of amyotrophic lateral sclerosis (ALS), often associated with Parkinsonism and/or Alzheimer-type dementia, used to be common in the Western Pacific. Affected populations included: (i) indigenous (Chamorro) people of Guam and Rota (Mariana Islands); (ii) Japanese residents of Kii peninsula (Honshu Island); and (iii) Auyu and Jaqai linguistic groups of Irian Jaya (west New Guinea, Indonesia). The apparent retreat of this clinical entity (usually called Western Pacific amyotrophic lateral sclerosis-Parkinsonism-dementia complex [ALS/PDC] or Guam's disease) is currently attributed to a non-transmissible environmental agent that is disappearing with the acculturation of the three affected population groups to a more modern lifestyle.[8,9]

Field studies in the disease geographical foci demonstrated an association between use (in food and traditional medicine) of the seed of *Cycas circinalis* (and related species) and the clinical appearance of ALS/PDC years or decades later.[8,9] Cycad plants are also known to induce a neuromuscular disease (cycadism) in ruminants.[10]

Two toxins contained in cycad seed have been indicated as possible etiological agents: the glycoside cycasin and the aminoacid β-N-methylamino-L-alanine (BMAA). This latter bears a close structural similarity to BOAA, the neurotoxin responsible for lathyrism (see above).[11] Spencer and colleagues[12] reported that *Cynomolgus* monkeys fed subconvulsive daily doses (250 mg/kg body weight) of synthetic BMAA developed neurological symptoms and pathological changes in motor neurons 2 to 12 weeks later.

The "cycad hypothesis" is, however, controversial. In particular, it has been pointed out that the amount of BMAA given to monkeys in Spencer's experiments is much higher than the quantity consumed by humans,[13,14] and that no unusual prevalence of ALS, Parkinsonism or dementia has ever been reported in many regions in the Asia-Pacific basin, where cycad has been traditionally eaten and used as a poultice.[15]

In addition to plant toxins, low calcium and magnesium in water and high aluminum concentrations in soil have also been proposed as a possible mechanism for ALS/PDC. The hypothesis is that chronic nutritional deficiences of calcium and magnesium combined with a basic defect in mineral metabolism induces a form of secondary hyperparathyroidism, leading to increased gastrointestinal absorption of aluminum and deposition of calcium, aluminum and silicon as hydroxyapatites in neurons.[9]

2.3   3-NITROPROPIONIC ACID

3-Nitropropionic acid (NPA) is the recognized cause of an extrapyramidal disorder characterized by bilateral putaminal necrosis and delayed dystonia in Chinese childern after consumption of mildewed sugar cane contaminated with a species of the fungus *Arthrinium*.[16] Within hours of ingesting the contaminated food, affected subjects present gastrointestinal disturbances followed by acute encephalopathy with convulsions and coma that may last as long as 20 days. After regaining consciousness, a number of patients develop the extrapyramidal disorder up to six weeks later. Although a contribution of hypoxia during coma cannot be ruled out, the primary trigger is NPA, which blocks energy transformation by acting as a suicide inhibitor of succinic dehydrogenase. While cases of the disease outside China are not reported, NPA is a widely distributed plant mycotoxin, with several possibilities of human contact.[16,17]

3   EXPOSURE TO CHEMICALS AND NEUROPSYCHIATRIC EFFECTS

3.1   MERCURY

The neuropsychiatric effects of occupational exposure to mercury (Hg) vapor have been recognized since the last century as a clinical disorder among felters who made hats and extensively employed Hg as a stiffener of the rabbit fur.[18,19] The psychiatric symptoms ("erethism") of chronic Hg poisoning probably contributed to the popular saying "mad as a hatter," and may have influenced Lewis Carrol's characterization of the hatter in *Alice in Wonderland*.[20] Nowadays, most reports of hazardous exposure come from mining and chloralkali industries.[19,21]

Mercurial erethism is characterized by severe behavioral and personality changes, increased excitabilty, loss of memory and insomnia, which may develop into depression. In severe cases, delirium and hallucination may occur. Parallel to the development of erethism, the characteristic mercurial tremor appears as fine trembling of the muscles interrupted by coarse shaking movements every few minutes. This begins in peripheral parts like fingers, eyelids and lips, and has the features of intentional tremor. It disappears during sleep. In progressive cases, it may develop into a generalized tremor involving the entire body, with violent chronic spasm of the extremities.[21] In addition to the CNS effects, patients may display stomatogingivitis with ptyalism.[22] Historically, a mercurial line (blue or dark brown) on the gums and loss of teeth was observed among heavily exposed workers.[22] Peripheral nerves and the kidney may also be affected.[21,23]

At lower exposure levels, the clinical features of the intoxication can together be characterized as an unspecific, asthenic-vegetative syndrome ("micromercurialism") involving symptoms like weakness, fatigue, anorexia, loss of weight, and gastrointestinal disturbances.[21]

Signs and symptoms of mercurialism seem to regress and disappear when exposure has ceased. However, in more severe cases persistent *sequelae* are common.[24] This is not surprising, since Hg tends to accumulate in the brain and other organs, where it can still be found several years after the end of exposure.[25]

Cases of manifest chronic Hg poisoning are currently rare, however, the possible CNS effects of long-term, low-level exposure are a source of concern. Several studies of workers with low exposure have shown both increased frequency of symptoms and disturbances in various psychological tests related to the degree of Hg exposure.[26-29] Fawer and collegues[30] reported a subclinical increase in forearm tremor frequency at an average exposure of about 26 $\mu g/m^3$. Another study[31] described increased finger tremor in otherwise asymptomatic Hg exposed workers. Recently, subclinical color vision loss, mainly in the blue-yellow range, was observed among workers engaged in the production of precision instruments.[32] However, other investigators exploring the subclinical effects of low-level Hg exposure reported negative findings.[33-35]

Recent neurobehavioral studies focused on dentists exposed to Hg vapor from dental amalgams. Dentists provide an excellent population for the study of threshold effects of Hg since they span the range of exposure to be evaluated and are more

homogenous than most exposed population with respect to several potential confounding variables.[36] The largest study,[37] conducted in Singapore, evaluated 98
dentists and 54 controls. Exposures to average Hg concentrations of 16.7 μg/m³ were
associated with poorer performance in a cognitive test battery. Similarly, Echeverria
and coworkers[36] provided evidence for subtle preclinical changes (poor mental
concentration, emotional and mood lability, somatosensory irritation) among dentists
with a mean urinary Hg level of 36 μg/l.

3.2 ORGANOTINS

While metallic tin has a low degree of toxicity, some organotin derivatives,
mainly trialkyltin compounds, are highly toxic. Human toxicity was dramatically
illustrated in 1953-1954 by an outbreak of medicinal poisoning in France.[38,39] A total
of 290 people were intoxicated, and 110 of them died after using an oral preparation
with the trade name *Stalinon* utilized to treat boils and other cutaneous infections.
The commercial product consisted of diethyltin (DET), but contained up to 10%
triethyltin (TET), a highly neurotoxic agent, as a contaminant. Signs and symptoms
appeared suddenly after a latent period of about four days and included headache,
altered consciousness, vertigo, nausea, vomiting, visual impairment (due to papilledema) with photophobia, sensory disturbances, hyporeflexia, and loss of sphincter
control. Convulsive manifestations were recorded in approximately 10% of the
victims. Flaccid paraplegia involving the abdominal and respiratory muscles
occurred in most cases. Death occurred after 4–10 days as the consequence of deep
coma or, more frequently, hyperacute intracranial hypertension. The cause of death
was attributed to brain edema. About one third of the patients recovered with
sequelae involving the psychic and intellectual sphere or the visual system.[38,39]

Presently, organotin compounds are mainly used as pesticides, antioxidants,
catalysts, lubricants, and polymer stabilizers. After the *Stalinon* catastrophy, other
cases of accidental acute organotin poisoning occurred in the occupational setting.
Signs and symptoms similar to those induced by TET (headache, vertigo, asthenia,
photophobia, nausea and vomiting), though less severe, appeared in two farmers a
few days after the exposure to a fungicide formulation containing triphenyltin
(TPT).[40] Rey and colleagues[41] reported six workers who were exposed to dimethyltin
(DMT) and trimethyltin (TMT) while cleaning a caldron. Once again, clinical manifestations appeared after a latency period (one–three days). They included headache,
tinnitus, deafness, impaired memory, disorientation, aggressiveness, psychotic
behavior, syncope, loss of consciousness and, in some cases, respiratory depression
requiring ventilatory assistance. One patient died. Postmortem examination revealed
cerebral edema with irreversible cell damage in the area of the amygdala. Two
patients suffered permanent *sequelae*. The other three victims recovered, but complained of memory loss for about six months.[41]

An acute limbic-cerebellar syndrome (with hearing loss and mild sensory disturbances) was described in six men who inhaled TMT while cleaning a tank.[42] This
episode resembled the collective intoxication described by Rey's group: clinical
features appeared after two–three days, one patient died and two remained seriously
disabled. The major neuropathological findings in the patient who died were neuronal

alterations, including necrosis, in the limbic structures, especially amigdala and cerebellar Purkinje's cells.[42]

In other reports of human organotin intoxication, exposure was chronic, but clinical manifestations were similar. Ross and associates[43] examined 22 chemical workers chronically exposed to organotins (DMT and TMT) and reported attacks of alternating rage and deep depression, forgetfulness, and loss of libido. Fortemps and collegues[44] described two chemists exposed for approximately three months to a mixture of DMT and TMT, who suddenly developed a status of mental confusion with tonic-clonic seizures. Before the acute episode, the patients had complained of headache, episodes of intense pain *sine materia* in various organs and psychic disturbances (memory defects, loss of vigilance, insomnia, anorexia, disorientation). Both patients recovered completely following removal from exposure.[44]

3.3 SOLVENTS

Acute solvent exposure is typically associated with manifestations of CNS disturbance. While there is some variation in signs and symptoms with different solvent structures, results of high-level exposure are very similar: euphoria, disorientation, giddiness and confusion, progressing to unconsciousness, paralysis, convulsion, and death are typically observed. These effects are due to the solvent itself and not to metabolites, and primarily result from a physical interaction with the neuronal cell membranes.[45]

Chronic organic solvent abuse is known to cause toxic encephalopathy with neurologic signs and symptoms including cognitive, pyramidal, cerebellar, and brainstem/cranial nerve findings.[46] Neuroimaging studies have revealed brain atrophy and diffuse white matter changes.[47,48] The abusers described in these studies had inhaled glues, spray lacquer, or spray paint containing some solvents, of which toluene was considered the main cause of the CNS damage.[47]

Several case reports and epidemiological studies suggest that chronic CNS effects, although less serious, may also occur in solvent-exposed workers. The studies covered such groups as housepainters, builders of fiberglass boats, workers in paint factories, and workers exposed to jet fuel. The compounds involved include aliphatic alcohols, trichloroethylene, turpentine, toluene, xylene, and styrene. Headaches, dizziness, concentration difficulties, memory impairment, fatigue, irritability, depression, alcohol intolerance, and personality changes were the most frequently reported symptoms. The psychometric tests revealed disturbances in memory and perception, prolonged reaction times and some loss of coordination. The symptomatology has been referred to as "chronic painter's syndrome" or "psycho-organic syndrome."[49,50] Occupational exposure to solvents was also found to be associated with an elevated incidence of early disability retirements due to neuropsychiatric diagnoses.[50-52] In Scandinavian countries, where most cases have been reported, the solvent-induced psycho-organic syndrome is a recognized compensable disorder (see Iregren in this book).

On the other hand, the results of the Scandinavian investigators could frequently not be verified in cross-sectional and epidemiological studies performed in other countries.[53-56] Criticism related to potential confounding factors has been raised. For example, it has been argued that housepainters could be characterized by higher ethanol consumption[57] and lower primary intellectual capacity,[58] than control populations. Nevertheless, recent studies taking account of these criticisms tend to support the conclusion that long-term occupational exposure to organic solvents may have chronic effects on the CNS.[59,60]

3.4 ORGANOPHOSPHATES

The effects of acute organophosphate (OP) exposure (nerve gas or pesticide poisoning) have been well known for many years. Acute OP poisoning is mainly due to inhibition of acetylcholinesterase (AChE) with subsequent rapid accumulation of acetylcholine (ACh) at cholinergic synapses. The signs and symptoms of the intoxication are usually classified into muscarinic (parasympathetic), nicotinic (sympathetic and motor) and CNS manifestations.[61]

The acute CNS effects of OPs may result from an action of ACh on brain muscarinic and nicotinic receptors, and from possible interactions with other neurotransmitter and neuropeptide systems. A wide variety of manifestations have been described including anxiety, excitability, lethargy, confusion, agitation, impaired memory and concentration, depression, schizophrenic reactions, seizure and coma.[61] These effects usually disappear within a few days or weeks in coincidence with the gradual recovery of AChE activity. Following acute poisoning, however, delayed neurological *sequelae*, such as motor and sensory disturbances, can be observed.[61,62] The survivors of acute OP poisoning have been reported to perform worse than contol subjects on neuropsychological tests, suggesting a persistent decline of cognitive functions.[63-65]

It appears that CNS effects may also result from chronic exposure. In 1961, Gershon and Shaw[67] described a series of neuropsychiatric disturbances in 16 workers exposed for 1.5 to 10 years to OP insecticides. The most persistent symptoms were schizophrenic and depressive reactions, severe memory impairment and difficulty in concentration. The authors claimed to have established a connection, if not a causal relationship, between OP exposure and the development of psychiatric disturbances, generating much concern and controversy. The main criticism to Gershon and Shaw's paper regarded a lack of information on the degree of the exposure. Additionally, the alleged conclusion that OPs activate a tendency towards schizophrenic and depressive disease was confuted, and clinical and epidemiological studies were reported, which showed no association between OP exposure and psychiatric disorders.[78-80] In more recent times, two neurobehavioral studies among workers with low-level OP exposure also yielded negative results.[81,82]

Although Gershon and Shaw might have overstated their conclusions, there are many other reports in the literature describing adverse effects of prolonged OP exposure on the CNS (Table 1). Recent studies have indicated that repeated low-level exposure may result in subtle changes which are unlikely to be manifest as clinical symptoms, but may be detected by accurate neurological examination and sensitive neuropsychological tests.[77,83]

The pathogenesis of the putative chronic effects of OPs is currently investigated. Animal studies have indicated that down-regulation of brain muscarinic receptors is the mechanism underlying, at least in part, the development of tolerance to OP

TABLE 1
**Effects of Chronic Occupational Exposure to Organophosphates
on the Central Nervous System**

Reference	Main clinical findings
Holmes and Gaon, 1957[66]	memory disturbances, irritability, EEG abnormalities
Gershon and Shaw, 1961[67]	schizophrenic and depressive reactions, severe memory impairment, difficulty in concentration
Dille and Smith, 1964[68]	anxiety, uneasiness, depression, dizziness, emotional lability, EEG abnormalities
Metcalf and Holmes, 1969[69]	fatigue, memory disturbances, EEG abnormalities
Ohto, 1974[70]	decreased retinal sensitivity
Levin et al., 1976[71]	anxiety
Korsak and Sato, 1977[72]	deficits in neuropsychological test performance
Davis et al., 1978[73]	Parkinsonism
Duffy et al., 1979[74]	EEG abnormalities (particularly increase in β activity), REM sleep disturbances
Perold and Beznidenhout, 1980[75]	depression, anxiety
Misra et al., 1982[76]	visual impairment, retinal macular degeneration
Stephens et al., 1995[77]	deficits in attention and information processing speed

toxicity (see Rerference 61 for review). In rodents, evidence has been provided that this decline in central muscarinic receptor may also underlie cognitive changes.[84,85] The relevance of these findings for humans needs to be assessed.

4 EXPOSURE TO CHEMICALS AND NEURODEGENERATIVE DISEASES

4.1 GENERAL REMARKS

Neurodegenerative diseases include a wide spectrum of disorders characterized by chronic loss of selected and functionally related groups of neurons within the CNS. Such pathogenic process is similar to that occurring in the normal aging brain, since certain groups of functionally related neurons tend to die preferentially with the passage of time. However, the neurodegenerative disorders can be distinguished from physiologic aging by the more rapid progress and more severe affliction that characterizes them.[3]

By far, the most common neurodegenerative disorders are Parkinson's disease (PD), Alzheimer's disease (AD), and amyotrophic lateral sclerosis (ALS). Other conditions are Huntington's chorea, the hereditary ataxias and less common diseases such as progressive supranuclear palsy, multiple system atrophy, Hallervorden Spatz disease, and Wilson's disease.

The etiopathogenesis of neurodegenerative diseases is far from being completely elucidated. However, it is clear that, for the group as a whole, both genetic and environmental factors may be involved in their causation, though to varying extents.[3]

TABLE 2

**Occupational Chemicals Possibly Contributing
to Neurodegenerative Disorders**

Chemical	Disease
aluminum	Alzheimer's disease
manganese	Parkinson's disease
mercury	Parkinson's disease, amyotrophic lateral sclerosis
lead	Alzheimer's disease, amyotrophic lateral sclerosis
MPTP	Parkinson's disease
carbon monoxide	Parkinson's disease
solvents	Alzheimer's disease, amyotrophic lateral sclerosis
pesticides	Parkinson's disease, Alzheimer's disease

The following sections provide an overview of the occupational toxicants that might contribute to the etiology of the three major neurodegenerative disorders, i.e., PD, AD, and ALS (Table 2). Parkinson-like syndromes secondary to chemical exposures are also considered. It is not the purpose of this chapter to review genetic factors, which certainly contribute to all three diseases, and seem to be of major importance in certain subcategories of each. Nor will we discuss the possible role of physical (e.g., trauma, electromagnetic fields) and biological (e.g., conventional or unconventional viral infections) environmental factors, which may also play an important role.

4.2 PARKINSON'S DISEASE AND PARKINSONISM

Parkinson's disease affects approximately 1% of the population over age 50. The clinical presentation of the disease includes the classic triad of tremor, rigidity, and bradykinesia. Other features are flexed posture, impaired righting reflexes, hypophonia, and a tendency to drool. In the late stages, dementia is common. Under the morphological and neurochemical point of view, PD is characterized by the degeneration of pigmented neurons in the substantia nigra with Lewy body formation, and dopamine depletion in the striatum, i.e., the brain area to which nigral neurons project their axons.[86]

The cause of death of dopaminergic neurons remains elusive, and PD is thus considered an idiopathic disease. However, both genetic and environmental factors have been implicated,[87-89] and different mechanisms of cytotoxicity (i.e., oxidative stress, impairment of mitochondrial respiratory activity, excitotoxicity) have been hypothesized.[90] In animal studies, the actions of three chemicals have become prototype models for investigating such neurotoxic mechanisms: namely, dopamine for oxidative stress, 1-methyl-4-phenyl-1,2,3,6-tetrahydropyridine (MPTP) for mitochondrial damage, and metamphetamine for excitotoxicity (see Reference 90 for review). A current hypothesis is that PD is not a single disease entity, but the reflection ot the brain's limited repertoire for the expression of nigrostriatal damage, irrespective of cause.[91]

Parkinsonism is a clinical syndrome mimicking PD which may be caused by any process that interferes with dopamine-mediated neurotransmission in the striatum. Depending on the nature and extent of such interference (anatomical damage or pharmacological interaction), the syndrome may be reversible (e.g., extrapyramidal side effects of neuroleptic drugs) or permanent and evolutive (e.g., traumatic, vascular, postencephalitic, or manganese-induced Parkinsonism). In the latter case the pathologic lesions are usually distinct (and more extensive) from those seen in PD.[86,87,91]

The following sections review the occupational chemical exposures which can cause Parkinsonism or might be related to the development of idiopathic PD.

4.2.1 MPTP and its analogues

The discovery that MPTP, a contaminant produced during the illicit manufacture of the synthetic street drug meperidine, causes a syndrome which is clinically almost identical to PD,[92] revolutionized research into the etiopathogenesis of PD and related disorders.

MPTP is metabolically activated to 1-methyl-4-phenylpyridinium ion (MPP+) by monoamino oxidase B (MAO-B). This reaction is likely to occur within glial cells. MPP+ is then released from glial cells and is actively accumulated by the catecholamine uptake system into nigrostriatal neurons, where it inhibits complex I of the respiratory chain in mitochondria, possibly by a free-radical mechanism (see Reference 90 for review). In contrast to most of the chemicals producing secondary Parkinsonism, such as those discussed later in this chapter, MPP+-induced injury specifically involves the dopaminergic neurons damaged in idiopathic PD.

MPTP-induced Parkinsonism has been mainly observed in addicts after intravenous injection. However, MPTP was also used in industry as a chemical intermediate and thus represented, though rarely, an occupational hazard. A chemist with chronic MPTP exposure developed Parkinsonism at an early age.[93] Subsequently, a clinical survey of workers with intermittent industrial exposure to the same compound found one with "early PD" and mild abnormalities suggestive of Parkinsonism in other exposed subjects.[94]

The striking similarity between the toxicant-induced Parkinsonism and the idiopathic disease suggests that environmental neurotoxins (e.g., tetrahydroisoquinolines, β-carbolines) with chemical structures similar to MPTP may be involved in the pathogenesis of PD. It has been suggested that genetic defects in enzymatic biotransformation processes (e.g., debrisoquine hydroxylation) could lead to potentiation of relatively low-level neurotoxic exposures in PD patients.[88,89,91]

Occupational agents resembling MPTP either structurally or in biologic activity include the pesticides paraquat and rotenone. These compounds are discussed below.

4.2.2 Pesticides

The commonly used herbicide paraquat is chemically similar to MPTP and its active metabolite MPP+, and two patients believed to suffer from paraquat-induced Parkinsonism have been described.[95] However, paraquat is a charged compound and, therefore, is not expected to readily cross the blood-brain barrier. Moreover, no cases of PD or Parkinsonism were found in clinical surveys of paraquat formulation workers[96] or in patients who had suffered acute paraquat poisoning.[97]

Of interest is that MPP+ had been developed as a potential herbicide under the name cyperquat, but was never marketed.[87] The insecticide of biological origin rotenone blocks mitochondrial respiration at the same site as MPP+, but it has not been reported to cause Parkinsonism.[87]

Davis and associates[73] reported the development of Parkinsonism in a 48-year-old patient with a history of chronic OP exposure (Table 1), and suggested the existence of a possible relationship between such exposure and alterations in central cholinergic or dopaminergic activity. The authors themselves, however, cautioned that the patient's syndrome and the toxic exposure may have been coincidental.

Permanent Parkinsonism was observed following chronic exposure to the fungicide maneb (manganese ethylene-bis-dithiocarbamate).[98,99] As discussed below, both manganese and ethylene-bis-dithiocarbamate are potential toxicants for the extrapyramidal system and may have had an etiological role in these cases. A toxic cooperative mechanism can also be hypothesized.

Several case-control studies indicated an association between PD and pesticide exposure,[100-104] but other investigations[105-108] did not find such a relationship. The patients who recalled specific exposures most frequently reported exposure to phenoxy herbicides, thiocarbamates, organochlorines, and alkylated phosphates.[102,104] Fleming et al.[109] examined brain tissue of a small sample of PD patients and control subjects for organochlorine pesticide residues and found dieldrin to be present significantly more frequently in brains of PD patients. Other studies investigated the relationship between PD and other (possibly) pesticide-related variables, such as rural living, farming, and well water consumption, once again with conflicting results (see Reference 91 for review).

The divergent findings of the epidemiological studies may be explained by methodological differences and small sample sizes, as well as by regional differences in agricultural practices. Moreover, most epidemiological studies solicited information about general, rather than product-specific, exposures to pesticides. Obviously, this could have masked the effect of compounds involved in the development of the disease.

4.2.3 Manganese

The neurotoxic effect of manganese (Mn) has been known for a long time. In 1837, only 20 years after James Parkinson published his "Essay on the Shaking Palsy," a syndrome similar to PD was described in five subjects who worked in a Mn ore-crushing plant.[110] Manganism has since been noted in hundreds of cases, particularly in miners, smelters, and workers involved in the manufacture of dry batteries.[91,111] The disorder starts gradually with nonspecific symptoms, often of a psychiatric nature ("manganese madness" or "locura manganica"), such as hallucinations, emotional lability, and compulsive and aberrant behavior. In a following stage, motor disturbances (generalized bradykinesia, rigidity, muscular weakness, clumsiness, impaired speech) somewhat similar to those of PD appear. Patients can also develop the motor deficits of manganism without having experienced any phase of manganese madness.[111,112]

Notably, idiopathic and Mn-induced Parkinsonism present several dissimilarities. Clinical features of manganism include more frequent dystonia, a particular propensity to fall backward, less frequent resting tremor, and failure to achieve a sustained therapeutic response to dopaminomimetics.[111,113,114] Histologically, the brunt of the pathologic changes in manganism is in the pallidum, where neuronal loss and gliosis occur, and not in the nigrostriatal pathway. Additionally, Lewy bodies are usually absent.[111,115]

Positron emission tomography (PET) also provides information of considerable value in discriminating between PD and manganism. Fluorodopa PET provides an index of the integrity of the dopaminergic nigrostriatal projection and is abnormal in PD, where a reduced uptake in the striatum can be demonstrated. In manganism, the fluorodopa scan is normal, compatible with the pallidum being the epicenter for Mn-induced damage.[111,116]

Much of the knowledge about Mn intoxication comes from high workplace exposures. However, several studies have indicated that lower exposures are associated with early signs and symptoms of CNS dysfunction. In a Mn oxide and salt producing plant, dose-response relationships were found between blood Mn concentration and impairment of eye-hand coordination and hand steadiness.[117] Following an outbreak of several cases of Parkinsonism in a ferromanganese plant in Taiwan,[112] examination of the 132 remaining workers in the plant revealed an increased frequency of extrapyramidal signs and other neurological symptoms with the degree of Mn exposure.[118] Low-level Mn exposure in foundry workers and steel smelters has been associated with slowed finger tapping, increased reaction time,[119,120] and a lower frequency of movement on diadochokinesometry.[120] Mn-exposed workers from a battery-plant likewise showed poorer performance on visual reaction time, eye-hand coordination, and hand steadiness, compared to a reference group.[121] In the latter study, the prevalence of abnormal results were related to the lifetime integrated exposure to Mn dust. Using an extensive battery of neurofunctional tests, manifestations of early CNS dysfunction associated with long-term exposure to low-level Mn dust (< 1 mg/m^3) have been described in workers employed in Mn alloy production.[122,123] Similar neuropsychological disturbances have been recently observed among welders exposed to Mn fumes.[124] Taken together, these findings indicate that manganism probably progresses subclinically on a continuum, and that early manifestations can be detected using sensitive testing methods. There is also concern that overexposure to Mn may accelerate neuronal loss, thereby augmenting the risk for developing PD later in life.[125] In contrast to other metals (e.g., lead), Mn is an essential, necessary element in small quantities. Thus, further research is required to determine the specific levels at which cumulative low-level exposure alters CNS functions.

4.2.4   Mercury

Parkinsonian features may be present in various types of mercury intoxication, though mercurialism is usually readily distinguishable from PD.[91] However, Finkelstein and associates[126] recently reported the case of a female dentist with elemental Hg intoxication who developed hemiParkinsonism in the absence of most classical neuropsychiatric signs of chronic mercurialism. Chelation therapy resulted in increased urinary Hg excretion and clinical improvement.

Two epidemiological studies have shown an association between Hg exposure and idiopathic PD,[105,127] and a positive association has been found between the number of dental amalgam fillings before illness onset and PD.[104] These provocative findings require confirmation.

4.2.5   Carbon monoxide

Carbon monoxide (CO) is a widespread environmental and occupational hazard and a leading cause of toxic accidents. Oxycarbonism is one of the few examples of acute poisoning that may cause permanent nervous system damage in a delayed fashion, in that a considerable proportion of CO-poisoned subjects have been shown to develop late neuropsychiatric abnormalities,[128,129] occasionally including a parkinsonian syndrome.[130,131] These complications, which usually appear after a period of apparent recovery, are indicated with the term "delayed neurological syndrome" (DNS) or "delayed encephalopathy." The syndrome usually appears within 40 days after the initial CO exposure,[128,132] but latencies as long as 8 months have also been reported.[133]

In 1926, Grinker[130] first described a Parkinson-like syndrome as a *sequela* of CO poisoning. It later became apparent that the clinical presentation of DNS may mimic almost every known neuropsychiatric syndrome: movements disorders (Parkinsonism, choreoathetosis), autonomic dysfunction reflected by urinary and fecal incontinence, gait disturbances, seizures, cortical blindness, symptoms resembling those of multiple sclerosis, peripheral neuropathy, personality changes, Wernicke's aphasia, Korsakoff's syndrome, agnosia, mutism, short-term memory deficits, dementia, psychosis, and maniac-depressive psychosis.[128,129,134] Additionally, more subtle cognitive and mental alterations, not evident by bedside neurological examination, can be detected through neuropsychologic tests.[135] Although a complete parkinsonian syndrome is uncommon, parkinsonian features are frequent delayed *sequelae* of CO poisoning. In a study of 65 DNS patients, more than one sign or symptom was present in over 80% of cases.[128]

The incidence of DNS has been variously estimated, depending on the duration and methods of follow-up. Most studies reported neuropsychiatric disturbances indicative of cerebral impairment in a number of cases ranging from 2 to 30%,[132,136] but the results of investigations using more rigorous methodology suggest that the frequency of the syndrome may be as high as 32–67%.[135]

The mechanisms of delayed CO neurotoxicity are a matter of debate.[129,135] Postulated pathogenic mechanisms include: (i) hypoxic-ischemic stress related to defective oxygen transport to the cells and to impaired cardiovascular function; (ii) CO interaction with intracellular targets (e.g., cytochromes) with impairment of mitochondral electron transport; (iii) brain lipid peroxidation with free radical injury; (iv) overstimulation of excitatory aminoacid receptors.

The development of DNS is unpredictable. Serum carboxyhemoglobin (COHb) levels do not reflect the severity of poisoning or the potential to develop delayed *sequelae*.[128,132] A series of risk factors has been identified, including age older than 40 years, initial coma, pre-existing cardiovascular diseases, and a type of occupation involving mental work.[128,137] A long period of CO exposure (one hour or more), coma of 2–3 days duration, abnormal EEG, and persistent dizziness and fatigue after regaining consciousness, as well as mental stimulation during the recovery phase have also been associated with higher risk, even though not invariably linked to unfavorable outcome.[128,136,137]

The DNS may resolve spontaneously, but significant improvement may require up to two years. Alternatively, the neurological damage may be permanent and severely affect the quality of life in the patients who survived the acute stage of CO poisoning.[128,136] Parkinsonism may also be persistent.[128]

Timely hyperbaric oxygen (HBO) treatment may interrupt the cascade of events leading to brain damage, and has been advocated to be effective in preventing DNS. However, clinical studies comparing the efficacy of hyperbaric vs. normobaric oxygen have generated conflicting results and controversies regarding both the causes of delayed CO neurotoxicity and the indications of HBO therapy.[129,135]

Recently, concern has been expressed that chronic low-dose CO exposure may be associated with delayed adverse cognitive effects.[87] A study has indicated an increased prevalence of PD in firefighters, possibly as the result of their chronic exposure to CO.[138] As yet, definitive evidence of an association between PD (or a parkinsonian syndrome) and chronic CO exposure has not been presented. Longitudinal evaluation of cohorts with long-term low-dose exposure will be important in clarifying this issue.

4.2.6   Carbon disulfide

Carbon disulfide (CS_2) was the first industrial solvent associated with neurotoxicity.[139] Early workers with massive continuous exposure to this compound exhibited behavioral changes reminiscent of those associated with acute Mn toxicity often followed by other signs of CNS injury, including tremor similar to that of PD. With long-term exposure, some subjects develop Parkinsonism.[91,140] Much of this experience comes from the viscose rayon industry, in which workers are exposed to both CS_2 and hydrogen disulfide (H_2S).

As in the case of Mn-induced Parkinsonism, the parkinsonian syndrome caused by CS_2 is characterized by pathological findings mostly localized in the basal ganglia. These features are distinct from those of idiopathic PD.[91]

The pathogenesis of CS_2-induced Parkinsonism is unclear. Postulated mechanisms include: (i) chelation of metal ions of biological importance (i.e., zinc and copper), that could decrease the activity of enzymes such as aromatic aminoacid decarboxylase which converts L-dopa to dopamine; (ii) inhibition of enzymes associated with catecholamine function, such as dopamine-β-hydroxylase or monoamino oxidase; (iii) interaction with the metabolism of vitamin B6, necessary for the function of aromatic aminoacid decarboxylase and for the activity of tyrosine hydroxylase which converts tyrosine to L-dopa.[91] CS_2 neurotoxicity is, at least in part, mediated by metabolic activation to H_2S and dithiocarbamates.[141]

4.2.7   Other compounds

Parkinsonism (predominantly due to putaminal lesions) has been described as a complication of acute methanol poisoning,[142-144] but no cases associated with work-place-level exposures are reported.

Pezzoli and associates[145,146] reported two cases of Parkinsonism apparently related to occupational exposure to *n*-hexane. In one of these patients, PET studies revealed regional abnormalities of the nigrostriatal dopaminergic system that were different from those of PD.[146] *n*-Hexane and its active metabolite 2,5-hexanedione have also been found to induce Parkinsonism in mice and rats, in which these compounds decrease striatal dopamine and homovanillic acid levels.[147]

It appears that other organic compounds may cause Parkinsonism in susceptible individuals. A case-report linked carbon tetrachloride to Parkinsonism in a chemist.[148] A partially reversible parkinsonian syndrome has been reported in a woman chronically exposed to an industrial solvent mixture based on trichlorotrifluoroethane and nitromethane.[149] Recently, Uitti and colleagues[150] presented the case of a woman who developed Parkinsonism acutely following repeated "huffing" (inhalation through the mouth) of lacquer thinners. The major constituent of the mixtures was toluene (approximately 60%) and the balance was made up of methanol, ethyl acetate, and methyl ethyl ketone. Similarly, Tetrud and associates[151] reported a worker who developed moderate, but persistent, Parkinsonism one week after accidentally ingesting a petroleum waste mixture. Some of the features of this case resembled the syndrome induced by MPTP. Mass spectrometry analysis of the waste mixture documented the presence of a large number of compounds, the majority of which were tentatively identified as hydrocarbons in the C-7 to C-16 range.

4.3   ALZHEIMER'S DISEASE

Alzheimer's disease accounts for about half of all causes of severe progressive dementia. It usually affects people above the age of 65 years (late onset AD), but can occur among individuals as young as 40 (early onset AD). Most cases are sporadic but some have a family history of dominant inheritance. Clinical features include profound memory loss, difficulty in learning, personality changes, impaired judgment, speech disturbances, increasing disorientation, and loss of everyday skills. The disease is relentlessly progressive and fatal in a few years. In an advanced state patients cannot talk and walk, becoming totally incapable of caring for themselves. No effective pharmacotherapy has yet been demonstrated.[152,153]

Diagnosis is made, with varying degrees of uncertainty, essentially on the basis of observed clinical symptoms and the exclusion of other causes of dementia. Confirmation is only possible at *postmortem* from examination of the neuropathological characteristics of brain tissue. The two major histological features are the presence of neurofibrillary tangles and neuritic (senile) plaques. Other neuropathological findings are neuronal loss (with cortical atrophy) and amyloid angiopathy.[152,153]

To date only increasing age and genetic predisposition are established etiological risk factors for AD, but environmental factors often have been suggested to play a role.[154-156] Occupational exposures which might be involved include aluminum, lead, and organic compounds.

4.3.1 Aluminum

A great deal of controversy exists over whether aluminum (Al) has a role in the development of AD.[157] The matter is of obvious importance since Al and its compounds are common contaminants of food and water, and are contained in several cosmetics and medications. Environmental and occupational exposure to Al by inhalation also is possible.[158,159]

Early evidence suggestive of a possible role for Al in AD came in the 1970s, when a neurological disease was reported among patients undergoing long-term dialysis. During the course of the disease — currently known as dialysis encephalopathy — patients developed a progressive state of dementia with symptoms similar to those of AD. Al from the dialysate water and from orally administered phosphate-binding gels was found to accumulate in the patient's brain and was thus identified as the causative factor.[160] It has been noted, however, that dialysis encephalopathy is relatively acute in nature, and that high Al doses are involved, much higher than those encountered in environmental exposure.[158]

Additional evidence for a link between Al and AD stems from reports indicating increased levels of this metal in the bulk brain tissue of Alzheimer's patients,[161-163] and the association of Al with the neurofibrillary tangles[164,165] and neuritic plaques.[166,167] However, the latter findings were not confirmed by other investigators using highly sensitive analytical techniques.[168-170] Al has also been associated with the neurofibrillary degeneration found among patients suffering from Guam's disease.[9] It is unclear whether the metal accumulates passively in neurons as a result of disrupted properties of the neural membrane, or whether it acts as a pathogenic agent.

Experimental animals injected intracranially or fed orally with Al show learning and memory disturbances,[171] as well as neurofibrillary degeneration and tangles similar to those of AD.[165,172,173] Moreover, alterations have been reported in the offspring of animals exposed to Al during pregnancy.[174,175] Evidence, however, indicates that AD and Al toxicity are different pathological processes with distinct clinical, microscopic, and neurochemical features.[176] The structure and antigenicity of neurofibrillary tangles seen in experimental animals differ from that found in AD patients, and the distribution of lesions is also quite different.

Epidemiological investigations[177-180] have indicated a relationship between AD and Al in drinking water. These studies are complicated by the fact that Al comes from a variety of dietary sources, among which water is only a minor contributor.[158] Moreover, criticism has been raised due to methodological limitations,[159] and negative results have also been published.[181,182] Studies that have considered the role of Al products such as antacids and antiperspirants have yielded equivocal results.[183,184] Support for the "aluminum hypothesis" comes from the finding of a slower progression of the disease in patients with AD treated with the chelator desferrioxamine.[185]

Occupational exposure to Al and its compounds is common. Evidence has been provided that the metal can be transported to the brain via the nasal-olfactory pathways[186] suggesting that it is possible for respirable Al dusts to circumvent certain natural barriers to uptake. Additionally, it has been shown that long-term exposure to Al by inhalation gives rise to its accumulation in the body and skeleton of healthy workers, and that the elimination of retained Al is very slow, in the order of several years.[187] However, relatively few studies have been carried out to investigate the possible relationship between occupational Al exposure and the occurrence of AD or long-term neurological disorders.

One case of severe occupational encephalopathy (associated with pulmonary fibrosis) due to Al has been reported in a man who had worked for 13 years as a ball-mill operator in an Al flake powder factory. Although the patient's memory became impaired during the course of the disease, AD was excluded. At autopsy, the Al content of the brain and other tissues was greatly raised.[188] A similar case was later reported in a Japanese man who had worked for 30 years in an Al refining plant.[189] Dementia was diagnosed when the patient was 60 years old, the first symptoms being noted at the age of 55. The patient died at 65. The neuropathological examination of his brain revealed cerebral atrophy, typical neurofibrillary tangles, and senile plaques. The latter were also present in the cerebellum which is not a target site in AD. X-ray microanalysis showed the presence of Al in the nuclei and cytoplasm of tangle-bearing neurons.[189] Signs of dementia have recently been observed in a patient more than 40 years after aluminosis was diagnosed in 1946.[190]

A few cohort studies on workers exposed to Al have been carried out. Rifat et al.[191] conducted a morbidity prevalence study among miners previously exposed to a mixture of finely ground Al and Al oxides. They did not find significant differences between exposed and non-exposed subjects in reported diagnosis of neurological disorders; however, exposed miners showed a lower level of cognitive functioning.[191] Cognitive deficits have also been reported following occupational exposure to Al fumes.[192] On the basis of the results of some neuroendocrine tests, it has been suggested that at the beginning of exposure to low Al concentrations there is an effect on the hypothalamus-pituitary axis.[193] This phenomenon appeared to be transient possibly because of an adaptation process.[193] Two Swedish studies[124,190] reported a higher prevalence of CNS symptoms among workers exposed to Al welding fumes, yet no specific reference was made to AD and the only conclusion was that Al is a neurotoxic metal. In another study conducted at a Norwegian Al plant,[194] subtle neuropsychological deficits were described among healthy workers at retirement age who had been chronically exposed to low Al levels. Salib and Hillier[195] found no association between having worked in an Al industry and the risk of developing AD later in life, although this may have partly been the result of random misclassification.

In conclusion, the combined experimental, clinical, and epidemiological evidence indicates that Al is neurotoxic in humans, and suggests, but does not prove, that the environmental or occupational exposure to this metal might be a cause of AD.

4.3.2 Lead

Since exposure to high levels of inorganic lead (Pb) causes acute lead enceph-alopathy,[196] it has been suggested that people with chronic exposure at low levels or those who have recovered from severe encephalopathy may be at risk for neu-robehavioral disturbances and mental deterioration.[155] Three cases appear to support this hypothesis: a survivor of childhood lead encephalopathy died at the age of 44 after a period of mental deterioration. The neuropathological examination revealed changes typical of AD with high Pb levels in brain tissue.[197] Two other cases were reported of workers with long-term occupational exposure to Pb. One was described as demented 15 years before death. In both cases neuropathological features of AD were observed at autopsy.[198] On the other hand, epidemiological studies do not indicate occupational exposure to inorganic Pb as a risk factor for AD.[199,200]

It has also been suggested that permanent brain damage may derive from expo-sure to organic Pb.[155] Cerebral atrophy and neuronal loss have been described in the brain of a gasoline sniffer with a history of hallucinations and poor memory possibly attributable to tetraethyl lead.[201] The same compound has been shown to induce neurofibrillary tangles in the rabbit brain.[198] To our knowledge, the possible associ-ation between occupational exposure to organic Pb and AD has not been investigated epidemiologically.

4.3.3 Solvents and Other Compounds

The debated role of organic solvents exposure in inducing chronic CNS damage is discussed above. With regard to AD, findings of epidemiological studies conducted to assess occupational exposure to solvents as a risk factor for the disease have been controversial. A meta-analysis involving the secondary evaluation of original data from three studies did not show a significant increased risk.[200] However, a recent case-control study indicated a statistically significant increase in risk of AD for men exposed to solvents such as benzene, toluene, phenols, alcohols, and ketones.[202] Studies on workers exposed to solvents have suggested that risk of neurological symptoms may be modified by heavy ethanol consumption;[203] this may represent an important confounding variable in these epidemiological studies.

In another case-control study,[184] an increased risk of AD was found for subjects exposed to glues and pesticides. These results await confirmation.

4.4 AMYOTROPHIC LATERAL SCLEROSIS

ALS is the major form of motor neuron disease (MND). This term, which also encompasses progressive bulbar palsy and spinal muscular atrophy, indicates a progressive affection of the CNS, characterized by a degeneration of spinal and/or bulbar motor neurons and corticospinal tracts. It usually develops in late adult life and affects males more often than females. MND has an invariably progressive and fatal course.

On the basis of genetic and epidemiologic features, three major types of ALS have been identified: (i) the classic, usually sporadic form, which accounts for 90%

of the patients; (ii) the familial, presumably hereditary form, which accounts for approximately 10% of the cases; and (iii) the Western Pacific form, often occurring in association with a parkinsonian-dementia complex (Guam's disease).[204]

The cause of the familial form of ALS is believed to be genetic: autosomal dominant inheritance with variable penetrance.[205,206] The cause of the Western Pacific form is currently under dispute, as discussed above. The etiology of the sporadic form is mostly unknown.[207] Knowledge on environmental and/or occupational chemical exposures that have been associated with sporadic ALS is summarized below.

4.4.1　Lead

The relationship between ALS and inorganic Pb exposure has received considerable attention.[208,209] In Aran's original description of the disease (dating back to 1850), three out of 11 patients had been in contact with Pb, and two of these had previously suffered from Pb poisoning. Aran, however, did not attribute any etiologic role to Pb exposure.[210] Cases of sporadic ALS associated with chronic Pb exposure were then described at the beginning of the century, but it was only in the 1960s that the possible role of Pb in the etiology of ALS started to be actively investigated. Simpson et al.[211] described a subject occupationally exposed to Pb for 20 years who presented with symptoms compatible with Pb poisoning and later developed a myelopathy with muscular atrophy resembling the early stages of ALS. Chelation therapy led to a surprising functional recovery. Another study, however, failed to demonstrate an association between ALS symptoms and the presence of Pb in body fluids and tissues.[212]

In 1970, Campbell et al. carried out a survey of 74 cases of ALS: 23 patients had a history of previous Pb exposure and were characterized by a more benign clinical course of the disease.[213] The authors emphasized that 25% of ALS patients had an anamnesis of bone fractures or skeleton diseases, while only 9% of control subjects had suffered this kind of pathology, and suggested that "the lead liberated from bone might affect the motor neuron." However, bone biopsy specimens in 25 random patients (exposed and non-exposed) showed no significant variations in Pb content compared to the control group.[213] The first study to report high Pb levels in the nerves, spinal cord, and cardiac and skeletal muscles in a patient with ALS came from Petkau et al.[214] Unlike the case described by Simpson's group,[211] this patient did not respond to chelating therapy.

Kurlander & Patten[215] found a higher average Pb content of spinal ventral horn tissue in seven patients who had died of MND compared with that of a control group. The same authors also found an association between MND and degenerative or traumatic skeletal pathologies, indicating a release of Pb from bone tissue.[215] In agreement with this study, ALS-affected patients were found to have higher Pb levels in cerebrospinal fluid (CSF), whole blood, and plasma (but not in the skeletal muscle).[216,217] By contrast, other investigators did not find significant differences for blood, plasma, CSF and nerve tissue Pb concentrations between ALS patients and controls.[218-221]

Epidemiological studies also provide contrasting results. For example, Gresham *et al.*[222] found no association between heavy metal exposure and ALS, while Armon *et al.*[204] suggested an association between ALS and occupational exposure to Pb vapors due to welding and soldering. Welding (together with work with electricity or solvents) also was indicated as a risk factor for MND in Swedish workers.[223] Studies conducted in Italy gave inconclusive[224] or negative[225] results. In the attempt to identify potential risk factors that can play a role in causing sporadic ALS, a case-control study was performed on a cluster of six patients in Two Rivers (Wisconsin).[226] A significant link was found with a history of physical trauma, frequent consumption of fresh fish (from Lake Michigan), and family history of cancer. Only one patient showed signs of Pb exposure, and blood Pb levels were normal in all patients.[226]

In conclusion, there are findings that suggest a link between Pb exposure and ALS, but none of them seem to offer conclusive proof, and caution should be used in making this association.

4.4.2 Mercury

The role of Hg in relation to ALS has not been pursued with the same enthusiasm as that of Pb. Both organic[227-229] and inorganic[230] Hg compounds, as well as elemental Hg,[231] have been reported to induce a syndrome closely resembling ALS. In some cases,[230,231] withdrawal from the contaminated environment led to the resolution of symptoms and signs. Recently, Schwarz and collegues[232] described the case of a female nurse who developed ALS three and a half years after accidental injection of Hg from a thermometer into her left hand. After the accident, particles of the metal had diffusely spread into the soft tissues of the palm and could not be removed completely by surgical means. Although the onset of the disease and the preceding event might have been coincidental, the authors speculated that Hg and trauma could have worked as synergic factors in this patient.[232]

On the other hand, neurological examinations and measurements of Hg contents in the blood, urine, and hair samples of ex-mercury miners, who were poisoned by or exposed to Hg vapor in Hokkaido (Japan), failed to disclose any pathogenicity of Hg to ALS.[233] Moreover, the epidemiological survey of MND in Hokkaido did not reveal an increased incidence of ALS.[234]

As with Pb, several tissue samples of ALS patients have been assayed to clarify the pathogenic relation of Hg to the disease: depending on the study, Hg levels have either been found increased,[235-237] decreased,[238] or unchanged.[239,240] Oral administration of a chelator (dimercaptosuccinic acid: DMSA) to MND patients did not result in a greater mobilization of Pb and Hg from peripheral depots than in control subjects.[241] Accordingly, chelation treatment has never been successful in MND.[232]

4.4.3 Solvents

In the United Kingdom, an association has been found between work in the leather industry and subsequent development of MND.[242,243] It has been hypothesized that the reason for this association may be occupational exposure to organic solvents, which may damage motoneurons either directly or through activation of latent viruses.[244] Suspected compounds include n-hexane , methyl n-butyl ketone (MnBK, 2-hexanone), toluene, and mehtyl ethyl ketone (MEK).[244] An association between MND and occupational exposure to solvents and glues has been suggested by Chiò et al.,[245] but the results of this study did not reach statistical significance due to the small number of exposed subjects. A statistically significant increased risk for MND in relation to occupational exposure to solvents was found by Gunnarson et al.[223] In this study solvents appeared to be a risk factor especially among subjects under 60 years of age or in combination with genetic disposition. Further investigation is needed to confirm or refute these results and to establish which compounds bring the greatest risk.

4.4.4   Unidentified chemicals

Epidemiological studies designed to investigate possible risk factors for MND have suggested the association with various jobs, including farming,[246-248] professional contacts with animal hides and carcasses,[249] and working in textile factories.[250] However, the specific etiologic agent(s) could not be identified, and other studies exploring the possible association of MND with the same occupations yielded negative results.[251-253]

5   CONCLUSIONS

Early knowledge on the acute neuropsychiatric effects of occupational exposures was derived from clinical observations. The observed effects primarily concerned exposure to metals (e.g., manganese and mercury), solvents (e.g., CS_2, trichloroethylene, styrene) and pesticides such as organophosphates. Over time, modern examination methods, such as neuropsychological testing, and epidemiological studies, have also revealed chronic and less evident neurotoxic effects, though at times the interpretation of the findings has been controversial. A major difficulty in interpreting chronic neurotoxic effects depends on both the diversity and vagueness of clinical features and the associated problem of defining a proper disease entity for conclusive epidemiologic studies.[254]

Clinicians should always be aware that neuropsychiatric signs and symptoms may derive from exposure to chemicals present in the workplace. A careful occupational anamnesis should always be collected. The identification of the causal link between a given clinical picture and a toxic exposure will lead to removal of the patient from the contaminated environment, and to preventive meaures in favor of other workers operating in the same place.

There is controversial evidence that occupational chemical exposures participate in the etiopathogenesis of some human neurodegenerative disorders (Table 2). MPTP is the best documented example of an exogenous compound leading directly to the loss of specific neurons resulting in a neurodegenerative disease, i.e., a Parkinson-like syndrome, closely resembling the idiopathic disease. Other chemicals able to induce secondary Parkinsonism (and possibly involved in the etiology of Parkinson's disease) include pesticides, metals (Mn, Hg), carbon monoxide, and solvents such as CS_2. Aluminum appears to play a role in the development of dialysis encephalopathy and Guam's disease. The possible involvement of this metal and of other chemicals (Pb, solvents) in Alzheimer's disease is an intensely debated issue. Similarly, uncertainity and controversy exist over whether metals (Pb, Hg) and industrial solvents have a role in the development of amyotrophic lateral sclerosis and other forms of motor neuron disease.

While neurodegenerative diseases have not been definitively demonstrated to arise from xenobiotic exposure, the possibility exists that exposure to toxins and toxicants could lead to neurological conditions after a period of clinical silence.[255] This is best illustrated by observations of Guam's disease: in some cases, decades elapsed between last exposure to the Guam environment and clinical appearance of disease.[256] Similarly, some immigrants to Guam who adopted the lifestyle of indigenous people (Chamorro) developed the disease after up to 26 year of residency.[257]

A prolonged period of latency (years or decades) between xenobiotic exposure and disease is a recognized phenomenon in oncology, but is a relatively new concept in neurotoxicology. It poses major challenges for epidemiological surveys searching for evidence of past chemical exposures, in that attention should be focused on the environment in early, rather than late, life. Indeed, as in the case of MPTP, exposure to a certain dose of a compound may rapidly induce a neurodegenerative disease (irreversible Parkinsonism), whereas other subjects (presumably with less severe exposure and/or a lower susceptibility) may develop a subclinical lesion.[92] These individuals might be at risk for developing the disorder later in life if the effects of toxic damage combine with age-related neuronal loss to overcome the brain functional reserve.[3,258]

The concept of an "event threshold" is best exemplified by the neuropathological course of Parkinson's disease. It has been estimated that loss of approximately 80% of nigrostriatal dopaminergic neurons is required before a patient becomes clinically parkinsonian. Once that level of loss is reached, signs and symptoms of disease appear, sometimes with astonishing rapidity and severity. This model may apply to other neurodegenerative diseases. On the other hand, it is also possible that silent neurotoxicity is detected only relatively infrequently, since compensatory and adaptive processes of the CNS can, under certain conditions, correct for substantial degrees of brain injury, and behavioral adaptations can obscure some motor and sensory deficits.[255]

Clearly, there is a need to expand research efforts in this field. Case reports, formal epidemiological studies, unconventional detective work, and basic laboratory research in animals and cultured cells represent complementary approaches which, taken together, hold the most promise. Clusters of psychiatric and neurodegenerative diseases (in space, in time, or by age) represent a valuable focus for intensive research. Increased communication across disciplinary fields, such as pharmacology, toxicology, epidemiology, occupational medicine, environmental health, neurology, and psychiatry, is the most essential ingredient for progress toward understanding causes and finding preventions and cures.

6   REFERENCES

1. Anthony, D. C., Montine, T.J., and Graham, D.G., "Toxic responses of the nervous system," in *Casarett and Doull's Toxicology. The Basic Science of Poisons,* Fifth Edition, Klassen, C. D., Ed., McGraw-Hill, New York, 1996, 463.
2. Norton, S., "The nervous system as a target for toxic agents," in *Recent Advances in Nervous System Toxicology,* Galli, C. L., Manzo, L., and Spencer, P. S., Eds., Plenum Press, New York, 1988, 3.
3. Calne, D. B., "Neurotoxins and degeneration in the central nervous system," *Neuro-Toxicology,* 12, 335, 1991.
4. Ludolph, A. C., Hugon, J., Dwivedi, M. P., Schaumburg, H. H., and Spencer, P. S., "Studies on the aetiology and pathogenesis of motor neuron diseases. 1. Lathyrism: clinical findings in established cases," *Brain,* 110, 149, 1987.
5. Spencer, P. S., Ludolph, A., Dwivedi, M. P., Roy, D. N., Hugon, J., and Schaumburg, H. H., "Lathyrism: evidence for role of the neuroexcitatory aminoacid BOAA," *Lancet,* 2, 1066, 1986.
6. Kessler, A., Lathyrismus, *Monatsschr. Psychiat. Neurol.,* 113, 76, 1947.
7. Cohn, D. F., Streifler, M., "Human neurolathyrism. A follow-up study of 200 patients. Part I: clinical investigation," *Schweiz Arch. Neurol. Neurochir. Psychiatry,* 128, 151, 1981.
8. Spencer, P. S., "Guam ALS/Parkinsonism-dementia: a long-latency neurotoxic disorder caused by 'slow toxin(s)' in food?," *Can J. Neurol. Sci.,* 14, 347, 1987.
9. Garruto, R. M., "Pacific paradigms of environmentally-induced neurological disorders: clinical, epidemiological and molecular perspectives," *NeuroToxicology,* 12, 347, 1991.
10. Whiting, M. G., "Toxicity of cycads," *Econ. Bot.,* 17, 271, 1963.
11. Kisby, G. E., Ross, S. M., Spencer, P. S., Gold, B. G., Nunn, P. B., and Roy, D. N., "Cycasin and BMAA: candidate neurotoxins for Western Pacific amyotrophic lateral sclerosis/Parkinsonism-dementia complex," *Neurodegeneration,* 1, 73, 1992.
12. Spencer, P. S., Nunn, P. B., Hugon, J., Ludolph, A. C., Ross, S. M., Roy, D. N., and Robertson, R. C., "Guam amyotrophic lateral sclerosis-Parkinsonism-dementia linked to a plant excitant neurotoxin," *Science,* 237, 517, 1987.
13. Duncan, M. W., Kopin, I. J., Garruto, R. M., Lavine, L., and Markey, S. P., "2-Amino-3 (methylamino)-propionic acid in cycad-derived foods is an unlikely cause of amyotrophic lateral sclerosis/Parkinsonism," *Lancet,* 2, 631, 1988.
14. Duncan, M. W., Steele, J. C., Kopin, I. J., and Markey, S. P., "2-Amino-3 (methylamino)-propanoic acid (BMAA) in cycad flour: an unlikely cause of amyotrophic lateral sclerosis and parkinsonism dementia of Guam," *Neurology,* 40, 767, 1990.
15. Gajdusek, D. C., "Cycad toxicity not the cause of high incidence amyotrophic lateral sclerosis/Parkinsonism dementia on Guam, Kii peninsula of Japan, or in West New Guinea," in *Amyotrophic Lateral Sclerosis: Concepts in Pathogenesis and Etiology,* Hudson, A. H., Ed., University of Toronto Press, Toronto, Canada, 1990, 317.
16. He, F., Zhang, S., and Zhang, C., "Extrapyramidal lesions induced by mildewed sugar cane poisoning. Three case reports," *China Med. J.,* 67, 395, 1987.
17. Ludolph, A. C., He, F., Spencer, P. S., Hammerstad, J., and Sabri, M., "3-Nitropropionic acid-exogenous animal neurotoxin and possible human striatal toxin," *Can. J. Neurol. Sci.,* 18, 492, 1991.
18. Hamilton, A., "Industrial diseases of fur cutters and hatters," *J. Ind. Hyg.,* 4, 219, 1922.

19. Candura, F., "The metallurgy of mercury," in: *Principles of Industrial Technology for Occupational Medicine Scholars' Use* (in Italian), Comet, Pavia, 1991, 431.

20. Goodman, J. C., "Biochemical explanation for folk tales. Australian madness and mad hatters: a unifying hypothesis," *Trends Biochem. Sci.*, 10, 406, 1985.

21. Berlin, M., "Mercury," in *Handbook on the Toxicology of Metals,* 2nd Edition, Friberg, L., Nordberg, G. F., and Vouk, V., Eds., Elsevier, Amsterdam, 1986, 387.

22. Gupta, B. N., "Occupational diseases of teeth," *J. Soc. Occup. Med.*, 40, 149, 1990.

23. Urban, P., Lukas, E., Benicky, L., and Moscovicova, E., "Neurological and electro-physiological examination on workers exposed to mercury vapors," *NeuroToxicology,* 17, 191, 1996.

24. O'Carrol, R. E., Masterton, G., Dougall, N., Ebmeier, K. P., and Goodwin, G. M., "The neuropsychiatric sequelae of mercury poisoning. The mad hatter's disease revisited," *Br. J. Psychiatry,* 167, 95, 1995.

25. Opitz, H., Schweinsberg, F., Grossmann, T., Wendtgallitelli, M. F., and Meyermann, R., "Demonstration of mercury in the human brain and other organs 17 years after metallic mercury exposure," *Clin. Neuropathol.*, 15, 139, 1996.

26. Roels, H. A., Lauwerys, R. R., Buchet, J. P., Bernard, A., Barthels, A., Oversteyns, M., and Gaussin, J., "Comparison of renal function and psychomotor performance in workers exposed to elemental mercury," *Int. Arch. Occup. Environ. Hlth.*, 50, 77, 1982.

27. Smith, P. J., Langolf, G. D., and Goldberg, J., "Effects of occupational exposure to elemental mercury on short-term memory," *Br. J. Ind. Med.,* 40, 413, 1983.

28. Soleo, L., Urbano, M., Petrera, V., and Ambrosi, L., "Effects of low exposure to inorganic mercury on psychological performance," *Br. J. Ind. Med.*, 47, 105, 1990.

29. Langworth, S., Almkvist, O., Söderman, E., and Wikström, B. O., "Effects of occupational exposure to mercury vapour on the central nervous system," *Br. J. Ind. Med.*, 49, 545, 1992.

30. Fawer, R. F., de Ribaupierre, Y., Guillemin, M. P., Berode, M., and Lob, M., "Measurement of hand tremor induced by industrial exposure to metallic mercury," *Br. J. Ind. Med.*, 40, 204, 1983.

31. Chapman, L. J., Sauter, S. L., Henning, R. A., Dodson, V. N., Reddan, W. G., and Metthews, C. G., "Differences in frequency of finger tremor in otherwise asymptomatic mercury workers," *Br. J. Ind. Med.*, 47, 838, 1990.

32. Cavalleri, A., Belotti, L., Gobba, F., Luzzana, G., Rosa, P., and Seghizzi, P., "Colour vision loss in workers exposed to elemental mercury vapour," *Toxicol. Lett.*, 77, 351, 1995.

33. Schuckmann, F., "Study of preclinical changes in workers exposed to inorganic mercury in chloralkali plants," *Int. Arch. Occup. Environ. Hlth.*, 44, 193, 1979.

34. Bunn, W. B., McGill, C. M., Barber, T. E., Cromer, J. W. J., and Goldwater, L. J., "Mercury exposure in chloralkali plants," *Am. Ind. Hyg. Assoc. J.*, 47, 249, 1986.

35. Piikivi, L., and Hänninen, H., "Subjective symptoms and psychological performance of chlorine-alkali workers," *Scand. J. Work Environ. Hlth.*, 15, 69, 1989.

36. Echeverria, D., Heyer, N. J., Martin, M. D., Naleway, C. A., Woods, J. S., and Bittner, A. C., "Behavioral effects of low-level exposure to Hg among dentists," *Neurotoxicol. Teratol.*, 17, 161, 1995.

37. Ngim, C. H., Foo, S. C., and Jeyaratnam, J., "Chronic neurobehavioral effects of elemental mercury in dentists," *Br. J. Ind. Med.*, 49, 782, 1992.

38. Alajouanine, T., Derobert, L., and Thieffry, S., "Etude clinique d'ensemble de 210 cas d'intoxication par les sels organiques d'etain," *Rev. Neurol.*, Paris, 98, 85, 1958.

39. Derobert, L., "Poisoning by an organic tin compound (diiodoethyl tin or Stalinon)," *J. Forensic Med.*, 7, 192, 1960.

40. Manzo, L., Richelmi, P., Sabbioni, E., Pietra, R., Bono, F., and Guardia, L., "Poisoning by triphenyltin acetate. Report of two cases and determination of tin in blood and urine by neutron activation analysis," *Clin. Toxicol.*, 18, 1343, 1981.

41. Rey, C., Reinecke, H. J., and Besser, R., "Methyltin intoxication in six men: toxicologic and clinical aspects," *Vet. Hum. Toxicol.*, 26, 121, 1984.

42. Besser, R., Krämer, G., Thümler, R., Bohl, J., Gutmann, L., and Hopf, H. C., "Acute trimethyltin limbic-cerebellar syndrome," *Neurology*, 37, 945, 1987.

43. Ross, W. D., Emmett, E. A., Steiner, J., and Tureen, R., "Neurotoxic effects of occupational exposure to organotins," *Am. J. Psychiatry*, 138, 1092, 1981.

44. Fortemps, E., Amand, G., Bomboir, A., Lauwerys, R., and Laterre, E. C., "Trimethyltin poisoning. Report of two cases," *Int. Arch. Occup. Environ. Hlth.*, 41, 1, 1978.

45. Snyder, R., and Andrews, L. S., "The toxic effects of solvents," in *Casarett and Doull's Toxicology. The Basic Science of Poisons, Fifth Edition*, Klassen, C. D., Ed., McGraw-Hill, New York, 1996, 739.

46. Hormes, J. T., Filley, C. M., and Rosenberg, N. L., "Neurologic sequelae of chronic solvent vapor abuse," *Neurology*, 36, 698, 1986.

47. Rosenberg, N. L., Kleinschmidt-DeMasters, B. K., Davis, K. A., Dreisbach, J. N., Hormers, J. T., and Filley, C. M., "Toluene abuse causes diffuse central nervous system white matter changes," *Ann. Neurol.*, 23, 611, 1988.

48. Yamanouchi, N., Okada, S., Kodama, K., Hirai, S., Sekine, H., Murakami, A., Komatsu, N., Sakamoto, T., and Sato, T., "White matter changes caused by chronic solvent abuse," *Am. J. Neuroradiol.*, 16, 1643, 1995.

49. Arlien-Søborg, P., Bruhn, P., Gyldensted, C., and Melgaard, B., "Chronic painter's syndrome: chronic encephalopathy in house painters," *Acta Neurol. Scand.*, 60, 149, 1979.

50. Valciukas, J. A., "Chronic toxic encephalopathies in solvent-exposed workers," in *Foundations of Environmental and Occupational Neurotoxicology*, Van Nostrand Reinhold, New York, 1991, 537.

51. Axelson, O., Hane, M., and Hogstedt, C., "A case-referent study on neuropsychiatric disorders among workers exposed to solvents," *Scand. J. Work Environ Hlth.*, 2, 14, 1976.

52. Brackbill, R. M., Maizlish, N., and Fischbach, T., "Risk of neuropsychiatric disability among painters in the United States," *Scand. J. Work Environ. Hlth.*, 16, 182, 1990.

53. Cherry, N., and Waldron, H. A., "The prevalence of psychiatric morbidity in solvent workers in Britain," *Int. J. Epidemiol.*, 13, 197, 1984.

54. Cherry, N., Hutchins, H., Pace, T., and Waldron, H. A., "Neurobehavioural effects of repeated occupational exposure to toluene and paint solvents," *Br. J. Ind. Med.*, 42, 291, 1985.

55. Triebig, G., Claus, D., Csuzda, I., Druschky, K. F., Holler, P., Kinzel, W., Lehrl, S., Reichwein, P., Weidenhammer, W., Weitbrecht, W. U., Weltle, D., Schaller, K. H., and Valentin, H., "Cross-sectional epidemiological study on neurotoxicity of solvents in paints and lacquers," *Int. Arch. Occup. Environ. Hlth.*, 60, 233, 1988.

56. Bolla, K., Schwartz, B., Agnew, J., Ford, P., and Bleeker, M., "Subclinical neuropsychiatric effect of low-level solvent exposure in U.S. paint manufactures," *J. Occup. Med.*, 32, 671, 1990.

57. Guberan, E., Usel, M., Raymond, L., Tissot, R., and Sweetnam, P. M., "Disability, mortality, and incidence of cancer among Geneva, Switzerland painters and electricians: a historical prospective study," *Br. J. Ind. Med.*, 46, 16, 1989.

58. Errebo-Knudsen, E. O., and Olsen, F., "Organic solvents and presenile dementia. The painters syndrome: a critical review of the Danish literature," *Sci. Total Environ.*, 48, 45, 1986.

59. Reinvang, I., Borchgrevink, H. M., Aaserud, O., Lie, V., Malt, U. F., Nalstad, P., Larsson, P. G., and Gierstad, L., "Neuropsychological findings in a non-clinical sample of workers exposed to solvents," *J. Neurol. Neurosurg. Psychiatry*, 57, 614, 1994.

60. Lundberg, I., Michélsen, H., Nise, G., Hogstedt, C., Högberg, M., Alfredsson, L., Almkvist, O., Gustavsson, A., Hagman, M., Herlofson, J., Hindmarsh, T., and Wennberg, A., "Neuropsychiatric function of housepainters with previous long-term heavy exposure to organic solvents," *Scand. J. Work Environ. Hlth.*, 21 (suppl. 1), 1, 1995.

61. Costa, L. G., "Organophosphorus compounds," in *Recent Advances in Nervous System Toxicology*, Galli, C. L., Manzo, L., and Spencer, P. S., Eds., Plenum Press, New York, 1988, 203.

62. Callender, T. J., Morrow, L., and Subramanian, K., "Evaluation of chronic neurological sequelae after acute pesticide exposure using SPECT brain scans," *J. Toxicol. Environ. Hlth.*, 41, 275, 1994.

63. Savage, E. P., Keefe, T. J., Mounce, L. M., Heaton, R. K., Lewis, J. A., and Burcar, P. J., "Chronic neurological sequelae of acute organophosphate pesticide poisoning," *Arch. Environ. Hlth.*, 43, 38, 1988.

64. Rosenstock, L., Keifer, M., Daniell, W. E., McConnell, Claypoole, K., and the Pesticide Health Effects Study Group, "Chronic central nervous system effects of acute organophosphate pesticide intoxication," *Lancet*, 338, 223, 1991.

65. Steenland, K., Jenkins, B., Ames, R. G., O'Malley, M., Chrislip, D., and Russo, J., "Chronic neurological sequelae to organophosphate pesticide poisoning," *Am. J. Public Hlth.*, 84, 731, 1994.

66. Holmes, J. H., and Gaon, M. D., "Observations on acute and multiple exposure to anticholinesterase agents," *Trans. Am. Clin. Climat. Ass.*, 68, 86, 1957.

67. Gershon, S., and Shaw, F. H., "Psychiatric sequelae of chronic exposure to organophosphorus insecticides," *Lancet*, 1, 1371, 1961.

68. Dille, J. R., and Smith, P. W., "Central nervous system effects of chronic exposure to organophosphate insecticides," *Aerospace Med.*, 35, 475, 1964.

69. Metcalf, R. L., and Holmes, J. H., "VII. Toxicology and physiology. EEG, psychological and neurological alterations in humans with organophosphorus exposure," *Ann. N. Y. Acad. Sci.*, 160, 357, 1969.

70. Ohto, K., "Long term follow up study of chronic organophosphate pesticide intoxication (Saku disease) with special reference with retinal pigmentary degeneration," *Acta Soc. Ophthalmol. Japon.*, 78, 237, 1974.

71. Levin, H. S., Rodnitzky, R. L., and Mick, D. L., "Anxiety associated with exposure to organophosphorus compounds," *Arch. Gen. Psychiat.*, 33, 225, 1976.

72. Korsak, R. J., and Sato, M M., "Effects of chronic organophosphate pesticide exposure on the central nervous system," *Clin. Toxicol.*, 11, 83, 1977.

73. Davis, K. L., Yesavage, J. A., and Berger, P. A., "Possible organophosphate-induced Parkinsonism," *J. Nerv. Ment. Dis.*, 166, 222, 1978.

74. Duffy, F. H., Burchfiel, J. L., Bartels, P. H., Gaon, M., and Sim, V. M., "Long term effects of an organophosphate upon the human electroencephalogram," *Toxicol. Appl. Pharmacol.*, 47, 161, 1979.

75. Perold, J. G., and Bezuidenhout, D. J. J., "Chronic organophosphate poisoning," *S. Afr. Med. J.*, 57, 7, 1980.

76. Misra, V. K., Nag, D., Misra, N. K., and Krishna Murti, C. R., "Macular degeneration associated with chronic pesticide exposure," *Lancet*, 1, 288, 1982.

77. Stephens, R., Spurgeon, A., Calvert, I. A., Beach, J., Levy, L. S., Berry, H., and Harrington, J. M., "Neuropsychological effects of long-term exposure to organophosphates in sheep dip," *Lancet*, 345, 1135, 1995.

78. Bidstrup, P. L., "Psychiatric sequelae of chronic exposure to organophosphorus insecticides," *Lancet*, 2, 103, 1961.

79. Stoller, A., Krupinski, J., Christophers, A. J., and Blanks, G. K., "Organophosphorus insecticides and major mental illness. An epidemiological investigation," *Lancet*, 1, 1387, 1965.

80. Durham, W. F., Wolfe, H. R., and Quinby, G. E., "Organophosphorus insecticides and mental alertness," *Arch. Environ. Hlth.*, 10, 55, 1965.

81. Rodnitzky, R. L., "Occupational exposure to organophosphate pesticides: a neurobehavioral study," *Arch. Environ. Hlth.*, 30, 98, 1975.

82. Maizlish, N., Schenker, M., Weisskopf, C., Seiber, J., and Samuels, S., "A behavioral evaluation of pest control workers with short term, low-level exposure to the organophosphate diazinon," *Am. J. Ind. Med.*, 12, 153, 1987.

83. Beach, J. R., Spurgeon, A., Stephens, R., Heafield, T., Calvert, I. A., Levy, L. S., and Harrington, J. M., "Abnormalities on neurological examination among sheep farmers exposed to organophosphorus pesticides," *Occup. Environ. Med.*, 53, 520, 1996.

84. McDonald, B. E., Costa, L. G., and Murphy, S. D., "Spatial memory impairment and central muscarinic receptor loss following prolonged treatment with organophosphates," *Toxicol. Lett.*, 40, 47, 1988.

85. Veronesi, B., Jones, K., and Pope, C., "The neurotoxicity of subchronic acetylcholinesterase inhibition in rat hippocampus," *Toxicol. Appl. Pharmacol.*, 104, 440, 1990.

86. Koller, W. C., "Classification of Parkinsonism," in *Handbook of Parkinson's Disease*, Koller, W. C., Ed., Marcel Dekker, New York, 1987, 51.

87. Tanner, C. M., "Occupational and environmental causes of Parkinsonism," *Occup. Med. (Philadelphia)*, 7, 503, 1992.

88. Gorrell, J. M., Di Monte D., and Graham, D., "The role of the environment in Parkinson's disease," *Environ. Hlth. Perspect.*, 104, 652, 1996.

89. Tanner, C. M., and Goldman, S. M., "Epidemiology of Parkinson's disease," *Neurol. Clin.*, 14, 317, 1996.

90. Dawson, R., Jr., Beal, M. F., Bondy, S. C., Di Monte, D. A., and Isom, G. E., "Excitotoxins, aging, and environmental neurotoxins: implications for understanding human neurodegenerative diseases," *Toxicol. Appl. Pharmacol.*, 134, 1, 1995.

91. Spencer, P. S., and Butterfield, P. G., "Environmental agents and Parkinson's disease," in *Etiology of Parkinson's Disease*, Ellenberg, J. H., Koller, W. C., and Langston, J. W., Eds., Marcel Dekker, New York, 1995, 319.

92. Langston, J. W., Ballard, P. A., Tetrud, J. W., and Irwin, I., "Chronic Parkinsonism in humans due to a product of meperidine-analog synthesis," *Science*, 219, 979, 1983.

93. Langston, J. W., and Ballard, P. A., "Parkinson's disease in a chemist working with 1-methyl-4-phenyl-1,2,3,6-tetrahydropyridine," *N. Engl. J. Med.*, 309, 310, 1983.

94. Barbeau, A., Roy, M., and Langston, J. W., "Neurological consequences of industrial exposure to 1-methyl-4-phenyl-1,2,3,6-tetrahydropyridine," *Lancet*, 1, 747, 1985.

95. Bocchetta, A. and Corsini, G. U., "Parkinson's disease and pesticides," *Lancet*, 2, 1163, 1986.

96. Howard, J. K., "A clinical survey of paraquat formulation workers," *Br. J. Ind. Med.*, 36, 220, 1979.

97. Zilker, T., Fogt, F., and von Clarmann, M., "Kein Parkinsonsyndrom nach akuter Paraquatintoxikation" [No Parkinsonian syndrome following acute paraquat poisoning], *Klin. Wochenschr.*, 66, 1138, 1988.

98. Ferraz, H. B., Bertolucci, P. H. F., Pereira, J. S., Lima, J. G. C., and Andrade, L. A. F., "Chronic exposure to the fungicide maneb may produce symptoms and signs of CNS manganese intoxication," *Neurology*, 38, 550, 1988.

99. Meco, G., Bonifati, V., Vanacore, N., and Fabrizio, E., "Parkinsonism after chronic exposure to the fungicide maneb (manganese ethylene-bis-dithiocarbamate)," *Scand. J. Work Envriron. Hlth.*, 20, 301, 1994.

100. Ho, S. C., Woo, J., and Lee, C. M., "Epidemiologic study of Parkinson's disease in Hong Kong," *Neurology*, 39, 1314, 1989.

101. Golbe, L. I., Farrell, T. M., and Davis, P. H., "Follow-up study of early life protective and risk factors in Parkinson's disease," *Mov. Disord.*, 5, 66, 1990.

102. Semchuk, K. M., Love, E. J., and Lee, R. G., "Parkinson's disease and exposure to agricultural work and pesticide chemicals," *Neurology*, 42, 1328, 1992.

103. Hertzman, C., Wiens, M., Snow, B., Kelly, S., and Calne, D. B., "A case-control study of Parkinson's disease in a horticultural region of British Columbia," *Mov. Disord.*, 9, 69, 1994.

104. Seidler, A., Hellenbrand, W., Robra, B.-P., Vieregge, P., Nischan, P., Joerg, J., Oertel, W. H., Ulm, G., and Schneider, E., "Possible environmental, occupational, and other etiologic factors for Parkinson's disease: a case-control study in Germany," *Neurology*, 46, 1275, 1996.

105. Ohlson, C.-G., and Hogstedt, C., "Parkinson's disease and occupational exposure to organic solvents, agricultural chemicals and mercury: a case-control study," *Scand. J. Work Environ. Hlth.*, 7, 252, 1981.

106. Stern, M., Dulaney, E., Gruber, S. B., *et al.*, "The epidemiology of Parkinson's disease: a case-control study of young-onset and old-onset patients," *Arch. Neurol.*, 48, 903, 1991.

107. Wong, G. F., Gray, C. S., Hassanein, R. S., and Koller, W. C., "Environmental risk factors in siblings with Parkinson's disease," *Arch. Neurol.*, 48, 287, 1991.

108. Jiménez-Jiménez, F. J., Mateo, D., and Gimenez-Roldan, S., "Exposure to well water and pesticides in Parkinson's disease: a case-control study in the Madrid area," *Mov. Disord.*, 7, 149, 1992.

109. Fleming, L., Mann, J. B., Bean, J., Briggle, T., and Sanchez-Ramos, J. R., "Parkinson's disease and brain levels of organochlorine pesticides," *Ann. Neurol.*, 36, 100, 1994.

110. Couper, J., "On the effects of black oxide of manganese when inhaled into the lungs," *Br. Ann. Med. Pharm.*, 1, 41, 1837.

111. Calne, D. B., Chu, N.-S., Huang, C.-C., Lu, C.-S., and Olanow, W., "Manganism and idiopathic Parkinsonism: similarities and differences," *Neurology*, 44, 1583, 1994.

112. Huang, C.-C., Chu, N.-S., Lu, C.-S., Wang. D.-J., Tsai, J.-L., Tzeng, J.-C., Wolters, E. C., and Calne, D. B., "Chronic manganese intoxication," *Arch. Neurol.*, 46, 1104, 1989.

113. Huang, C.-C., Lu, C.-S., Chu, N.-S., Hochberg, F., Lilienfeld, D., Olanow, W., and Calne D. B., "Progression after chronic manganese exposure," *Neurology*, 43, 1479, 1993.

114. Lu, C.-S., Huang, C.-C., Chu, N.-S., and Calne, D. B., "Levodopa failure in chronic manganism," *Neurology*, 44, 1600, 1994.

115. Yamada, M., Ohno, S., Okayasu, I., Okeda, R., Hatakeyama, S., Watanabe, H., Ushio, K., and Tsukagoshi, H., "Chronic manganese poisoning: a neuropathological study with determination of manganese distribution in the brain," *Acta Neuropathol. (Berl.)*, 70, 723, 1986.

116. Wolters, E. C., Huang, C.-C., Clark, C., Peppard, R. F., Okada, J., Chu, N. S., Adam, M. J., Ruth, T. J., Li, D., and Calne D. B., "Positron emission tomography in manganese intoxication," *Ann. Neurol.*, 26, 647, 1989.

117. Roels, H. A., Lauwerys, R. R., Buchet, J.-P., Genet, P., Sarhan, M. J., Hanotiau, I., de Fayes, M., Bernard, A., and Stanescu, D., "Epidemiological survey among workers exposed to manganese: effects on lung, central nervous system, and some biological indicators," *Am. J. Ind. Med.*, 11, 307, 1987.

118. Wang, J.-D., Huang, C.-C., Hwang, Y. H., Chiang, J. R., Lin, J. M., and Chen, J. S., "Manganese induced Parkinsonism: an outbreak due to an unrepaired ventilation control system in a ferromanganese smelter," *Br. J. Ind. Med.*, 46, 856, 1989.

119. Iregren, A., "Psychological test performance in foundry workers exposed to low levels of manganese," *Neurotoxicol. Teratol.*, 12, 673, 1990.

120. Wennberg, A., Iregren, A., Struwe, G., Cizinsky, G., Hagman, M. M., and Johansson, L., "Manganese exposure in steel smelteries. A health hazard to the nervous system," *Scand. J. Work Environ. Hlth.*, 17, 255, 1991.

121. Roels, H. A., Ghyselen, P., Buchet, J.-P., Ceulemans, E., and Lauwerys, R. R., "Assessment of the permissible exposure level to manganese in workers exposed to manganese dioxide dust," *Br. J. Ind. Med.*, 49, 25, 1992.

122. Mergler, D., Huel, G., Bowler, R., Iregren, A., Bélanger S., Baldwin, M., Tardif, R., Smargiassi, A., and Martin, L., "Nervous system dysfunction among workers with long-term exposure to manganese," *Environ. Res.*, 64, 151, 1994.

123. Lucchini, R., Selis, L., Folli, D., Apostoli, P., Mutti, A., Vanoni, O., Iregren, A., Alessio, L., "Neurobehavioral effects of manganese in workers from a ferroalloy plant after temporary cessation of exposure," *Scand. J. Work Environ. Hlth.*, 21, 143, 1995.

124. Sjögren, B., Iregren, A., Frech, W., Hagman, M., Johansson, L., Tesarz, M., and Wennberg, A., "Effects on the nervous system among welders exposed to aluminium and manganese," *Occup. Environ. Med.*, 53, 32, 1996.

125. Mergler, D., "Manganese: the controversial metal. At what levels can deleterious effects occur?" *Can. J. Neurol. Sci.*, 23, 93, 1996.

126. Finkelstein, Y., Vardi, J., Kesten, M. M., and Hod, I., "The enigma of Parkinsonism in chronic borderline mercury intoxication, resolved with challenge with penicillamine," *NeuroToxicology*, 17, 291, 1996.

127. Ngim, C. H., and Devathasan, G., "Epidemiologic study on the association between body burden mercury level and idiopathic Parkinson's disease," *Neuroepidemiology*, 8, 128, 1989.

128. Choi, I. S., "Delayed neurologic sequelae in carbon monoxide intoxication," *Arch. Neurol.*, 40, 433, 1983.

129. Butera, R., Candura, S. M., Locatelli, C., Varango, C., Li, B., Manzo, L., "Neurological sequelae of carbon monoxide poisoning: role of hyperbaric oxygen," *Indoor Environ.*, 4, 134, 1995.

130. Grinker, R. R., "Parkinsonism following carbon monoxide poisoning," *J. Nerv. Ment. Dis.*, 64, 18, 1926.

131. Klawans, H. L., Stein, R. W., Tanner, C. M., Goetz, C. G., "A pure parkinsonian syndrome following acute carbon monoxide intoxication," *Arch. Neurol.*, 39, 302, 1982.

132. Myers, R. A., Snyder, S. K., Emhoff, T. A., "Subacute sequelae of carbon monoxide poisoning," *Ann. Emerg. Med.*, 14, 1163, 1985.

133. Yang, Z. D., "Observation of hyperbaric oxygen in 160 patients with later manifestations after acute carbon monoxide poisoning," *J. Hyperbar. Med.*, 1, 188, 1986.

134. Min, S. K., "A brain syndrome associated with delayed neuropsychiatric sequelae following acute carbon monoxide intoxication," *Acta Psychiatr. Scand.*, 73, 80, 1986.

135. Seger, D., and Welch, L., "Carbon monoxide controversies: neuropsychologic testing, mechanism of toxicity, and hyperbaric oxygen," *Ann. Emerg. Med.*, 24, 242, 1994.
136. Smith, J. S., and Brandon, S., "Morbidity from acute carbon monoxide poisoning at three-year follow-up," *Br. Med. J.*, 1, 318, 1973.
137. He, F., Qin, J., Chen, S., Pan, X., Xu, G., Zhang, S., and Fang, G., "Risk factors of development of delayed encephalopathy following acute carbon monoxide poisoning," *Indoor Environ.*, 1, 268, 1992.
138. Minerbo, G. M., and Jankovic, J., "Prevalence of Parkinson's disease among firefighters," *Neurology*, 40 (suppl.), 348, 1990.
139. Delpech, A., "Accidents que developp, chez les ouvriers en caoutchoue, l'inhalation du CS$_2$ en vapeur," *Union Med.*, 10, 265, 1856.
140. Vigliani, E. C., "Carbon disulphide poisoning in viscose rayon workers," *Br. J. Ind. Med.*, 11, 325, 1954.
141. Bus, J., "The relationship of carbon disulfide metabolism to development of toxicity," *NeuroToxicology*, 4, 73, 1985.
142. Guggenheim, M. A., Couch, J. R., Weinberg, W., "Motor dysfunction as a permanent complication of methanol ingestion," *Arch. Neurol.*, 24, 550, 1971.
143. McClean, D. R., Jacobs, H., and Mielke, B. W., "Methanol poisoning: a clinical and pathological study," *Ann. Neurol.*, 8, 161, 1980.
144. LeWitt, P. A., and Martin, S. D., "Dystonia and hypokinesis with putaminal necrosis after methanol intoxication," *Clin. Neuropharmacol.*, 11, 1612, 1988.
145. Pezzoli, G., Barbieri, S., Ferrante, C., Zecchinelli, A., and Foà, V., "Parkinsonism due to *n*-hexane exposure," *Lancet*, 2, 874, 1989.
146. Pezzoli, G., Antonini, A., Barbieri, S., Canesi, M., Perbellini, L., Zecchinelli, A., Mariani, C. B., Bonetti, A., and Leenders, K. L., "*n*-Hexane-induced Parkinsonism: pathogenetic hypotheses," *Mov. Disord.*, 10, 279, 1995.
147. Pezzoli, G., Ricciardi, S., Masotto, C., Mariani, C. B., and Carenzi, A., "*n*-Hexane induces Parkinsonism in rodents," *Brain Res.*, 531, 355, 1990.
148. Melarned, E., and Lavy, S., "Parkinsonism associated with chronic inhalation of carbon tetrachloride," *Lancet*, 1, 1015, 1977.
149. Sandyk, R., and Gillman, M. A., "Motor dysfunction following chronic exposure to a fluoroalkane solvent mixture containing nitromethane," *Eur. Neurol.*, 23, 479, 1984.
150. Uitti, R. J., Snow, B. J., Shinotoh, H., Vingerhoets, F. J. G., Hayward, M., Hashimoto, S., Richmond, J., Markey, S. P., Markey, C. J., and Calne, D. B., "Parkinsonism induced by solvent abuse," *Ann. Neurol.*, 35, 616, 1994.
151. Tetrud, J. W., Langston, J. W., Irwin, I., and Snow, B., "Parkinsonism caused by petroleum waste ingestion," *Neurology*, 44, 1051, 1994.
152. McKhann, G., Drachman, D., Folstein, M., Katzman, R., Price, D., and Stadlan, E. M., "Clinical diagnosis of Alzheimer's disease: report of the NINCDS-ADRDA Work Group under the auspices of Department of Health and Human Services Task Force on Alzheimer's Disease," *Neurology*, 34, 939, 1984.
153. Cumlings, C. L., "The failing brain," *Lancet*, 345, 1481, 1995.
154. Mozar, H. N., Bal, D. G., and Howard, J. T., "Perspectives on the etiology of Alzheimer's disease," *JAMA*, 257, 1503, 1987.
155. Gautrin, D., and Gauthier, S., "Alzheimer's disease: Environmental factors and etiologic hypotheses," *Can. J. Neurol. Sci.*, 16, 375, 1989.
156. van Duijn, C. M., "Epidemiology of the dementias: recent developments and new approaches," *J. Neurol. Neurosurg. Psychiatry*, 60, 478, 1996.

157. Savory, J., Exley, C., Forbes, W. F., Huang, Y., Joshi, J. G., Kruck, T., Crapper McLachlan, D. R., and Wakayama, I., "Can the controversy of the role of aluminum in Alzheimer's disease be resolved? What are the suggested approaches to this controversy and methodological issues to be considered?" *J. Toxicol. Environ. Hlth.*, 48, 615, 1996.

158. Copestake, P., "Aluminum and Alzheimer's disease. An update," *Food Chem. Toxicol.*, 31, 679, 1993.

159. Doll, R., "Review: Alzheimer's disease and environmental aluminium," *Age Ageing*, 22, 138, 1993.

160. Alfrey, A. C., LeGendre G. R., and Kaehney, W. D., "The dialysis encephalopathy syndrome: possible aluminum intoxication," *N. Engl. J. Med.*, 294, 184, 1976.

161. Crapper, D. R., Krishnan, S. S., and Dalton, A. J., "Brain aluminium distribution in Alzheimer's disease and experimental neurofibrillary degeneration," *Science*, 185, 511, 1973.

162. Crapper, D. R., Krishnan, S. S., and Quittkat, S., "Aluminum neurofibrillary degeneration in Alzheimer's disease," *Brain*, 99, 67, 1976.

163. Trapp, G. A., Miner, G. D., Zimmerman, R. L., Mastri, A. R., and Heston, L. L., "Aluminum levels in brain in Alzheimer's disease," *Biol. Psychiatry*, 13, 709, 1978.

164. Perl, D. P., and Brody, A. R., "Alzheimer's disease: X-ray spectrometric evidence of aluminum accumulation in neufibrillary tangle-bearing neurons," *Science*, 208, 297, 1980.

165. Perl, D. P., and Good, P. F., "The association of aluminum, Alzheimer's disease and neurofibrillary tangles," *J. Neural Transm.*, 24, 205, 1987.

166. Masters, C. L., Multhaup, G., Simms, G., Pottgiesser, J., Martins, R. N., and Beyreuther, K., "Neuronal origin of a cerebral amyloid: neurofibrillary tangles of Alzheimer's disease contain the same protein as the amyloid of plaque cores and blood vessels," *EMBO J.*, 4, 2757, 1985.

167. Candy, J. M., Klinowski, J., Perry, R. H., Perry, E. K., Fairbairn, A., Oakley, A. E., Carpenter, T. A., Atack, J. R., Blessed, G., and Edwardson, J. A., "Aluminosilicates and senile plaques formation in Alzheimer's disease," *Lancet*, 1, 354, 1986.

168. McDermott, J. R., Smith, A. I., Iqbal, K., and Wisniewski, H. M., "Brain aluminum in aging and Alzheimer disease," *Neurology*, 29, 809, 1979.

169. Markesbery, W. R., Ehmann, W. D., Hossain, T. I. M., Alaudin, M., Goodin, D. T., "Instrumental neutron activation analysis of brain aluminum in Alzheimer disease and aging," *Ann. Neurol.*, 10, 511, 1981.

170. Landsberg, J. P., McDonald, B., and Watt, F., "Absence of aluminum in neuritic plaque cores in Alzheimer's disease," *Nature*, 360, 65, 1992.

171. Connor, D. J., Jope, R. S., and Harrell, L. E., "Chronic, oral aluminum administration to rats: cognition and cholinergic parameters," *Pharmacol. Biochem. Behav.*, 31, 467, 1988.

172. Klatzo, I., Wisniewski, H. M., and Streicher, E., "Experimental production of neurofibrillary degeneration: I. Light microscopic observations," *J. Neuropathol. Exp. Neurol.*, 24, 187, 1965.

173. Varner, J. A., Huie, C., Horvath, W., Jensen, K. F., and Isaacson, R. L., "Chronic AlF_3 administration: II. Selected histological observations," *Neurochem. Res. Commun.*, 13, 99, 1993.

174. Bernuzzi, V., Desor, D., and Lehr, P. R., "Developmental alterations in offspring of female rats orally intoxicated by aluminum chloride or lactate during gestation," *Teratology*, 40, 21, 1989.

175. Golub, M. S., Keen, C. L., and Gershwin, M. E., "Neurodevelopmental effect of aluminum in mice: fostering studies," *Neurotoxicol. Teratol.*, 14, 177, 1992.

176. Hamdy, R. C., "Aluminum toxicity and Alzheimer's disease. Is there a connection?" *Postgrad. Med.*, 88, 239, 1990.
177. Martyn, C. N., Barker, D. J. P., Osmond, C., Harris, E. C., Edwardson, J. A., and Lacey, R. F., "Geographical relation between Alzheimer's disease and aluminum in drinking water," *Lancet*, 1, 59, 1989.
178. Flaten, T. P., "Geographical association between aluminum in drinking water and death rates with dementia (including Alzheimer's disease), Parkinson's disease, and amyotrophic lateral sclerosis in Norway," *Environ. Geochem. Hlth.*, 12, 152, 1990.
179. Neri, L. C., and Hewitt, D., "Aluminium, Alzheimer's disease, and drinking water," *Lancet*, 338, 390, 1991.
180. Crapper McLachlan, D. R., Bergeron, M. D., Smith, J. E., Boomer, D., and Rifat, S. L., "Risk for neuropathologically confirmed Alzheimer's disease and residual aluminum in municipal drinking water employing weighted residential histories," *Neurology*, 46, 401, 1996.
181. Wettstein, A., Aeppli, J., Gautschi, K., and Peters, M., "Failure to find a relationship between mnestic skills of octogenarians and aluminum in drinking water," *Int. Arch. Occup. Environ. Hlth.*, 63, 97, 1991.
182. Forster, D. P., Newens, A. J., Kay, D. W., and Edwardson, J. A., "Risk factors in clinically diagnosed presenile dementia of the Alzheimer type: a case-control study in northern England," *J. Epidemiol. Comm. Hlth.*, 49, 253, 1995.
183. Li, G., Shen, Y. C., Li, Y. T., Chen, C. H., Zhau, Y. W., and Silverman, S. M., "A case-control study of Alzheimer's disease in China," *Neurology*, 42, 1481, 1992.
184. The Canadian Study of Health and Aging, "Risk factors for Alzheimer's diseae in Canada," *Neurology*, 44, 2073, 1994.
185. Crapper McLachlan, D. R., Dalton, A. J., Kruck, T. P. A., Bell, M. Y., Smith, W. L., Kalow, W., and Andrews, D. F., "Intramuscular desferrioxamine in patients with Alzheimer's disease," *Lancet*, 337, 1304, 1991.
186. Perl, D. P., and Good, P. F., "Uptake of aluminium into central nervous system along nasal-olfactory pathways," *Lancet*, 1, 1028, 1987.
187. Elinder, C.-G., Ahrengart, L., Lidums, V., Pettersson, E., and Sjögren, B., "Evidence of aluminium accumulation in aluminium welders," *Br. J. Ind. Med.*, 48, 735, 1991.
188. McLaughlin, A. I. G., Kazantis, G., King, E., Teare, D., Porter, R. J., and Owen, R., "Pulmonary fibrosis and encephalopathy associated with the inhalation of aluminium dust," *Br. J. Ind. Med.*, 19, 253, 1962.
189. Kobayashi, S., Hirota, N., Saito, K., and Utsuyama, M., "Aluminum accumulation in tangle-bearing neurons of Alzheimer's disease with Balint's syndrome in a long term aluminum refiner," *Acta Neuropathol. (Berl.)*, 74, 47, 1987.
190. Sjögren, B., Ljunggren, K. G., Almkvist, O., Frech, W., and Basun, H., "Aluminosis and dementia," *Lancet*, 344, 1154, 1994.
191. Rifat, S. L., Eastwood, M. R., Crapper McLachlan, D. R., and Corey, P. N., "Effect of exposure of miners to aluminium powder," *Lancet*, 336, 1162, 1990.
192. White, D. W., Longstreth, W. T., Jr., Rosenstock, L., Claypoole, K. H. J., Brodkin, C. A., and Townes, B. D., "Neurologic syndrome in 25 workers from an aluminum smelting plant," *Arch. Int. Med.*, 152, 1443, 1992.
193. Alessio, L., Apostoli, P., Ferioli, A., Di Sipio, I., Mussi, I., Rigosa, C., and Albertini, A., "Behaviour of biological indicators of internal dose and some neuro-endocrine tests in aluminium workers," *Med. Lav.*, 80, 290, 1989.
194. Bast-Pettersen, R., Drabløs, P. A., Goffeng, L. O., Thomassen, Y., and Torres, C. G., "Neuropsychological deficit among elderly workers in aluminum production," *Am. J. Ind. Med.*, 25, 649, 1994.

195. Salib, E., and Hillier, V., "A case-control study of Alzheimer's disease and aluminum occupation," *Br. J. Psychiatry*, 168, 244, 1996.

196. Castellino, P., Anzelmo, V., Bianco, P., Mattei, O., and Castellino, N., "Acute encephalopathy," in *Inorganic Lead Exposure*, Castellino, N., Castellino, P., and Sannolo, N., Eds., Lewis Publishers, Boca Raton, FL, 1995, 298.

197. Niklowitz, W. J., and Mandybur, T. I., "Neurofibrillary changes following childhood lead encephalopathy," *J. Neuropathol. Exp. Neurol.*, 34, 445, 1975.

198. Niklowitz, W. J., "Neurofibrillary changes after acute experimental lead poisoning," *Neurology*, 25, 927, 1975.

199. Shalat, S. L., Seltzer, B., Baker, E. L., Jr., "Occupational risk factors and Alzheimer's disease: a case-control study," *J. Occup. Med.*, 30, 934, 1988.

200. Graves, A. B., van Duijn, C. M., Chandra, V., Fratiglioni, L., Heyman, A., Jorm, A. F., Kokmen, E., Kondo, K., Mortimer, J. A., Rocca, W. A., *et al.*, "Occupational exposure to solvents and lead as risk factors for Alzheimer's disease: a collaborative re-analysis of case-control studies," *Int. J. Epidemiol.*, 20 (suppl. 2), S58, 1991.

201. Cassels, D. A. K., and Dodds, E. C., "Tetraethyl lead poisoning," *Br. J. Med.*, 2, 681, 1946.

202. Kukull, W. A., Larson, E. B., Bowen, J. D., McCormick, W. C., Teri, L., Pfanschmidt, M. L., Thompson, J. D., O'Meara, E. S., Brenner, D. E., and van Belle, G., "Solvent exposure as a risk factor for Alzheimer's disease: a case-control study," *Am. J. Epidemiol.*, 141, 1059, 1995.

203. Edling, C., Ekberg, K., Ahlborg, G., Jr., Alexandersson, R., Barregard, L., Ekenvall, L., Nilsson, L., and Svensson, B. G., "Long-term follow up of workers exposed to solvents," *Br. J. Ind. Med.*, 47, 75, 1990.

204. Armon, C., Kurland, L. T., Daube, J. R., and O'Brien, P., "Epidemiologic correlates of sporadic amyotrophic lateral sclerosis," *Neurology*, 41, 1077, 1991.

205. Mulder, D. W., Kurland, L. T., Offord, K. P., and Beard, C. M., "Familial adult motor neuron disease: amyotrophic lateral sclerosis," *Neurology*, 36, 511, 1986.

206. Chiò A., Brignolio, F., Meineri, P., and Schiffer, D., "Phenotypic and genotypic heterogeneity of dominantly inherited amyotrophic lateral sclerosis," *Acta Neurol. Scand.*, 75, 277, 1987.

207. Tandan, R., and Bradley, W. G., "Amyotrophic lateral sclerosis: Part 2. Etiopathogenesis," *Ann. Neurol.*, 18, 419, 1985.

208. Mitchell, J. D., "Heavy metals and trace elements in amyotrophic lateral sclerosis," *Neurol. Clin.*, 5, 43, 1987.

209. Castellino, P., Anzelmo, V., Bianco, P., Mattei, O., and Castellino, N., "Motor neuron disease and amyotrophic lateral sclerosis (ALS)," in *Inorganic Lead Exposure*, Castellino, N., Castellino, P., and Sannolo, N., Eds., Lewis Publishers, Boca Raton, FL, 1995, 330.

210. Aran, F. A., "Réchèrches sur une maladie non encore décrite du système musculaire (atrophie musculaire progressive)," *Arch. Gén. Méd.*, 24, 4, 1850.

211. Simpson, J. A., Seaton, D. A., and Adams, J. F., "Response to treatment with chelating agents of anemia, chronic encephalopathy, and myelopathy due to lead poisoning," *J. Neurol. Neurosurg. Psychiatry*, 27, 536, 1964.

212. Currier, R. D., and Haerer, A. F., "Amyotrophic lateral sclerosis and metallic toxins," *Arch. Environ. Hlth.*, 17, 712, 1968.

213. Campbell, A. M. G., Williams, E. R., and Baltrop, D., "Motor neuron disease and exposure to lead," *J. Neurol. Neurosurg. Psychiatry*, 33, 877, 1970.

214. Petkau, A., Sawatzky, A., Hillier, C. R., and Hoogstraten, J., "Lead content of neuro-muscular tissue in amyotrophic lateral sclerosis: case report and other considerations," *Br. J. Ind. Med.*, 31, 275, 1974.

215. Kurlander, H. M., and Patten, B. M., "Metals in spinal cord tissue of patients dying of motor neuron disease," *Ann. Neurol.*, 6, 21, 1979.

216. Conradi, S., Ronnevi, L. O., and Vesterberg, O., "Increased plasma levels of lead in patients with amyotrophic lateral sclerosis compared with control subjects as determined by flameless atomic absorption spectrophotometry," *J. Neurol. Neurosurg. Psychiatry*, 41, 389, 1978.

217. Conradi, S., Ronnevi, L. O., and Vesterberg, O., "Lead concentration in skeletal muscle in amyotrophic lateral sclerosis patients and control subjects," *J. Neurol. Neurosurg. Psychiatry*, 41, 1001, 1978.

218. Manton, W. I., and Cook, J. D., "Lead content of cerebrospinal fluid and other tissue in amyotrophic lateral sclerosis (ALS)," *Neurology*, 29, 611, 1979.

219. Stober, T., Stelte, W., and Kunze, K., "Lead concentrations in blood, plasma, erythrocytes, and cerebrospinal fluid in amyotrophic lateral sclerosis," *J. Neurol. Sci.*, 61, 21, 1983.

220. Cavalleri, A., Minoia, C., Ceroni, M., and Poloni, M., "Lead in cerebrospinal fluid and its relationship to plasma lead in humans," *J. Appl. Toxicol.*, 4, 63, 1984.

221. Kapaki, E., Segditsa, J., Zournas, Ch., Xenos, D., and Papageorgiou, C., "Determination of cerebrospinal fluid and serum lead levels in patients with amyotrophic lateral sclerosis and other neurological diseases," *Experientia*, 45, 1108, 1989.

222. Gresham, L. S., Molgaard, C. A., Golbeck, A. L., and Smith, R., "Amyotrophic lateral sclerosis and occupational heavy metal exposure: a case-control study," *Neuroepidemiology*, 5, 29, 1986.

223. Gunnarson, L.-G., Bodin, L., Söderfeldt, B., and Axelson, O., "A case-control study of motor neurone disease: its relation to heritability, and occupational exposures, particularly to solvents," *Br. J. Ind. Med.*, 49, 791, 1992.

224. Scarpa, M., Colombo, A., Panzetti, P., and Sorgato, P., "Epidemiology of amyotrophic lateral sclerosis in the province of Modena, Italy. Influence of environmental exposure to lead," *Acta Neurol. Scand.*, 77, 456, 1988.

225. De Domenico, P., Malara, C. E., Marabello, L., Puglisi, R. M., Meneghini, F., Serra, S., Gallitto, G., and Musolino, R., "Amyotrophic lateral sclerosis: an epidemiological study in the province of Messina, Italy, 1976–1985," *Neuroepidemiology*, 7, 152, 1988.

226. Sienko, D. G., Davis, J. P., Taylor, J. A., and Brooks, B. R., "Amyotrophic lateral sclerosis. A case-control study following detection of a cluster in a small Wisconsin community," *Arch. Neurol.*, 47, 38, 1990.

227. Brown, I. A., "Chronic mercurialism: a cause of the clinical syndrome of amyotrophic lateral sclerosis," *Arch. Neurol. Psychiatry*, 72, 674, 1954.

228. Kantarjiam, A. D., "A syndrome clinically resembling amyotrophic lateral sclerosis following chronic mercurialism," *Neurology*, 11, 639, 1961.

229. Shirakawa, K., Yuasa, K., Hirota, T., Tsubaki, T., and Hoshi, M., "A case of methylmercury poisoning with onset of a clinical syndrome resembling amyotrophic lateral sclerosis," *Neurol. Med. (Tokyo)*, 4, 58, 1976.

230. Barber, T. E., "Inorganic mercury intoxication reminiscent of amyotrophic lateral sclerosis," *J. Occup. Med.*, 20, 667, 1978.

231. Adams, C. R., Ziegler, D. K., and Lin, J. T., "Mercury intoxication simulating amyotrophic lateral sclerosis," *JAMA*, 250, 642, 1983.

232. Schwarz, S., Husstedt, I. W., Bertram, H. P., and Kuchelmeister, K., "Amyotrophic lateral sclerosis after accidental injection of mercury," *J. Neurol. Neurosurg. Psychiatry*, 60, 698, 1996.

233. Moriwaka, F., Tashiro, K., Doi, R., Satoh, H., and Fukuchi, Y., "A clinical evaluation of the inorganic mercurialism. Its pathogenic relation to amyotrophic lateral sclerosis," *Clin. Neurol. (Tokyo)*, 31, 885, 1991.

234. Okumura, H., Moriwaka, F., Tashiro, K., Hamada, A., Matsumoto, A., Matsumoto, H., Itoh, N., Shindo, R., Takahata, N. and ALS study group, "Epidemiological study of motor neuron disease in Hokkaido Island. Its incidence, prevalence and regional distribution," *Brain Nerve (Tokyo)*, 44, 727, 1992.

235. Mano, Y., Takayanagi, T., Ishitani, A., and Hirota, T., "Mercury in hair of patients with ALS," *Clin. Neurol. (Tokyo)*, 29, 844, 1989.

236. Mano, Y., Takayanagi, T., Abe, T., and Takizawa, Y., "Amyotrophic lateral sclerosis and mercury. A preliminary report," *Clin. Neurol. (Tokyo)*, 30, 1257, 1990.

237. Khare, S. S., Ehmann, W. D., Kasarskis, E. J., and Markesbery, W. R., "Trace element imbalances in amyotrophic lateral sclerosis," *NeuroToxicology*, 11, 521, 1990.

238. Moriwaka, F., Satoh, H., Ejima, A., Watanabe, C., Tashiro, K., Hamada, T., Matsumoto, A., Shima, K., Yanagihara, T., Fukazawa, T., Okumura, H., Maruo, Y., Itoh, K., and Yoshida, K., "Mercury and selenium contents in amyotrophic lateral sclerosis in Hokkaido, the northernmost island of Japan," *J. Neurol. Sci.*, 118, 38, 1993.

239. Pierce-Ruhland, R., and Patten, B. M., "Muscle metals in motor neuron disease," *Ann. Neurol.*, 8, 193, 1980.

240. Oishi, M., Takasu, T., and Tateno, M., "Hair trace elements in amyotrophic lateral sclerosis," *Trace Elements Med.*, 7, 182, 1990.

241. Louwerse, E. S., Buchet, J. P., Van Dijk, M. A., de Jong, V. J., and Lauwerys, R. R., "Urinary excretion of lead and mercury after oral administration of meso-2,3-dimercaposuccinic acid in patients with motor neurone disease," *Int. Arch. Occup. Environ. Hlth.*, 67, 135, 1995.

242. Hawkes, C. H., and Fox, A. J., "Motor neuron disease in leather workers," *Lancet*, 1, 507, 1981.

243. Buckley, J., Warlow, C., Smith, P., Hilton-Jones D., Irvine S., and Tew, J. R., "Motor neuron disease in England and Wales 1959–1979," *J. Neurol. Neurosurg. Psychiatry*, 46, 197, 1983.

244. Hawkes, C. H., Cavanagh, J. B., and Fox, A. J., "Motoneuron disease: a disorder secondary to solvent exposure?" *Lancet*, 2, 73, 1989.

245. Chiò A., Tribolo, A., and Schiffer, D., "Motoneuron disease and solvent and glue exposure," *Lancet*, 2, 921, 1989.

246. Granieri, E., Carreras, M., Tola, R., Paolino, E., Tralli, G., Eleopra, R., and Serra, G., "Motor neuron disease in the province of Ferrara, Italy, in 1964–1982," *Neurology*, 38, 1604, 1988.

247. Chiò, A., Meineri, P., Tribolo, A., and Schiffer, D., "Risk factors in motor neuron disease: a case-control study," *Neuroepidemiology*, 10, 174, 1991.

248. Kalfakis, N., Vassilopoulos, D., Voumvourakis, C., Ndjeveleka, M., and Papageorgiou, C., "Amyotrophic lateral sclerosis in southern Greece: an epidemiologic study," *Neuroepidemiology*, 10, 170, 1991.

249. Hanish, R., Dworsky, R. L., Henderson, B. E., "A search for clues to the cause of amyotrophic lateral sclerosis," *Arch. Neurol.*, 33, 456, 1976.

250. Abarbanel, J. M., Herishanu, Y. O., Osimani, A., and Frisher, S., "Motor neuron disease in textile factory workers," *Acta Neurol. Scand.*, 79, 347, 1989.

251. Kurtzke, J. F., and Beebe, G. W., "Epidemiology of amyotrophic lateral sclerosis. 1. A case-control comparison based on ALS deaths," *Neurology*, 30, 453, 1980.
252. Deapen, D. M., and Henderson, B. E., "A case-control study of amyotrophic lateral sclerosis," *Am. J. Epidemiol.*, 123, 790, 1986.
253. Imaizumi, Y., "Mortality rate of amyotrophic lateral sclerosis in Japan: effects of marital status and social class, and geographical variation," *Jap. J. Hum. Genet.*, 31, 101, 1986.
254. Axelson, O., "Where do we go in occupational neuroepidemiology?" *Scand. J. Work Environ. Hlth.*, 22, 81, 1996.
255. Reuhl, K. R., "Delayed expression of neurotoxicity: the problem of silent damage," *NeuroToxicology*, 12, 341, 1991.
256. Garruto, R. M., Gajdusek, D. C., and Chen, K. M., "Amyotrophic lateral sclerosis among Chamorro migrants from Guam," *Ann. Neurol.*, 8, 612, 1980.
257. Garruto, R. M., Gajdusek, D. C., and Chen, K. M., "Amyotrophic lateral sclerosis and Parkinsonism-dementia among Filipino migrants to Guam," *Ann. Neurol.*, 10, 341, 1981.
258. Calne, D. B., Esen, A., McGeer, E., and Spencer, P., "Alzheimer's disease, Parkinson's disease, and motoneurone disease: abiotrophic interaction between ageing and environment," *Lancet*, 2, 1067, 1986.

8 Sensory System Alterations Following Occupational Exposure to Chemicals

Donald A. Fox

CONTENTS

1 INTRODUCTION

Occupational exposure to toxic chemicals frequently results in structural and functional alterations in the auditory, olfactory, somatosensory, vestibular and/or visual system.[1] It has been estimated that almost half of all neurotoxic chemicals affect some aspect of sensory function.[2] In many cases, disturbances in sensory system function are the first symptoms noted or reported following chemical exposure.[3-5] Equally, and possibly more, important is the fact that these alterations are often present in the absence of any clinical signs of poisoning.[1,5] This suggests that sensory systems may be especially vulnerable to toxic insult. In fact, alterations in the structure and/or function of the eye or central visual system is one of the criteria utilized for setting permissible exposure levels for 33 different chemicals in the United States.[6] Overall, the most frequently reported sensory system alterations

occur in the visual system.[1,2] Finally, sensory system impairments, especially auditory and visual, can lead to increased occupational injuries as well as decrease the
overall quality of life.

The aim of this chapter is to review the clinical and corresponding experimental
data regarding the structural and functional retinal and central visual system deficits
commonly produced by a few selected neurotoxic chemicals found in the workplace:
carbon disulfide, trichloroethylene, organic solvent mixtures and inorganic lead.
These specific industrial chemicals were chosen as prototypical examples because
of their divergent structural, functional and behavioral alterations. That is, carbon
disulfide and trichloroethylene appear to primarily affect central retinal ganglion
cells and the optic nerve resulting in the loss of cone-dominated (or photopic) vision
whereas inorganic lead primarily affects rod photoreceptors resulting in the rod (or
scotopic) vision deficits (see references below). Based on their clinical effects, the
complex mixtures of organic solvents appear to have several target sites: retinal and
optic nerve. It is recognized, however, that other industrial chemicals produce selective visual system disturbances. For example, occupational exposure to methylmercury produces a severe loss of peripheral vision and scotopic deficits. These are
thought to primaril results from selective lesions in the calcarine fissure, as opposed
to the occipital pole, of the visual cortex.[7-9] However, there are direct effects of
methylmercury on the retina.[10-12] In addition, some industrial neurotoxicants (e.g.,
acrylamide, n-hexane) have been utilized in animal models to examine the pathogenesis of various neuronal and axonal diseases as well as basic functions of the
retinocortical pathways.[13-15]

Several review chapters and books have addressed various aspects related to
ocular and visual system toxicology.[16-19] For example, the acute and chronic adverse
effects of industrial chemicals on the external adnexa of the eye (i.e., conjunctiva,
cornea) and the lens have been presented by several authors.[17,20-23] The neurobehavioral, electrophysiological and morphological methods for assessing visual system
alterations also have been discussed in detail.[24-29] In contrast, the important areas of
auditory, olfactory and gustatory toxicity following exposure to industrial neurotoxicants have not yet received as much attention as they deserve.[30-34]

2   CARBON DISULFIDE

2.1   HUMAN STUDIES

Carbon disulfide is an important industrial solvent that has been in use since the
1850s. Occupational exposure to this chemical is a potential health hazard world-
wide: most notably in the viscose rayon industry. More than 20,000 workers in the
United States are exposed daily to this industrial chemical.[35] Neurological alterations,
central and peripheral distal axonopathies, and vascular alterations in the retina,
kidney and cardiovascular system commonly occur following exposure. The most
common retinal alterations include retrobulbar neuritis and pre-senile arteriosclerotic
changes in the fundus.[36,37] In addition, retinal microaneurysms and small hemorrhages, delayed peripapillary filling, and irregularities in the caliber of the retinal

arterioles and veins have been observed during fluorescein angiography.[36,38,39] In the only follow-up study (4 year) conducted to date, the percent of subjects with fundus alterations increased and the severity of the pigmentary and vascular lesions increased in most cases.[39] Many of these patients had no exposure during the intervening four-year period.

Alterations in visual function following carbon disulfide exposure were first observed during ophthalmological examinations.[40,41] These old clinical reports described the severe loss of central vision (central scotoma), depressed sensitivity in the peripheral visual field (rod scotoma) and accompanying optic atrophy. In addition, blurred vision, impaired color vision and pupillary changes were reported.[41-43] Moreover, electrophysiological analysis revealed that the light/dark ratio of the electrooculogram (EOG) was decreased and that the amplitude of the dark-adapted (rod or scotopic) and light-adapted (cone or photopic) electroretinograms (ERGs) were decreased with only a slight decrease in latency.[39,43-45] Long-term alterations in both the EOG and ERG were still observable in a four-year follow-up study.[39] Moreover, high frequency hearing loss has been found in workers chronically exposed to carbon disulfide.[46]

2.2 ANIMAL STUDIES

Mice, rabbits, rats and/or monkeys exposed to carbon disulfide have retinal ganglion cell loss and neurofilamentous axonopathy in the optic nerve fibers.[47-50] The most thorough analysis of the effects of carbon disulfide on visual system histopathology, retinal vasculature and visual function was conducted using non-human primates.[50,51] Monkeys were exposed to carbon disulfide via inhalation and then either sacrificed immediately or 12–18 months later. The monkey sacrificed immediately had numerous swellings in the distal optic tract which were irregular in their distribution.[50] This is consistent with the distal filamentous axonopathies observed in the peripheral nervous system and/or central visual pathway of humans and experimental animals.[37,47-49] In the monkeys examined at least one year following exposure, marked retinal ganglion cell degeneration located almost exclusively in the central retinal region was present.[50] Moreover, these same animals had diminished horseradish peroxidase transport from the eye to the lateral geniculate and vision cortical areas associated with the central retinal ganglion cells as well as decreased cytochrome oxidase activity in these same areas.[50] These results are consistent with the clinical reports of optic atrophy, a central scotoma, altered color vision and altered photopic ERG.[39,40-45] However, they do not adequately explain the rod deficits: that is, peripheral scotoma and altered scotopic ERG.[39,43-45] Moreover, and in contrast to the pathology observed in workers exposed to carbon disulfide,[36,38,39] the marked long-term visual system injury observed in the monkeys occurred in the total absence of retinal vascular abnormalities.[50,51]

All monkeys exposed to carbon disulfide exhibited severe and permanent reductions in visual acuity and contrast sensitivity.[51] This is consistent with clinical reports of decreased visual sensitivity and impaired color vision.[41-43] Merigan and co-workers suggested that the color vision alterations could result from the adverse effects of carbon disulfide on the central retinal ganglion cells projecting to the parvocellular layers of the lateral geniculate nucleus: the area most adversely affected by carbon disulfide exposure in their monkeys.[50,51] Furthermore, they suggested that color deficits may be a good indicator of mild carbon disulfide exposure. In contrast, and interestingly, flicker resolution did not exhibit a long-term decrease following carbon disulfide exposure.[51] This does not correlate with the clinically observed peripheral scotoma[40,41] since flicker resolution is dependent upon intact peripheral retinal function.[52] No explanation was offered for these dissimilarities.

3   TRICHLOROETHYLENE

3.1   HUMAN STUDIES

In the United States, more than 100,000 workers are exposed daily to the halogenated hydrocarbon, trichloroethylene.[53] Acute exposure to high concentrations of trichloroethylene is most frequently associated with cranial nerve palsies: especially of the trigeminal (V) and to a lesser extent the optic (II), facial (VII) and auditory (VIII) nerves. Chronic exposure produces similar effects as well as cerebellar dysfunction.[54] Damage to the fifth nerves results in the loss of most sensory modalities over the entire distribution of the fifth nerve and weakness of the muscles of mastication (Feldman, 1979). Damage to the other nerves may result in the loss of vision due to optic nerve atrophy or retrobulbar neuritis,[55,56] loss of taste and facial weakness due to seventh nerve damage[54,57] or hearing loss due to eighth nerve damage.[58] Although the cellular mechanisms responsible for these alterations are unknown, they are consistent with demyelinating neuropathy.[54]

Workers acutely or chronically exposed to trichloroethylene exhibit altered visual evoked potentials.[59,60] Long-term follow-up (16–18 years) of a patient following a single acute exposure to trichlorethylene revealed residual occulomotor and ciliary reflex dysfunction, impaired neuropsychological performance, but no residual changes in the visual evoked potential.[61] This is consistent with experimental data in human volunteers showing vestibulo-occulomotor, but not flicker fusion or form perception, abnormalities following acute inhalation exposure to trichlorethylene.[62,63]

3.2   ANIMAL STUDIES

Electrophysiological studies following the acute and subchronic effects of trichloroethylene on the rabbit and rat visual systems have revealed equivocal results.[64-66] The b-wave, but not the a-wave, of four light-adapted rabbits acutely injected with trichloroethylene was decreased for at least four hours post-injection.[66] Rabbits subchronically exposed (via inhalation) to low and high levels of trichloroethylene exhibited decreases and increases, respectively, in the amplitude of selected waveforms of the light-adapted visual evoked potential.[65] Six weeks following exposure, some recovery was evident. In contrast, rats exposed (via inhalation) to much higher levels of trichloroethylene exhibited no changes in the light-adapted visual evoked potential or somatosensory evoked potential, but instead exhibited a high-frequency hearing loss as assessed by the brainstem auditory evoked potential.[66]

4 OTHER ORGANIC SOLVENTS

4.1 HUMAN STUDIES

Color vision loss (acquired dyschromatopsia) and decreases in the contrast sensitivity function have been reported by Mergler and co-workers to occur in workers exposed to high concentrations of a mixture of organic solvents such as trichlorethylene, alcohols, xylene, toluene and others.[67-70] These color vision losses appear to exhibit dose-response relationships since other investigators have found no such alterations in workers exposed to lower concentrations of similar chemicals.[71,72] Using the Lanthony D-15 desaturated color arrangement panel as an assessment tool, Mergler and co-workers reported that a large percentage of solvent-exposed workers in microelectronic plants, print shops and paint manufacturing facilities had acquired dyschromatopsia in the absence of observable clinical abnormalities even though the exposures exceeded the threshold limit values.[67-70] Ocular examinations including biomicroscopy, funduscopy and peripheral visual field tests revealed no abnormalities. These color vision losses were mainly blue-yellow losses although the more severe red-green losses also were reported. Acquired blue-yellow losses generally result from lens opacification or retinal alterations whereas red-green losses are associated with retrobular or central visual pathway alterations.[80] Moreover, these occupationally-exposed workers also exhibited intermediate spatial frequency losses in their contrast sensitivity function, which reflect alterations in neural function.[52] As noted above, other investigators used the Lanthony color test and found no significant differences in color discrimination between solvent-exposed print shop workers and controls.[71,72]

Determining which solvent or solvents in mixtures may be responsible for the reported adverse effects of solvent mixtures on vision[67-70] is not an easy task. A detailed reading and interpretation of the literature suggests that this effort may be further complicated by possible gender differences in these effects. For example, one recent comprehensive color vision study on male workers exposed to only toluene shows that even high concentration of this solvent had no effect on color vision.[73] However, one study on female workers, where a single measure was used to assess color vision (Lanthony D-15 desaturated test), shows a trend toward increased prevalence of color vision impairment following exposure to low to moderate concentrations of toluene.[74] Clearly, more detailed and well-designed studies are needed to determine which solvent(s) cause alterations in color vision, if there are gender differences in these effects and whether these effects are reversible.

Five independent studies report that workers exposed to mean atmospheric concentrations of styrene ranging from 20–70 ppm exhibit concentration-dependent alterations in color vision.[75-79] A combined data analysis from two of the above studies[75,79] suggests that the threshold for color visual impairments is ≤4 ppm styrene.[80] This is well below the threshold limit value-time weighted average (TLV-TWA) value for any country: range 20–50 ppm. The findings of similar blue-yellow color vision deficits by five different groups of investigators in different countries argues convincingly for the reproducibility and validity of these styrene-induced color vision deficits. The reversibility of these impairments is unknown, however, in one study no recovery was found after a one month period of no exposure.[75] The findings reveal the sensitivity of the visual system and especially the photoreceptors to toxicant exposure. Moreover, the results demonstrate that the Lanthony D-15 desaturated test is a sensitive and reliable test for detecting color vision abnormalities in solvent-exposed workers. Finally, and importantly, the results suggest that the TLV-TWA for styrene be reevaluated.

Overall, the evidence indicates that organic solvents can produce color vision deficits in occupationally-exposed workers. In addition to the exposure differences in the above cited studies, there are also some important methodological problems with and differences between these studies. Several of these issues were discussed by Baird *et al.*[72] These relate to lighting sources, reliability of tests due to lack of test-retest protocols and the establishment of proper and valid criteria for identifying deficits. However, other important factors such as lack of pre-screening ophthalmo-logical examinations, appropriate statistical analysis, short term changes due to acute or confounding effects, and long-term residual effects were not discussed. All of these factors deserve further discussion, investigation and validation.

5   INORGANIC LEAD

5.1   HUMAN STUDIES

Although occupational lead poisoning occurs much less frequently today than in the past, visual[5,81-83] and auditory[94-98] system processing deficits and alterations are observed in battery factory workers, workers still handling lead-based chemicals and products, and workers exposed to lead dust. Several cases of retrobulbar neuritis with consecutive optic nerve atrophy have been observed following chronic lead exposure (mean blood lead, BPb, 40–60 µg/dl) or acute high level lead expo-sure.[83,99-103] Most of these cases presented with fundus lesions, peripheral or para-central scotomas while the most severe cases also had a central scotoma. Generally, the scotomas were not observed until approximately 5 years of continuous lead exposure. Interestingly, the earliest observable scotomas were not detected under standard photopic (cone-mediated) viewing conditions, but became evident only under mesopic (rod- and cone-mediated) viewing conditions. Consistent with these findings is the frequent complaint made by lead-exposed workers that they have difficulty walking and seeing in the dark.[93]

These ophthalmological and neuropsychological findings correlate directly with the critical flicker fusion threshold data observed in similarly lead-exposed workers. No alterations in the critical flicker fusion threshold were observed when the test was conducted under photopic conditions or when using red lights but consistent decreases were observed when the test was conducted under scotopic conditions or when green lights were used.[83-87] Moreover, in occupationally lead-exposed workers with no visual deficits or no observable alterations following ophthalmological examination, the a-wave and b-wave amplitudes of the dark-adapted ERG were decreased.[81] In summary, these results suggest that occupational lead exposure pro-duces concentration- and time-dependent alterations in the retina such that higher levels of lead directly and adversely affect both the retina and optic nerve/tract whereas lower levels of lead appear to primarily affect the rod photoreceptors and the rod pathway. Interestingly, these latter clinical findings showing a preferential lead-induced rod-selective deficits are observed in both experimental *in vivo* and *in vitro* animal studies (see Section 5.2 below).

In addition to the retinal deficits, several clinical studies have documented occulomotor deficits in chronically lead-exposed workers who have no observable ophthalmological deficits (mean BPb: 56–67 μg/dl). For example, in a prospective study Baloh and co-workers found that the mean accuracy of saccadic eye movements was lower in lead-exposed workers at both baseline and during follow-up examination.[90-92] In two other studies, each on a different lead-exposed population, the mean accuracy of saccadic eye movements also were reduced and the number of overshoots were increased.[5,89] Moreover, in the studies cited above the decreases in saccade maximum velocity in lead-exposed workers were almost significantly different.[5,90-92] Although the site and mechanism of action underlying these alterations are unknown, they most likely result from CNS-mediated deficits.[104]

These retinal and occulomotor alterations were, in most cases, correlated with the blood lead levels and, equally important, occurred in the absence of observable ophthalmological changes, CNS symptoms and abnormal performance test scores. Thus, these measures of visual function may be among the most sensitive for the early detection of the neurotoxic effect of inorganic lead.

5.2   ANIMAL STUDIES

Acute and subchronic exposure of adult experimental animals to lead produce retinal deficits, rod degeneration and apoptotic rod cell death[105-108] as well as auditory system dysfunction.[109-111] *In vitro* biochemical and electrophysiological experiments with lead, using visual system tissues obtained from adult animals, produce alterations similar to those observed following *in vivo* lead exposure.[112-123] Recently, new information regarding the mechanisms of action accounting for the retinal deficits and rod degeneration has been presented.[119-124]

Experimental support for a selective rod photoreceptor site of action was first suggested by the findings of Fox and Sillman[112] and later by Sillman and co-workers.[113-115] These investigators demonstrated that micromolar concentrations of lead chloride selectively depressed the amplitude and sensitivity of the rod, but not cone, photoreceptor potential in isolated, perfused bullfrog retinas.[112-115] Intracellular recordings from isolated axolotl rods[116] and from mudpuppy horizontal cells[117,118] exposed to lead chloride validated these original findings.

More recently, quantitative histological, ultrastructural, biochemical and ERG studies have shown that moderate level lead exposure only during adulthood (BPb: 59 μg/dl) produces selective rod photoreceptor deficits and apoptotic cell death in rods and bipolar cells.[108] These results are consistent with neurobehavioral and ERG studies conducted in lead-exposed workers (vide supra) and similar to studies conducted in developing rats exposed to lead only during development.[125,126] That is, the rod-mediated, but not cone-mediated, ERG showed decreases in a-wave and b-wave amplitude, sensitivity, critical flicker fusion threshold and dark adaptation. These results suggest that lead produces similar functional alterations in the mature and developing visual system: albeit with different degrees of sensitivity and severity.

In vivo and *in vitro* data strongly suggest that lead-induced alterations in retinal cGMP metabolism provides a common biochemical mechanism underlying the ERG deficits and the rod and bipolar apoptotic cell death. For example, adult lead exposure results in decreased retinal cGMP phosphodiesterase (PDE) activity and corresponding increases in retinal cGMP levels.[108] Moreover, the inhibition of cGMP PDE can be produced in adult control retinas or isolated rod enzyme following *in vitro* exposure to picomolar to micromolar concentrations of lead.[108,119-124] The *in vivo* and *in vitro* data suggest that lead exposure in humans may produce similar retinal deficits. Additionally, if rods and blue-sensitive cones in humans exhibit the same sensitivity to a lead-induced inhibition of cGMP-PDE, as they do to the drug-induced inhibition of cGMP-PDE[127,128] blue-cone color vision deficits, in addition to scotopic vision deficits, may result from lead exposure. Visual testing of occupationally-exposed lead workers for blue-yellow color vision impairments using the Lanthony D-15 desaturated color arrangement panel would appear to be useful and worthwhile (e.g., see Section 4 on organic solvents).

Recently, the kinetics of the rod cGMP PDE as well as the molecular mechanism accounting for the lead-induced inhibition of rod cGMP PDE was elucidated.[119,123,124] Picomolar to nanomolar concentrations of Pb^{2+} directly inhibited the rod cGMP PDE by competing with millimolar concentrations of the co-factor Mg^{2+} for the metal site and this inhibition was proportional to the cGMP concentration.[123,124] Thus, as the rod intracellular concentration of Mg^{2+} decreased or cGMP increased, so did the inhibition. The relevance and implication of these interactions deserve further investigation, especially since tissue magnesium is decreased in bone and liver following lead exposure.[129,130] Moreover, and consistent with these findings, Mg^{2+} reversed the Pb^{2+}-induced inhibition of PDE.[123,124] The similarity of findings in these *in vivo* and *in vitro* ERG and biochemical studies in lead-exposed humans and/or animals argues strongly that the ERG deficits result from the direct effect of lead on cGMP PDE. In addition, these findings provide a novel mechanism for understanding the Pb^{2+}-induced inhibition of cGMP PDE. Furthermore, these results may have implications for other enzymes using Mg^{2+} as a co-factor and suggest that Mg^{2+} may be useful in these situations for reversing the inhibition by Pb^{2+}.

Retinal histopathological alterations due to moderate- or high-level lead exposure (BPb of 40–60 µg/dl and ≥60 µg/dl, respectively) have been observed in adult animals.[108,125,126] As expected, the degree and extent of damage depended upon the tissue level of lead and duration of exposure. For example, chronic high-level lead exposure resulted in focal necrosis of the rod inner and outer segments, necrosis in the inner nuclear layer and Muller cells, and lysosomal inclusions in the retinal pigment epithelium.[105-107] In contrast, subacute moderate-level lead exposure resulted in selective rod and bipolar cell death.[108] At the ultrastructural level, all dying cells exhibited the classical morphological features of apoptotic cell death. That is, they had compaction and segregation of the nuclear chromatin into sharply defined areas that abutted the nuclear membrane and condensation of the cytoplasm.[131] These results reveal that the histopathological aspects of lead exposure during adulthood are similar to those observed during low and moderate level developmental lead exposure and thus by extension, they strongly suggest that they share common underlying biochemical alterations (e.g., inhibition of cGMP PDE). An alternative, although not mutually exclusive, hypothesis is that lead produces its selective rod degeneration and functional alterations via an inhibition of the retinal Na^+,K^+-ATPase activity.[126,132] This latter hypothesis is consistent with the observed retinal degeneration that occurs following the intraocular injection of ouabain[133,134] with the rod-selective alterations in amplitude, sensitivity and kinetics following inhibition of retinal Na,K-ATPase by ouabain or strophanthidin[135,136] and with the observed alteration in retinal Na^+,K^+-ATPase activity following *in vivo* or *in vitro* lead exposure.[132]

6   CONCLUSIONS

Sensory system toxicology is a research area of immense importance and opportunity. Subtle alterations in visual or auditory processing can have profound immediate, long-term and maybe even delayed effects on the mental, social, and physical health and performance of an individual. In addition, such alterations can markedly affect the quality of life. The determination and understanding of cellular and molecular mechanisms underlying the toxicant-induced retinal and visual system disorders produced by various neurotoxicants are just now beginning. However, the adverse effects of neurotoxicants on other sensory modalities are relatively unexplored and important areas of investigation. The data cited in this brief review also show that *in vivo* and *in vitro* experimental models, using tissue from a wide variety of species, can faithfully and reliably reproduce most of the structural, functional and/or behavioral deficits in the visual system produced by neurotoxicant chemicals — a point not stressed often enough. Finally, these data suggest the need for simple routine scotopic visual system evaluation and color vision testing in workers occupationally exposed to neurotoxic chemicals as early and sensitive screening measures for detecting neurotoxic insult.

7   ACKNOWLEDGMENTS

Original work reported herein from Dr. Donald A. Fox's laboratory was supported in part by an NIH Grant RO1 ES 03183 from the National Institute of Environmental Health Sciences.

8   REFERENCES

1. Anger, W. K. and Johnson, B. L., "Chemicals affecting behavior," in *Neurotoxicity of Industrial and Commercial Chemicals*, Vol. I, O'Donoghue, J. L., Ed., CRC Press, Boca Raton, FL, 1985, Chapter 3.
2. Crofton, K. M. and Sheets, L. P., "Evaluation of sensory system function using reflex modification of the startle response," *J. Am. Coll. Toxicol.*, 8, 199, 1989.
3. Damstra, T., "Environmental chemicals and nervous system dysfunction," *Yale J. Biol. Med.*, 51, 457, 1981.

4. Hanninen, H., Nurminen, M., Tolonen, M. and Martelin, T., "Psychological tests as indicators of excessive exposure to carbon disulfide," *Scand. J. Work Environ.* 19, 163, 1978.

5. Baker, E. L., Feldman, R. G., White, R. A., Harley, J. P., Niles, C. A., Dinse, G. E., and Berkey, C., "Occupational lead neurotoxicity: A behavioral and electrophysiological evaluation," *Br. J. Ind. Med.*, 41, 352, 1984.

6. Anger, W. K., "Neurobehavioral testing on chemicals: Impact on recommended standards," *Neurobehav. Toxicol. Teratol.*, 6, 147, 1984.

7. Berlin, M., Grant, C. A., Hellber, J., Hellstrom, J., and Schutz, A., "Neurotoxicity of methylmercury in squirrel monkeys," *Arch. Environ. Hlth.*, 30, 340, 1975.

8. Evans, H. L., Laties, V. G., and Weiss, B., "Behavioral effects of mercury and methylmercury," *Fed. Proc.* 34, 1858, 1975.

9. Merigan, W. H., Maurrisen, J. P.J., Weiss, B., Eskin, T., and Lapham, L. W., "Neurotoxic actions of methylmercury on the primate visual system," *Neurobehav. Toxicol. Teratol.*, 5, 649, 1983.

10. Hansson, H. A., "Selective effects of metabolic inhibitors on retinal culture," *Exp. Eye Res.*, 5, 335, 1966.

11. Braekevelt, C. R., "Morphological changes in rat retinal photoreceptors with acute methylmercury intoxication," in *The Structure of The Eye*, Hollyfield, J. G., Ed., Elsevier Biomedical, New York, 1982, Chapter 13.

12. Yonemaya, M., Sharma, R. P., and Kleinschuster, S. J., "Methylmercury and organogenesis *in vitro*: Inhibition of glutamine synthetase induction and alteration of selected cellular enzymes in aggregation of dissociated embryonic chick retinal cells," *Arch. Environ. Contam. Toxicol.* 12, 157, 1983.

13. Merigan, W. H., Barkdoll, E., Maurissen, J. P. J., Eskin, T. A., and Lapham, L. W., "Acrylamide effects on the macaque visual system. I. Psychophysics and electrophysiology," *Investig. Ophthalmol. Vis. Sci.*, 26, 309, 1985.

14. Eskin, T. A., Lapham, L. W., Maurissen, J. P. J., and Merigan, W. H., "Acrylamide effects on the macaque visual system. II. Retinogeniculate morphology," *Investig. Ophthalmol. Vis. Sci.*, 26, 317, 1985

15. Lynch, J. J., Silveira, L. C. L., Perry, V. H., and Merigan, W. H., "Visual effects of damage to P ganglion cells in macaques," *Vis. Neurosci.*, 8, 575, 1992.

16. Merigan, W. H. and Weiss, B., *Neurotoxicity of the Visual System*, Academic Press, New York, 1980.

17. Grant, W. M., *Toxicology of the Eye*, 3rd edition, Charles C. Thomas, Springfield, IL, 1986.

18. Chiou, G. C. Y., *Ophthalmic Toxicology*, Raven Press, New York, 1992.

19. Hockwin, O., Green, K. and Rubin, L. F., *Manual of Oculotoxicity Testing of Drugs*, Gustav Fisher Verlag, Stuttgart, 1992.

20. Basu, P. K., "Air pollution and the eye," *Surv. Ophthalmol.*, 17, 78, 1972.

21. Potts, A. M., "Toxic Responses of the Eye," in *Casarett and Doull's Toxicology: The Basic Science of Poisons*, 4th edition, Amdur, M. O., Doull, J., and Klaassen, C. D., Eds., Pergamon Press, New York, 1991, 521.

22. Way, J. L., Baxter, L., Cannon, E. P., and Pei, L., "Air pollution and ocular toxicology," in *Ophthalmic Toxicology*, Chiou, G. C. Y., Ed., Raven Press, New York, 1992, Chapter 8.

23. Way, J. L., Baxter, L., Cannon, E. P., and Pei, L., "Occupational hazards and ocular toxicology," in *Ophthalmic Toxicology*, Chiou, G. C. Y., Ed., Raven Press, New York, 1992, Chapter 9.

24. Evans, H. L., "Assessment of vision in behavioral toxicology," in *Nervous System Toxicology*, Mitchell, C. L., Ed., Raven Press, New York, 1982, 81.
25. Fox, D. A., Lowndes, H. E., and Bierkamper, G. G., "Electrophysiological techniques in neurotoxicology," in *Nervous System Toxicology*, Mitchell, C. L., Ed., Raven Press, New York, 1982, 299.
26. Chang, L. W., "Selected histopathological and histochemical methods for neurotoxicity assessment," in *Neurotoxicology: Approaches and Methods*, Chang, L. W. and Slikker, W., Jr., Eds., Academic Press, New York, 1995, Chapter 1.
27. Herr, D. W. and Boyes, W. K., "Electrophysiological analysis of complex brain systems: Sensory-evoked potentials and their generators," in *Neurotoxicology: Approaches and Methods*, Chang, L. W. and Slikker, W., Jr., Eds., Academic Press, New York, 1995, Chapter 9.
28. Maurissen, J. P. J., "Neurobehavioral methods for the evaluation of sensory functions," in *Neurotoxicology: Approaches and Methods*, Chang, L. W. and Slikker, W., Jr., Eds., Academic Press, New York, 1995, Chapter 11.
29. Mergler, D., "Behavioral neurophysiology: Quantitative measures of sensory toxicity," in *Neurotoxicology: Approaches and Methods*, Chang, L. W. and Slikker, W., Jr., Eds., Academic Press, New York, 1995, Chapter 47.
30. Prosen, C. A. and Stebbins, W. C., "Ototoxicity," in *Experimental and Clinical Neurotoxicology*, Spencer, P. S. and Schaumburg, H. H., Eds., Williams & Wilkins, Baltimore, MD, 1980, Chapter 5.
31. Crofton, K. K., "Reflex modification and the assessment of sensory dysfunction," in *Neurotoxicology*, Tilson, H. and Mitchell, C. L., Eds., Raven Press, New York, 1992, 181.
32. Schiffman, S. S. and Nagle, H. T., "Effect of environmental pollutants on taste and smell," *Otolaryngol. Head Neck Surg.* 106, 693, 1992.
33. Shusterman, D. J. and Sheedy, J. E., "Occupational and environmental disorders of the special senses," *Occup. Med.*, 7, 515, 1992.
34. Henley, C. M. and Rybak, L. P., "Ototoxicity in developing mammals," *Brain Res. Rev. 20*, 68, 1995.
35. National Institute of Occupational Safety and Health, "Occupational Exposure to Carbon Disulfide." Dept. HEW, Publ. No. 77-156, Washington, DC, 1977.
36. Raitta, C., Tolonen, M., and Nurminen, M., "Microcirculation of ocular fundus in viscose rayon workers exposed to carbon disulfide," *Arch. Klin. Exp. Ophthalmol.*, 191, 151, 1974.
37. Seppalainen, A. M. and Haltia, M., "Carbon disulfide," in *Experimental and Clinical Neurotoxicology*, Spencer, P. S. and Schaumburg, H. H., Eds., Williams & Wilkins, Baltimore, MD, 1980, Chapter 25.
38. Hotta, R. and Gotto, S., "A fluorescein angiographic study on micro-angiopathia sulfocarbonica," *Jpn. J. Ophthalmol.*, 15, 132, 1971.
39. De Rouck, A., De Laey, J. J., Van Hoorne, M., Pahtak, A., and Devuyst, A., "Chronic carbon disulfide poisoning: A 4 year follow-up study of the ophthalmological signs," *Int. Ophthalmol.*, 9, 17, 1986.
40. McDonald, R. W., "Carbon disulfide poisoning," *Arch. Ophthalmol.*, 20, 839, 1938.
41. Gordy, S. T. and Trumper, M., "Carbon disulfide poisoning," *Ind. Med.*, 9, 231, 1940.
42. Savic, S. M., "Influence of carbon disulfide on the eye," *Arch. Environ. Hlth.*, 14, 325, 1967.
43. Raitta, C., Teir, H., Nurminen, M., Helpiuo, E., and Malmstrom, S., "Impaired color discrimination among viscose rayon workers exposed to carbon disulfide," *J. Occup. Med.*, 23, 189, 1981.

44. Palacz, O., Szymanska, K., and Czepita, D., "Studies on the ERG in subjects with chronic exposure to carbon disulfide: I. Assessment of the condition of the visual system taking into account ERG findings depending on the duration of exposure to CS_2," *Klin. Oczna*, 82, 65, 1980.

45. Palacz, O., Karczewicz, D., and Dzikowski, J., "Studies on the ERG in subjects with chronic exposure to carbon disulfide: II. Effects of Cynarex administration in ERG curves," *Klin. Oczna*, 82, 69, 1980.

46. Zenk, H., "CS_2 effects upon olfactory and auditory functions of employees in the synthetic fiber industry," *Int. Arch. Arbeitmed.*, 27, 210, 1970.

47. Ide, T., "Histopathological studies on retina, optic nerve and arachnoidal membrane of mouse exposed to carbon disulfide poisoning," *Acta Soc. Ophthalmol. Jpn.*, 62A, 85, 1958.

48. World Health Organization Environ. *Criteria 10, Carbon disulfide*, WHO, Geneva, Switzerland, 1979.

49. Papolla, M., Monaco, S., Weiss, H., Miller, C., Saheuk, Z, Autilo-Gambetti, L., and Gambetti, P., "Slow axonal transport in carbon disulfide giant axonopathy," *J. Neuropathol. Exp. Neurol.* 43, 305, 1984.

50. Eskin, T. A., Merigan, W. H., and Wood, R. W., "Carbon disulfide effects on the visual system. II. Retinogeniculate degeneration," *Investig. Ophthalmol. Vis. Sci.*, 29, 519, 1988.

51. Merigan, W. H., Wood, R. W., Zehl, D., and Eskin, T. A., "Carbon disulfide effects on the visual system. I. Visual thresholds and ophthalmoscopy," *Investig. Ophthalmol. Vis. Sci.*, 29, 512, 1988.

52. Woodhouse, J. M. and Barlow, H. B., "Spatial and temporal resolution and analysis," in *The Senses*, Barlow, H. B. and Mollon, J. D., Eds., Cambridge Univ. Press, Cambridge, UK, 1982, Chapter 8.

53. National Institute for Occupational Safety and Health (NIOSH), *Criteria for a Recommended Standard to Occupational Exposure to Trichloroethylene*, Dept. HEW, Washington, DC, 1973.

54. Feldman, R. G., "Trichloroethylene," in *Handbook of Clinical Neurology, Intoxications of the Nervous System*, Vol. 36, Part I, Vinken, P. J. and Bruyn, G. W., Eds., North-Holland Pub. Co., Amsterdam, 1979, 457.

55. Grandjean, E., Munchinger, R., Turrian, V., Haas, P. A., Knoepfel, H. K., and Rosemund, H., "Investigations into the effects of exposure to trichloroethylene in mechanical engineering," *Br. J. Ind. Med.*, 12, 131, 1955.

56. Tabbacchi, G., Corsico, R., and Gallinelli, R., "Retrobulbar neuritis caused by suspected chronic trichloroethylene poisoning," *Ann. Otalmol. Clin. Ocul.*, 92, 787, 1966.

57. Mitchell, A. B. S. and Parsons-Smith, B. G., "Trichloroethylene neuropathy," *Br. Med. J.*, 1, 422, 1969.

58. Lawrence, W. H. and Partyka, E. K., "Chronic dysphagia and trigeminal anesthesia after trichloroethylene exposure," *Ann. Intern. Med.*, 95, 710, 1981.

59. Barret, L., Arsac, P., Vincent, M., Faure, J., Garrel, S. and Reymond, F., "Evoked trigeminal potential in chronic trichloroethylene exposure," *J. Toxicol. Clin. Toxicol.*, 19, 419, 1982.

60. Winneke, G., "Acute behavioral effects of exposure to some organic solvents," *Acta Neurobiol. Exp.*, 66, 117, 1982.

61. Feldman, R. G., White, R. F., Currie, J. N., Travers, P. H., and Lessell, S., "Long-term follow-up after single toxic exposure to trichloroethylene," *Am. J. Ind. Med.*, 8, 119, 1985.

62. Vernon, R. J. and Ferguson, K., "Effects of trichloroethylene on visual-motor performance," *Arch. Environ. Hlth.*, 18, 894, 1969.

63. Larsby, B., Tham, R., Eriksson, B., Hyden, D., Odkvist, L., Liedgren, C., and Bunnfors, I., "Effects of trichloroethylene on the human vestibulo-occulomotor system," *Acta Otololaryngol.*, Stockholm, 101, 193, 1986.

64. Blain, L., Lachapelle, P., and Molotchnikoff, S., "The effect of acute trichloroethylene exposure on electroretinogram components," *Neurotoxicol. Teratol.* 12, 633, 1990.

65. Blain, L., Lachapelle, P., and Molotchnikoff, S., "Evoked potentials are modified by long-term exposure to trichloroethylene," *Neurotoxicology*, 13, 203, 1992.

66. Rebert, C. S., Day, V. L., Matteucci, M. J., and Pryor, G. T., "Sensory-evoked potentials in rats chronically exposed to trichloroethylene: Predominant auditory discrimination," *Neurotoxicol. Teratol.* 13, 83, 1991.

67. Mergler, D., Blain, L., and Lagace, J. P., "Solvent related colour vision loss: An indicator of damage?" *Int. Arch. Occup. Environ, Hlth.*, 59, 313, 1987.

68. Mergler, D., Belanger, S. De Grosbois, S., and Vachon, N., "Chromal focus of acquired chromatic discrimination loss and solvent exposure among printshop workers," *Toxicology*, 49, 341, 1988.

69. Mergler, D., Huel, G., Bowler, R., Frenette, B., and Cone, J., "Visual dysfunction among former microelectronics assembly workers," *Arch. Environ, Hlth.*, 46, 326, 1991.

70. Frenette, B., Mergler, D., and Bowler, R., "Contrast-sensitivity loss in a group of former microelectronics workers with normal visual acuity," *Optom. Vis. Sci.*, 68, 556, 1991.

71. Nakatsuka, H., Watanabe, T., Takeuchi, Y., Hisanga, N., Shibatta, E., Suzuki, H. Huang, M. Y., Chen, Z., Qu, Q. S., and Ikeda, M., "Absence of color vision loss among workers exposed to toluene or tetrachlorethylene, mostly at levels below occupational exposure limits," *J. Occup. Env. Hlth.*, 64, 113, 1990.

72. Baird, B., Camp, J., Daniell, W., and Antonelli, J., "Solvents and color discrimination ability," *J. Occup. Med.*, 36, 747, 1994.

73. Muttray, A., Wolters, V., Mayer-Popken, O., Schicketanz, K. H., and Konietzko, J., "Effect of subacute occupational exposure to toluene on color vision," *Int. J. Occup. Med. Environ. Hlth.*, 8, 339-345, 1995.

74. Zavalic, M., Turk, R., Bogadi-Sare, A., and Skender, L., "Colour vision impairment in workers exposed to low concentrations of toluene," *Arch. Hig. Rada Toksikol.*, 47, 167-175, 1996.

75. Gobba, F., Galassi, C., Imbriani, M., Ghittori, S., and Cavalleri, A., "Acquired dyschromatopsia among styrene-exposed workers," *J. Occup. Med.*, 33, 761-765, 1991.

76. Fallas, C., Fallas, J., Maslard, P., and Dally, S., "Subclinical impairment of colour vision among workers exposed to styrene," *Br. J. Ind. Med.*, 49, 679-682, 1992.

77. Chia, S. E., Jeyaratnam, J., Ong, C. N., Ng, T. P., and Lee, H. S., "Impairment of color vision among workers exposed to low concentrations of styrene," *Am. J. Ind. Med.*, 26, 481-488, 1994.

78. Eguchi, T., Kishi, R., Harabuchi, I., Yuasa, J., Arata, Y., Katakura, Y., and Mitake, H., "Impaired colour discrimination among workers exposed to styrene: relevance of a urinary metabolite," *Occup. Environ. Med.*, 52, 534-538, 1995.

79. Campagna, D., Mergler, D., Huel, G., Belanger, S., Truchon, G., Ostiguy, C., and Drolet, D., "Visual dysfunction among styrene-exposed workers," *Scand. J. Work Environ. Hlth.*, 21, 382-390, 1995.

80. Campagna, D., Gobba, F., Mergler, D., Moreau, T., Galassi, C., Cavalleri, A., and Huel, G., "Color vision loss among styrene-exposed workers: neurotoxicological threshold assessment," *Neurotoxicology*, 17, 367-373, 1996.

81. Porkony, J., Smith, V. S., Verriest, G., and Pinckers, A., *Congenital and Acquired Color Vision Defects*, Grune & Stratton, New York, 1979.

82. Guguchkova, P. T., "Electroretinographic and electrooculographic examinations of persons occupationally exposed to lead," *Vestnik. Oftalmol.*, 85, 60, 1972

83. Signorino, M., Scarpino, O., Provincialli, L., Marchesi, G. F., Valentino, M., and Governa, M., "Modification of the electroretinogram and of different components of the visual evoked potentials in workers exposed to lead," *Ital. Electroenceph. J.*, 10, 51, 1983.

84. Cavalleri, A., Trimarchi, F., Gelmi, C., Baruffini, A., Minoia, C., Biscaldi, G., and Gallo, G., "Effects of lead on the visual system of occupationally exposed subjects," *Scand. J. Work Environ. Hlth.*, 8 (Suppl. 1), 148, 1982.

84. Betta, A., De Santa, A., Savonitto, C., and D'Andrea, F., "Flicker fusion test and occupational toxicology performance evaluation in workers exposed to lead and solvents," *Human Toxicol.*, 2, 83, 1983.

86. Campara, P., D'Andrea, F., Micciolo, R., Savonitto, C., Tansella, M., and Zimmer-mann-Tansella, C., "Psychological performance of workers with blood-lead concentration below the current threshold limit value," *Int. Arch. Occup. Environ. Hlth.*, 53, 233, 1984.

87. Jeyaratnam, J., Boey, K. W., Ong, C. N., Chia, C. B., and Phoon, W. O., "Neuropsychological studies on lead workers in Singapore," *Br. J. Ind. Med.*, 43, 626, 1986.

88. Williamson, A. M. and Teo, R. K. C., "Neurobehavioral effects of occupational exposure to lead," *Br. J. Ind. Med.*, 43, 374, 1986.

89. Sborgia, G., Assennato, G., L'Abbate, N., DeMarinis, L., Paci, C., DeNicolo, M., DeMarinis, G., Montrone, N., Ferrannini, E., Specchio, L., Masi, G., and Olivieri, G., "Comprehensive neurophysiological evaluation of lead-exposed workers," in *Neurobehavioral Methods in Occupational Hlth.*, Gilioli, R., Cassito, M., and Foà, V., Eds., Pergamon Press, New York, 1983, 283.

90. Specchio, L. M., Bellomo, R., Dicuonzo, F., Assennato, G., Federici, A., Miscaigna, G., and Puca, F. M., "Smooth pursuit eye movements among storage battery workers," *Clin. Toxicol.*, 18, 1269, 1981.

91. Baloh, R. W., Spivey, G. H., Brown, C. P., Morgan, D., Campion, D. S., Browdy, B. L., Valentine, J. L., Gonick, H. C., Massey, F. J., and Culver, B. D., "Subclinical effects of chronic increased lead absorption — a prospective study. II. Results of baseline neurologic testing," *J. Occup. Med.*, 21, 490, 1979.

92. Baloh, R. W., Langhofer, L., Brown, C. P., and Spivey, G. H., "Quantitative tracking tests in lead workers," *Am. J. Med.*, 1, 109, 1980.

85. Spivey, G. H., Baloh, R. W., Brown, C. P., Browdy, B. L., Campion, D. S., Valentine, J. L., Morgan, D., and Culver, B. D., "Subclinical effects of chronic increased lead absorption — a prospective study. III. Neurologic findings at follow-up examination," *J. Occup. Med.*, 22, 607, 1980.

93. Hanninen, H., Mantere, P., Hernberg, S., Seppalainen, A. M., and Kock, B., "Subjective symptoms in low-level exposure to lead," *Neurotoxicology*, 1, 333, 1979

94. Balzano, I., "Chronic lead poisoning and changes in the internal ear," *Rass. Med. Ind.*, 21, 320, 1952.

95. Koch, C. and Serra, M., "The effects of tetraethyllead poisoning on hearing and vestibular systems," *Acta Med. Ital. Med. Trop. Subtrop. Gastroenterol.*, 17, 77, 1962.

96. Gammarrota, M. and Bartoli, E., "Considerations on relations between lead intoxication and cochlear defect," *Clin. Otorhinolaringoratrica*, 16, 136, 1964.

97. Valcie, I. and Monojlovic, C., "Results obtained in examination of a group of workers in the chemical industry in regard to effects of hearing," in *XVI Int. Congress Occupational Health*, Tokyo, Japan, 1969.

98. Repko, J. D., Morgan, B. B., Jr., and Nicholson, J. A., "Behavioral effects of occupational exposure to lead," in *NIOSH Research Report Hlth., Education, and Welfare Publication*, (NIOSH). 1975, 75.

99. Sherer, J. W., "Lead amblyopia with cataract from the same source," *J. Missouri State Med. Assoc.*, 32, 275, 1935.

100. Belova, S. F., "The toxic effect of lead on the optic nerve," *Vestnik. Oftalmol.*, 78, 43, 1965.

101. Baghdassarian, S. A., "Optic neuropathy due to lead poisoning," *Arch. Ophthalmol.*, 80, 721, 1968.

102. Karai, I., Horiguchi, S. H., and Nishikawa, N., "Optic atrophy with visual field defect in workers occupationally exposed to lead for 30 years," *J. Toxicol. Clin. Toxicol.*, 19, 409, 1982.

110. Gruszczynska, M., "Effect of lead on the visual system," *Klin. Oczna.*, 91, 76, 1989.

104. Alpern, M., "Eye movements and strabismus," in *The Senses*, Barlow, H. B. and Mollon, J. D., Eds., Cambridge Univ. Press, Cambridge, 1982, Chapter 11.

105. Hass, G. M., Brown, D. V. L., Eisenstein, R., and Hemmens, A., "Relation between lead poisoning in rabbit and man," *Am. J. Pathol.*, 45, 691, 1964.

106. Brown, D. V. L., "Reaction of the rabbit retinal pigment epithelium to systemic lead poisoning," *Tr. Am. Ophthalmol. Soc.*, 72, 404, 1974.

107. Hughes, W. F. and Coogan, P. S., "Pathology of the pigment epithelium and retina in rabbits poisoned with lead," *Am. J Pathol.*, 77, 237, 1974.

108. Fox, D. A., Campbell, M. L., and Blocker, Y. S., "Functional alterations and apoptotic cell death in the retina following developmental or adult lead exposure," *Neurotoxicology*, 18, 645-665, 1997.

109. Gozdzik-Zolnierkiewicz, T. and Moszynski, B., "Eighth nerve in experimental lead poisoning," *Acta Otolaryngol.*, 68, 85, 1969.

110. Van Gelder, G. A., Carson, T., Smith, R. M., and Buck, W. B., "Behavioral toxicologic assessment of the neurologic effect of lead in sheep," *Clin. Toxicol.*, 6, 405, 1973.

111. Yamamura, K., Terayama, K., Yamamoto, N., Kohyama, A., and Kishi, R., "Effects of acute lead acetate exposure on adult guinea pigs electrophysiological study of the inner ear," *Fund. Appl. Toxicol.*, 13, 509, 1989.

112. Fox, D. A. and Sillman, A. J., "Heavy metals affect rod, but not cone photoreceptors," *Science*, 206, 78, 1979.

113. Sillman, A. J., Bolnick, D. A., Bosetti, J. B., Haynes, L. W. and Walter, A. E., "The effects of lead and cadmium on the mass photoreceptor potential the dose-response relationship," *Neurotoxicology*, 3, 179, 1982.

114. Sillman, A. J., Bolnick, D. A., Bosetti, J. B., Haynes, L. W., and Walter, A. E., "The effect of lead on photoreceptor response amplitude — influence of the light stimulus," *Exp. Eye Res.*, 39, 183, 1984.

115. Sillman, A. J., Bolnick, D. A., Bosetti, J. B., Haynes, L. W., and Walter, A. E., "The effect of lead on photoresponse amplitude — influence of removing external calcium and bleaching rhodopsin," *Neurotoxicology*, 7, 1, 1986.

116. Tessier-Lavigne, M., Mobbs, P., and Attwell, D., "Lead and mercury toxicity and the rod light response," *Invest. Ophthalmol. Vis. Sci.*, 26, 1117, 1985.

117. Frumkes, T. E. and Eysteinsson, T., "The cellular basis for suppressive rod-cone interaction," *Vis. Neurosci.*, 1, 263, 1988.

118. Eysteinsson, T. and Frumkes, T. E., "Physiological and pharmacological analysis of suppressive rod-cone interaction in Necturus retina," *J. Physiol.*, 61, 866, 1989.

119. Fox, D. A., Srivastava, D., Hurwitz, R. L., "Lead-induced alterations in rod-mediated visual functions and cGMP metabolism: New insights," *Neurotoxicology*, 15, 503, 1994.

120. Medrano, C. J. and Fox, D. A., "Substrate-dependent effects of calcium on rat retinal mitochondria respiration: Physiological and toxicological studies," *Toxicol. Appl. Pharmacol.* 125, 309, 1994.

121. Medrano, C. J. and Fox, D. A., "Oxygen consumption in rat outer and inner retina: Light- and pharmacologically-induced inhibition," *Exp. Eye Res.*, 61, 273, 1995.

122. Srivastava, D., Fox, D. A., and Hurwitz, R. L., "Effects of magnesium on cyclic GMP hydrolysis by the bovine retinal rod cyclic GMP phosphodiesterase," *Biochem. J.*, 308, 653, 1995.

123. Srivastava, D., Hurwitz, R. L., and Fox, D. A., "Lead- and calcium-mediated inhibition of bovine rod cGMP phosphodiesterase: Interactions with magnesium," *Toxicol. Appl. Pharmacol.*, 134, 43, 1995.

124. Fox, D. A. and Srivastava, D., "Molecular mechanisms of the lead-induced inhibition of rod cGMP phosphodiesterase," *Tox. Letts.*, 82/83, 263-270, 1995.

125. Fox, D. A., "Visual and auditory system alterations following developmental or adult lead exposure: A critical review," in *Human Lead Exposure*, H. L. Needleman, Ed., CRC Press, Boca Raton, FL, 1992., Chapter 6.

126. Otto D. and Fox, D. A., "Auditory and visual dysfunction following lead exposure," *Neurotoxicology*, 14, 191, 1993.

127. Zrenner, E. and Gouras, P., "Blue-sensitive cones of the cat produce a rod like electroretinogram," *Invest. Ophthalmol. Vis. Sci.*, 18, 1076, 1979.

128. Zrenner, E., Kramer, W., Bittner, C., Bopp, M., and Schlepper, M., "Rapid effects on colour vision following intravenous injection of a new, non-glycoside positive ionotropic substance (AR-L 115 BS)," *Doc. Ophthalmol. Proc. Ser.*, 33, 493, 1982.

129. Dowd, T. L., Rosen, J. F., and Gupta, R. K., "^{31}P NMR and saturation transfer studies of the effect of Pb^{2+} on cultured osteoblastic bone cells," *J. Biol. Chem.*, 256, 20833, 1990.

130. Flora, S. J. S., Kumar, D., Sachan, S. R. S., and Gupta, S. D., "Combined exposure to lead and ethanol on tissue concentration of essential metals and some biochemical indices in rat," *Biol. Trace Elem. Res.*, 28, 157, 1991.

131. Wyllie, A. H., Kerr, J. F. R., and Currie, A. R., "Cell death: The significance of apoptosis," *Int. Rev. Cytol.*, 68, 251, 1980.

132. Fox, D. A., Rubinstein, S. D., and Hsu, P., "Lead inhibits rat retinal, but not kidney, Na$^+$ K$^+$-ATPase," *Toxicol. Appl. Pharmacol.*, 109, 482, 1991.

133. Wolburg, H., "Time- and dose-dependent influence of ouabain on the ultrastructure of optic neurones," *Cell Tissue Res.*, 164, 503, 1975.

134. Wolburg, H., "Axonal transport degeneration and regeneration in the visual system of the goldfish," *Adv. Anat. Embryol. Cell. Biol.*, 67, 1, 1981.

135. Frank, R. N. and Goldsmth, T. H., "Effects of cardiac glycosides on electrical activity in isolated retina of the frog," *J. Gen. Physiol.*, 50, 1585, 1967.

136. Torre, V., "The contribution of the electrogenic sodium-potassium pump to the electrical activity of toad rods," *J. Physiol.*, 333, 315, 1982.

9 Electrophysiological Approaches to Occupational Neurotoxicology

Anna Maria Seppäläinen

CONTENTS

1 INTRODUCTION

Electrophysiological methods are sensitive indicators of nervous system damage and defective function. They are mostly noninvasive and well tolerated. Usually the subject cannot voluntarily affect the results, which thus can be regarded as objective and quite often quantitative measures. Mild neurotoxic effects can be overlooked in purely clinical examination, although the subject may complain of various subjective symptoms. The symptoms, on the other hand, may be very nonspecific. As neurotoxic effects can be regarded adverse even when they are mild, it is important to reveal them at a stage when they still are reversible. Electrophysiological methods can be applied to groups of exposed workers, and thus are suitable for epidemiological studies to reveal exposure–effect relationships as well as to reveal mild adverse effects in a certain work-site or at certain exposure levels.

Various electrophysiological methods have been used in epidemiological studies among groups of workers with a specific exposure to reveal neurotoxic effects as a group level. These types of studies have helped in determining safe exposure levels, and have offered guidelines for threshold limit values. Electrophysiological methods have been used in individual diagnosis of occupational diseases similarly as in the assessment of other diseases and injuries of the nervous system. Some authors have suggested electrophysiological methods as screening methods at work sites for early detection of neurotoxicity.

Electroencephalography (EEG) was the earliest electrophysiological test to be applied in industrial medicine, and in the beginning to ascertain acute neurotoxicity.[1-2] The next step was to reveal neurotoxicity in the peripheral nervous system by nerve conduction studies in occupational[3] and experimental[4] lead poisoning as well as in carbon disulfide exposure.[5] Neuropathy at higher exposure levels was revealed by reduced motor conduction velocities (MCVs). Higher sensitivity was reached by measuring the conduction velocity of slower motor fibers with a collision technique.[6] Electromyography has been applied mainly in clinical cases of occupational diseases, and later on sensory conduction velocity (SCV) studies have been found more sensitive in revealing especially axonal neuropathy.[7] In recent years various evoked or event-related potentials have been applied in occupational neurotoxicology.[8]

2 NEUROTOXICITY STUDIES

2.1 SOLVENTS

Many hydrocarbon and chlorinated solvents have nervous system effects. Often symptoms, which as such are nonspecific, refer to the central nervous system, but the peripheral nervous system is affected as well.

2.1.1 Carbon Disulfide

With long-term exposure at low concentrations, carbon disulfide (CS_2) predominantly causes neuropathy, although various psychic symptoms also have been usual. Several papers appeared from 1969 on reporting slowed nerve conduction velocities among viscose workers exposed to CS_2 and impaired MCVs.[5,9-12] Neuropathological evidence of axonal degeneration with axonal swellings filled with neurofilaments was also shown in experimental exposure of rats and rabbits.[9] Workers with long-term exposure to concentrations of 20 to 80 ppm CS_2 have slower MCVs and SCVs, especially in the distal parts, as well as electromyographic signs of axonal neuropathy. When 118 viscose workers exposed for median 15 years (for the last 10 years mostly to less than 10 ppm, and before that at times to 20 ppm of CS_2), were compared to 100 paper mill workers, the MCVs especially of the leg nerves were significantly slower.[10] The conduction velocity of the slower motor fibers was more

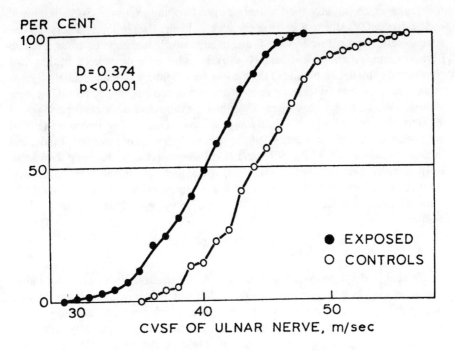

FIGURE 1 Cumulative percentage distribution of the conduction velocity of slower motor fibers (CVSF) of the ulnar nerve among workers with long-term occupational exposure to carbon disulfide and age-matched controls. (From Seppäläinen AM and Tolonen M. *Work Environ Health* 11, 145, 1974. With permission.)

sensitive in showing the difference in the ulnar nerve as well (Figure 1).[10] Small nervous system effects may occur after 40 years of exposure to concentrations below 10 ppm as revealed by conduction velocity of the slow fibers measured by using the refractory period method.[12] Nonspecific EEG abnormalities have been reported among workers with chronic CS_2 as well as impairment of the neuromuscular junction.[9] Recently, prolongation was found in the main components and interpeak latencies of brainstem auditory evoked potentials (BAEP) among current spinning workers with CS_2 exposure for more than 240 months.[13]

2.1.2 Hexacarbons

Hydrocarbon solvents with six carbon atoms, like *n*-hexane and methyl-*n*-butyl ketone are neurotoxicants. Neuropathologically, they cause focal axonal swelling associated with accumulations of neurofilaments, especially in the distal regions of long peripheral and central nerve fibers.[14] Clinical symptoms are motor weakness accentuated in the distal parts of the limbs and sensory deficits. The first reports of electromyographic examinations came from Japan[15] followed by MCV slowing in American factory workers.[16] Hexacarbon neuropathy has not disappeared. Serious polyneuropathy cases were reported from Taiwan. These patients regained full motor functions, which initially even worsened after stopping exposure. SCVs improved earlier than MCVs.[17-18] In a study of 56 workers in an offset printing factory, 20 had symptomatic neuropathy, 26 were found to have subclinical neuropathy as revealed by nerve conduction studies, and 10 workers, who were considered healthy had significantly smaller amplitude of the median nerve sensory action potentials. Symptomatic subjects had motor action potentials of reduced amplitude and MCVs were slowed.[19] Workers, who had been occasionally exposed to peak concentrations of 2000 to 3000 ppm of *n*-hexane, had low-amplitude visual evoked potentials (VEPs) with delayed early components.[20] Pattern reversal VEP was of lower amplitude, and interpeak latencies of VEPs, SEPs and BAEPs were prolonged among Taiwanese workers with *n*-hexane polyneuropathy and also among subclinical cases.[17] Similarly, short-latency components of median nerve somatosensory evoked potentials (SEP) were delayed and cortical peaks of SEPs were of low amplitude among Italian shoe-workers.[21]

2.1.3 Toluene and Xylene

Toluene and xylene are often used together in various thinners and other industrial solvents. They both are considered rather safe solvents, and central nervous system damage has occurred mainly after abuse, which may have started in occupational exposure. Excessive slow activity either diffusely or episodically in the EEG of hospitalised patients has been reported after toluene abuse.[22] Car painters exposed in the mean 14.8 yrs to mixtures of organic solvents, the greatest component of which was toluene, did not differ as a group level from their referents concerning EEG abnormalities, MCVs or SCVs, although abnormally slow nerve conduction velocities were detected only among a few solvent exposed workers, not among referents.[7] MCV and SCV of the median nerve in the palm was affected only among 10 printers exposed to toluene. These printers had a lower coefficient of variation in electrocardiographic R-R intervals indicating autonomous system effects.[23] Rotogravure printers exposed to toluene had prolonged latencies of P1 of pattern reversal VEP as well as VEPs of reduced amplitude.[24] Similarly, eight out of 36 printers with severe exposure to toluene had some abnormalities in SEPs at peripheral or spinal and cortical levels (three among them having prolonged central conduction times with normal peripheral conduction).[25]

Xylene is rarely used alone, usually only as a component of solvent mixtures. Only experimental electrophysiological human studies exist, and those report on minor changes during short-term exposure. An increased amount of scattered slow waves in the posterior regions were detected in the EEG during exposure peaks of 400 ppm of xylene combined with periods of ergometer exercise. This was interpreted as a slight lowering of vigilance.[26] A more recent study with automatic analysis of EEG in experimental xylene exposure showed that exercise increased delta and theta power, but this increase was partly counteracted by xylene towards the end of peak exposure to 400 ppm. Increased percentage and higher frequencies of alpha activity toward the later phases of exposure seemed to be the only changes attributable to xylene exposure.[27]

When nine young men were exposed to constant concentrations of 150 and 300 ppm of xylene with or without simultaneous oral alcohol intake, alcohol increased latencies P50, N60, P115 and N155 of the flash-VEP, but xylene seemed to counteract these alcohol effects. On the other hand, 300 ppm of xylene together with alcohol intake reduced the amplitude N60-P115.[28] The latency N135 of the pattern reversal VEP decreased after exposure to peaks of 400 ppm of xylene combined with exercise, and the latency P210 of the flash-VEP decreased in both stable exposure to 200 ppm and fluctuating peak exposure conditions with exercise.[29] No changes were noted in SEPs or BAEPs during these exposure conditions.[28-29]

2.1.4   Styrene

Styrene is used especially in the reinforced plastic industry. Styrene seems to affect mainly the central nervous system. Abnormal EEGs — mainly mild slow wave abnormalities either locally or diffusely — were found among 23 out of 96 male workers exposed to styrene, and the mandelic acid concentration in the urine was significantly higher for those with abnormal EEGs. A certain threshold for EEG response was noted at the level of 700 mg/g^3 creatinine of mandelic acid.[30] Quantitative EEG studies of 99 workers occupationally exposed to styrene showed a significant increase of abnormal EEG classifications (Neurometrics statistical normative database comparisons) in workers with higher exposure. Those with higher subject exposure had higher absolute EEG power in alpha and beta bands in the frontotemporal regions.[31]

Slowed MCVs of radial and peroneal nerves were found among few styrene exposed workers,[32] and some workers with very low long-term exposure had sensory action potentials of low amplitude.[33] Mild sensory nerve conduction deficits (SCVs slower than mean — 1 SD of controls) were found in 23% in those exposed to less than 50 ppm and in 71% in those exposed to more than 100 ppm of styrene.[34] No definitely abnormal SCVs were detected in that study. SEP studies showed significantly prolonged latencies both peripherally and centrally among 20 workers with severe styrene exposure, two of them having only central delays.[25]

2.1.5   Halogenated hydrocarbons

Many halogenated hydrocarbons have been employed for anesthesia and analgesia. Trichloroethylene has caused trigeminal neuropathy after anesthesia, and abnormalities have been reported in trigeminal SEPs.[35-36] Abnormalities were latency delays and increased stimulus thresholds for obtaining responses.

During the workshift, trichloroethylene induced slow waves in the EEG measured telemetrically, and there also was some alpha activation.[37] Alpha percentage increased in experimental exposure to 95 ppm of trichloroethylene after one and two hours of exposure.[38]

Methylene chloride induced EEG abnormalities more pronounced in the frontal and precentral areas. Abnormalities were noted after a weekend without exposure.[39] Changes in VEPs after one hour of exposure to methylene chloride were attributed

TABLE 1
Abnormal EEGs (Percentages) among
Solvent Exposed Workers in Various
Studies Conducted at the Institute of
Occupational Health in Helsinki

Number of workers	%	Reference
72 painters in the building industry	17	44
102 car painters	31	7
107 patients with solvent poisoning	65	42
87 patients with solvent poisoning	67	43

to carbon monoxide, which is one metabolic product of methylene chloride.[40] The same explanation could apply for EEG abnormalities as well.

1,1,1-Trichloroethane did not cause neuropathy according to previous reports,[8] but a recent paper reported on low-amplitude sensory action potentials of the sural nerve in a woman exposed at work to 1,1,1-trichloroethane. The symptoms of neuropathy disappeared quickly after removal from work and findings partly improved with time.[41]

2.1.6   Solvent mixtures

Occupational exposure to solvents occurs most usually with various solvent mixtures. The components of the mixtures vary, some contain hexacarbons, some mainly aromatic hydrocarbons, at times also halogenated hydrocarbons. The concentrations in the inhaled air vary depending on ventilation and use of protective masks.

Symptoms among workers with solvent exposure often refer to the central nervous system. Several EEG studies have been reported at times combined to psychological tests.[8] In solvent poisoning — referred to often as psycho-organic syndrome — EEG abnormalities were frequently seen. For example, among 107 patients with a diagnosis of solvent poisoning, 65% had an abnormal EEG, usually consisting of mildly excessive amount of slow waves, and 54% of them exhibited excessive beta activity.[42] EEG abnormalities tended to improve in some patients after removal from exposure, although also new, especially paroxysmal abnormalities appeared in a follow-up study.[43] EEG abnormalities seem to appear in increasing number, when exposure is more intensive (Table 1). House painters showed only in 17% abnormal EEGs, which does not differ from EEG abnormalities in general populations.[44]

Computerized EEG frequency analysis showed no statistically significant differences when comparing car and industrial painters exposed to 30% of the contemporary TLV to a reference group; some tendencies to higher alpha power, smaller alpha bandwidth, lower delta power, and larger delta bandwidth were noted among exposed workers.[45] Power spectrum analysis of EEGs of 50 paint production workers showed significant differences in comparison to 50 sugar refinery workers; the power of the beta band was greater among the exposed subjects.[46]

Jonkman *et al.*[47] used the Neurometrics method, which takes the person's age into account, to study quantitative EEG in workers exposed to solvents or pesticides. Groups of young painters (30 to 40 years of age), old painters (55 to 72 years of age) and bulb growers (30 to 55 years of age) were compared to normal groups. Only old painters differed significantly from controls in detailed visual analysis, and they were also the only ones to differ significantly in multivariate measures of quantitative EEG. They showed in several leads less alpha power and more theta and beta power than the normal controls. Young painters had less delta in fronto-temporal leads and in some derivations higher coherences than the normal controls. Bulb-growers again had an increased amount of beta activity in the central and parieto-occipital leads.

Solvent mixtures which contain *n*-hexane especially cause neuropathy. MCVs and SCVs are abnormally slow among workers who complain of mild symptoms like paresthesiae or muscular pain.[42] Fibrillations are at times found in electromyography, and loss of motor units is noted during maximal voluntary contraction. Italian authors showed exposure-response relationships between glue exposure and polyneuropathy, and between exposure indexes and nerve conduction velocities.[48] In a Swedish epidemiological study[45] on 80 car and industrial spray painters with no relevant *n*-hexane exposure, the mean nerve conduction velocities of several nerves were slower and the sensory action potential of the sural nerve was lower among the solvent exposed than among the referents.

With time, nerve conduction velocities tend to improve when the exposure is stopped. This is achieved more easily if exposure has been short. With long-term exposure and when symptoms of solvent poisoning have developed, improvement is not a rule. On the contrary, in a follow-up study[43] neuropathic electromyographic signs were detected in 74% of 87 patients who had been diagnosed on the average 5.9 years earlier as having solvent poisoning. Most of them had been studied also around the time of the diagnosis, and some impairment in electroneuromyographic findings were detected as often as some improvement.

Evoked potential studies of shoe-workers with minimal polyneuropathy have shown decreased amplitudes of pattern reversal VEPs, of median nerve SEPs (Erb's point and N20) and often the amplitude ratio V/I of BAEP between 0.5 and 1.0.[49] Among workers exposed to a mixture of solvents for 1 to 37 years, but without clinical neurological symptoms, in 95% of 18 women and in 75% of 28 men at least one abnormality was found in three types of evoked potentials.[50] In SEPs the most frequent abnormalities were a delayed P22 component and a decrease in the peripheral conduction velocity. At group level, the amplitude of N20-P22 was significantly increased among women, but no significant differences were noted for pattern reversal VEPs or BAEPs. Another study also reported higher amplitude N20-P22 of the median nerve SEP among 50 patients, for whom a solvent induced psychoorganic syndrome was suspected.[51]

Event-related potentials of 12 individuals with a history of organic solvent exposure showed significantly delayed latencies of N250 and P300, when compared to 19 nonexposed controls. Moreover, the latencies of P300 were longer in the solvent-exposed group when compared to clinically stable outpatient schizophrenics.[52]

Sleep apnoea has recently been under keen interest. In 39% of 51 patients referred for evaluation of possible organic solvent encephalopathy, and 31% of 16 house painters had pathological sleep apnea (apnea index greater than 5). Twelve of the patients with abnormal apnea index were retested after 2 or more weeks without solvent exposure, and they showed a significant drop in the index.[53]

2.2 METALS

2.2.1 Lead

Lead poisoning has been recognized for hundreds of years. In children the main symptoms come from the central nervous system, convulsions being the most prominent features. A sign of occupational lead poisoning was lead palsy resulting for example in "drop hand." Slowing in a part of peroneal nerve fibers was reported by Catton *et al.*[54] Seppäläinen and Hernberg[6] found slowing in the slower motor fibers of the ulnar nerve, and significantly slower MCVs of the median, ulnar and peroneal nerves in a group of subjects including lead poisoning patients and subjects with excessive lead absorption. Since that time several research groups have shown exposure response relationships with long-term lead absorption measures and slowing of MCVs and SCVs.[8] Arm nerves seem to be affected earlier than leg nerves; the median and ulnar nerves already in the first year of lead exposed work, if the blood lead rose over 30 mg/100 mL according to a prospective study.[55] Some authors have suggested that SCVs are earlier affected than MCVs in lead exposure: others have found no changes in SCVs, while MCVs have slowed.[8] Negative reports about nerve conduction velocities and lead exposure have also been published. Autonomous nervous system effects of lead have been studied as well, and coeffecients of variation in the R-R interval were found significantly reduced among workers with elevated lead, zinc, and copper absorption. MCVs and SCVs of the median nerve were also slowed among these workers. It was speculated that lead can affect parasympathetic nerves and that zinc may antagonize this autonomic system dysfunction.[56]

BAEPs of workers, with blood lead concentrations around 50 μg/100 mL at the study time displayed significantly longer interpeak I-V latencies.[57-58] The mean interpeak III-V latency of BAEP was significantly prolonged among 15 lead workers compared to controls.[59] In the same study, latency N145 of VEP showed a positive correlation to duration of lead exposure, and negative correlations were found between MCV and SCV of the radial nerve and blood lead measures.

2.2.2 Mercury

Although 10 of 21 workers exposed to metallic mercury or its compounds had abnormally slow MCVs or SCVs (eight of those with abnormal values belonging to the highest exposure group) no significant differences were found at the group level in comparison to controls.[60] Another report described abnormal conduction velocities in subjects also drawing correlations between electrophysiological measures and mercury exposure indexes.[61] Distal motor latencies and reduced sensory response amplitudes have also been associated to mercury exposure.[8] Among 77 chloralkali workers, the SCV of the median nerve and the amplitude of the sural nerve were associated with measures of cumulative mercury exposure.[62]

Visually interpreted EEGs showed only a tendency towards an increased number of EEG abnormalities among 41 workers with long-term low-level exposure to mercury vapor in a chloroalkali plant. However, computerized EEG with FFT showed lower total power especially in alpha frequency and in the occipital areas.[63] Brainstem auditory evoked potentials of eight workers exposed to mercury showed longer I-V interpeak latencies of BAEP than among controls.[58]

2.2.3   Other metals

In a group of 70 workers in a copper smelting factory producing arsenic trioxide as a byproduct, there were subjects with reduced nerve conduction velocities and sensory amplitudes.[64] Thirty men at steel smelting working with low level exposure to manganese were compared to 60 nonexposed referents; no differences were found in EEG or BAEP, but in event related auditory evoked potential there was a tendency towards prolonged P300.[65]

2.3   Insecticides

Although low amplitude motor responses to nerve stimulation have been connected to exposure to organophosphorous compounds, single-fibre EMG did not reveal any defect in the neuromuscular transmission.[8] Exposure to chlorinated insecticides in a production plant caused epileptic seizures and EEG abnormalities, which disappeared within 2 to 6 months after exposure was terminated.[2,8]

More recently slightly decreased MCVs of the median and peroneal nerves and SCVs of the median and sural nerves as well as increased refractory periods in the leg nerves were found in 131 flower bulb farmers exposed to mixed pesticides and compared to 67 well-matched controls.[66] P300 to tone bursts was of longer latency for a group of 31 workers spraying an organophosphate pesticide as compared to age- and sex-matched hospital workers. Only one exposed worker had an abnormal P300.[67]

2.4   Selected Chemicals

Acrylamide monomer is neurotoxic. However, MCVs and SCVs of workers handling acrylamide have been reportedly normal, and only some low amplitude sensory responses have been found.[8] Methyl methacrylate caused local slowing of SCVs in the fingers of dental technicians.[8] Polychlorinated biphenyls (PCB) caused slowed MCVs and SCVs after ingestion of cooking oil contaminated with PCBs.[8] Reversible slowing of SCVs and decrease in the sensory amplitude were noted after extensive accidental exposure of firemen and cleaning workers, who after explosion of capacitors were exposed to PCBs and their degradation products.[68]

3   CONCLUSIONS

Clinical EEG studies have produced nonspecific evidence of encephalopathy especially in exposure to various solvents and solvent mixtures. Quantitative EEG studies have shown differences at the group level. Often solvent exposed workers have presented increases in beta activity and changes in alpha activity. Nerve conduction studies have given more precise information on peripheral nervous disturbances. Mostly neuropathy is axonal in type, and the resulting decrease in MCVs and SCVs is small, often detectable only at the group level. Recently, evoked potential studies have offered evidence of disturbances in the central nervous system, although SEP delays may reflect also peripheral neuropathy. Interestingly, some authors report on higher amplitude of cortical SEP components among exposed.

Electrophysiological methods do not offer an etiological diagnosis, although they give objective evidence of disturbances. Well designed epidemiological studies using these methods can reveal etiological factors, and some electrophysiological studies can also be applied to experimental studies on humans and animals.

4   REFERENCES

1. Chalupa, B., Synková, J., and Sevcik, M., "The assessment of electroencephalographic changes and memory disturbances in acute intoxications with industrial poisons," *Br. J. Ind. Med.*, 17, 238, 1960.
2. Hoogendam, I., Versteeg, J. P. J., and De Vlieger, M., "Electroencephalograms in insecticide toxicity," (in Italian) *Arch. Environ. Health*, 10, 441, 1965.
3. Sessa, T., Ferrari, E., and Colucci D'Amato, C., "Nerve conduction velocity in occupational lead poisoning," *Fol. Med.*, 48, 658, 1965.
4. Fullerton, P. M., "Chronic peripheral neuropathy produced by lead poisoning in guinea-pigs," *J. Neuropathol. Exp. Neurol.*, 25, 214, 1966.
5. Lukás, E., "Leitgeschwindigkeit peripherer Nerven bei Schwefelkohlenstoff ausgesetzten Personen," *Int. J. Clin. Pharmacol. Ther. Toxicol.*, 2, 354, 1969.
6. Seppäläinen, A. M., and Hernberg, S., "Sensitive technique for detecting subclinical lead neuropathy," *Br. J. Ind. Med.*, 29, 443, 1972.
7. Seppäläinen, A. M., Husman, K., and Mårtenson, C., "Neurophysiological effects of long-term exposure to a mixture of organic solvents," *Scand. J. Work Environ. Health*, 4, 304, 1978.
8. Seppäläinen, A. M., "Neurophysiological approaches to the detection of early neurotoxicity in humans," *CRC Crit. Rev. Toxicol.*, 18, 245, 1988.
9. Seppäläinen, A. M., and Haltia, M., "Carbon disulfide," in *Experimental and Clinical Neurotoxicology*, Spencer, P. S., and Schaumburg, H. H., Eds., Williams & Wilkins, Baltimore, MD, 1980, 356.
10. Seppäläinen, A. M. and Tolonen, M., "Neurotoxicity of long-term exposure to carbon disulfide in the viscose rayon industry. A neurophysiological study," *Work Environ. Health*, 11, 145, 1974.
11. Corsi, G., Maestrelli, P., Picotti, G., Manzoni, S., and Megrin, P., "Chronic peripheral neuropathy in workers with previous exposure to carbon disulphide," *Br. J. Ind. Med.*, 40, 209, 1983.
12. Ruijten, M. W., Salle, H. J., Verberk, M. M., and Muijser, H., "Special nerve functions and colour discrimination in workers with long term low level exposure to carbon disulphide," *Br. J. Ind. Med.*, 47, 589, 1990.
13. Hirata, M., Ogawa, Y., Okayama, A., and Goto, S., "A cross-sectional study on the brainstem auditory evoked potential among workers exposed to carbon disulfide," *Int. Arch. Occup. Environ. Health*, 64, 321, 1992.

14. Spencer, P. S. and Schaumburg, H. H., "Experimental neuropathy produced by 2,5-hexanedione — a major metabolite of the neurotoxic industrial solvent methyl n-butyl ketone," *J. Neurol. Neurosurg. Psychiatry,* 38, 771, 1975.

15. Iida, M., Yamamura, Y., and Sobue, I., "Electromyographic findings and conduction velocity on n-hexane polyneuropathy," *Electromyogr. Clin. Neurophysiol.,* 9, 247, 1969.

16. Allen, N., Mendell, J. R., Billmaier, D. J., Fontaine, R. E., and O'Neill, J., Jr., "Toxic polyneuropathy due to methyl n-butyl ketone," *Arch. Neurol.,* 32, 209, 1975.

17. Chang, Y. C., "Neurotoxic effects of n-hexane on the human central nervous system: evoked potential abnormalities in n-hexane polyneuropathy," *J. Neurol. Neurosurg. Psychiatry,* 50, 269, 1987.

18. Chang Y. C., "Patients with n-hexane induced polyneuropathy: a clinical follow up." *Br. J. Ind. Med.,* 47, 485, 1990.

19. Chang C. M., Yu, C. W., Fong, K. Y., Leung, S. Y., Tsin, T. W., Yu, Y. L., Cheung, T. F., and Chan, S. Y., "N-Hexane neuropathy in offset printers," *J. Neurol. Neurosurg. Psychiatry,* 56, 538, 1993.

20. Seppäläinen, A. M., Raitta, C., and Huuskonen, M. S., "n-Hexane-induced changes in visual evoked potentials and electroretinograms of industrial workers." *Electroencephalogr. Clin. Neurophysiol.,* 47, 492, 1979.

21. Mutti, A., Ferri, F., Lommi, G., Lotta, S., and Franchini, I., "n-Hexane-induced changes in nerve conduction velocities and somatosensory evoked potentials," *Int. Arch. Occup. Environ. Health,* 51, 45, 1982.

22. Knox, J. W. and Nelson, J. R., "Permanent encephalopathy from toluene inhalation," *N. Engl. J. Med.,* 275, 1494, 1966.

23. Murata, K., Araki, S., Yokoyama, K., Tanigawa, T., Yamashita, K., Okajima, F., Sakai, T., Matsunaga, C., and Suwa, K., "Cardiac autonomic dysfunction in rotogravure printers exposed to toluene in relation to peripheral nerve conduction," *Ind. Health,* 31, 79, 1993.

24. Urban, P. and Lukás, E., "Visual evoked potentials in rotogravure printers exposed to toluene," *Br. J. Ind. Med.,* 47, 819, 1990.

25. Stetkarova, I., Urban, P., Prochazka, B., and Lukás, E., "Somatosensory evoked potentials in workers exposed to toluene and styrene," *Br. J. Ind. Med.,* 50, 520, 1993.

26. Savolainen, K., Riihimäki, V., Seppäläinen, A. M., and Linnoila, M., "Effects of short-term n-xylene exposure and physical exercise on the central nervous system," *Int. Arch. Occup. Environ. Health,* 45, 105, 1980.

27. Seppäläinen, A. M., Laine, A., Salmi, T., Verkkala, E., Riihimäki, V., and Luukkonen, R., "Electroencephalographic findings during experimental human exposure to n-xylene," *Arch. Environ. Health,* 46, 16, 1991.

28. Seppäläinen, A. M., Savolainen, K., and Kovala, T., "Changes induced by xylene and alcohol in human evoked potentials," *Electroencephalogr. Clin. Neurophysiol.,* 51, 148, 1981.

29. Seppäläinen, A. M., Laine, A., Salmi, T., Riihimäki, V., and Verkkala, E, "Changes induced by short-term xylene exposure in human evoked potentials," *Int. Arch. Occup. Environ. Health,* 61, 443, 1989.

30. Seppäläinen, A. M. and Härkönen, H., "Neurophysiological findings among workers occupationally exposed to styrene," *Scand. J. Work Environ. Health,* 2, 140, 1976.

31. Matikainen, E., Forsman–Grönholm, L., Pfäffli, P., and Juntunen, J., "Nervous system effects of occupational exposure to styrene; a clinical and neurophysiological study," *Environ. Res.,* 61, 84, 1993.

32. Lilis, R., Lorimer, W. V., Diamond, S., and Selikoff, I. J., "Neurotoxicity of styrene in production and polymerization workers," *Environ. Res.,* 15, 133, 1978.

33. Rosén, I., Haeger-Aronsen, B., Rehnström, S., and Welinder, H., "Neurophysiological observations after chronic styrene exposure," *Scand. J. Work Environ. Health*, 4 (Suppl. 2), 184, 1978.

34. Cherry, N. and Gautrin, D., "Neurotoxic effects of styrene," *Br. J. Ind. Med.*, 47, 29, 1990.

35. Barret, L., Arsac, P., Vincent, M., and Faure, J., "Evoked trigeminal nerve potential in chronic trichloroethylene intoxication," *J. Toxicol. Clin. Toxicol.*, 19, 419, 1982.

36. Dogui, M., Mrizak, N., Yacoubi, M., Ali, B. B., and Paty, J., "Trigeminal somatosensory evoked potentials in workers handling trichloroethylene," *Neurophysiol. Clin.*, 21, 95, 1991.

37. Konietzko, H., Elster, I., Schomann, P., and Weichardt, H., "Felduntersuchungen in Lösungsmittelbetrieben. V. Mitteilung," *Zentralbl. Arbeitsmed. Arbeitsschutz*, 26, 60, 1976.

38. Konietzko, H., Elster, I., Bencsath, A., Drysch, K., and Weichart, H., "EEG-Veränderungen unter definierter Trichloräthylen-Exposition," *Int. Arch. Occup. Environ. Health*, 35, 257, 1975.

39. Hanke, C., Ruppe, K., and Otto, J., "Untersuchungsergebnisse zur Toxischen Wirkung von Dichlormethan bei Fussbodenlegern," *Zentralbl. Gesamt. Hyg. Grenzgeb.*, 20, 81, 1974.

40. Stewart, R. D., Fisher, T. N., Hosko, M. J., Peterson, J. E., Baretta, E. D., and Dodd, H. C., "Experimental human exposure to methylene chloride," *Arch. Environ. Health*, 25, 342, 1972.

41. House, R. A., Liss, G. M, and Wills, M. C., "Peripheral sensory neuropathy associated with 1,1,1-trichloroethane," *Arch. Environ. Health*, 49, 196, 1994.

42. Seppäläinen, A. M., Lindström, K., and Martelin, T., "Neurophysiological and psychological picture of solvent poisoning," *Am. J. Ind. Med.*, 1, 31, 1980.

43. Seppäläinen, A. M. and Antti-Poika, M., "Time course of electrophysiological findings for patients with solvent poisoning, *Scand. J. Work Environ. Health*, 9, 15, 1983.

44. Seppäläinen, A. M. and Lindström, K., "Neurophysiological findings among house painters exposed to solvents," *Scand. J. Work Environ. Health*, 8 (Suppl. 1), 131, 1982.

45. Elofsson, S. A., Gamberale, F., Hindmarsch, T., Iregren, A., Isaksson, A., Johnsson. I., Knave, B., Lydahl, E., Mindus, P., Persson, H. E., Philipson, B., Steby, M., Struwe, G., Söderman, E., Wennberg, A., and Widén, A., "Exposure to organic solvents: a cross-sectional epidemiologic investigation on occupationally exposed car and industrial spray painters with special reference to the nervous system," *Scand. J. Work Environ. Health*, 6, 239, 1980.

46. Orbaek, P., Risberg, J., Rosén, I., Haeger-Aronsen, B., Hagstadius, S., Hjorstberg, U., Regnell. G., Rehnström, S., Svensson, K., and Welinder, H., "Effects of long-term exposure to solvents in the paint industry: a cross-sectional epidemiologic study with clinical and laboratory methods," *Scand. J. Work Environ. Health*, 11 (Suppl. 2), 1, 1985.

47. Jonkman, E. J., De Weerd, A. W., Poortvliet, D. C. J., Veldhizen, R. J., and Emmen, H., "Electroencephalographic studies in workers exposed to solvents or pesticides," *Electroencephalogr. Clin. Neurophysiol.*, 82, 438, 1992.

48. Buiatti, E., Cecchini, S., Ronchi, O., Dolara, P., and Bulgarelli, G., "Relationship between clinical and electromyographic findings and exposure to solvents in shoe and leather workers," *Br. J. Ind. Med.*, 35, 168, 1978.

49. Nolfe, G., Palma, V., Guadagnino, M., Serra, L. L., and Serra, C., "Evoked potentials in shoe-workers with minimal polyneuropathy," *Electromyogr. Clin. Neurophysiol.*, 31, 157, 1991.

50. Lille, F., Margules, S., Mallet, A., Deschamps, D., Garnier, R., and Dally, S., "Evoked potentials in workers occupationally exposed to organic solvents," *Electromyogr. Clin. Neurophysiol.,* 33, 279, 1993.

51. Deschamps, D., Garnier, R., Lille, F., Tran-Dinh, Y., Bertaux, L., Reygagne, A., and Dally S., "Evoked potentials and cerebral blood flow in solvent induced psycho-organic syndrome," *Br. J. Ind. Med.,* 50, 325, 1993.

52. Morrow, L. A., Steinhauer, S. R., Hodgson, M. J., "Delay in P300 latency in patients with organic solvent exposure," *Arch. Neurol.,* 49, 315, 1992.

53. Monstad, P., Mellgren, S. I., and Sulg, A., "The clinical significance of sleep apnoea in workers exposed to organic solvents: implications for the diagnosis of organic solvent encephalopathy," *J. Neurol.,* 239, 195, 1992.

54. Catton, M. J., Harrison, M. J. G., Fullerton, P. M., and Kazantzis, G., "Subclinical neuropathy in lead workers," *Br. Med. J.,*2, 80, 1970.

55. Seppäläinen,. A. M., Hernberg, S., Vesanto, R., and Kock, B., "Early effects of occupational lead exposure: a prospective study," *Neurotoxicology,* 4, 181, 1983.

56. Murata, K and Araki, S., "Autonomic nervous system dysfunction in workers exposed to lead, zinc, and copper in relation to peripheral nerve conduction: a study of R-R interval variability," *Am. J. Ind. Med.,* 20, 663, 1991.

57. Discalzi, G. L., Capellaro, F., Bottalo, L., Fabbro, D., and Mocellini, A., "Auditory brainstem evoked potentials (BAEPs) in lead-exposed workers," *Neurotoxicology,* 13, 207, 1992.

58. Discalzi, G., Fabbro, D., Meliga, F., Mocellini, A., and Capellaro, F., "Effects of occupational exposure to mercury and lead on brainstem auditory evoked potentials," *Int. J. Psychophysiol.,* 14, 21, 1993.

59. Hirata, M. and Kosaka, H., "Effects of lead exposure on neurophysiological parameters," *Environ. Res.,* 63, 60, 1993.

60. Triebig, G., Grobe, T., Saure, E., Schaller, K.-H., Weltle, D., and Valentin, H., "Neurotoxicity of chemical substances in the workplace. VI Longitudinal study of persons occupationally exposed to mercury," *Int. Arch. Occup. Environ. Health,* 55, 19, 1984.

61. Levine, S. P., Cavender, D., Langolf, G. D., and Albers, J. W., "Elemental mercury exposure: peripheral neurotoxicity," *Br. J. Ind. Med.,* 39, 136, 1982.

62. Ellingsen, D. G., Morland, T., Andersen, A., and Kjuus, H., "Relation between exposure related indices and neurological and neurophysiological effects in workers previously exposed to mercury vapour," *Br. J. Ind. Med.,* 50, 736, 1993.

63. Piikivi, L. and Tolonen, U., "EEG findings in chloralkali workers subjected to low long-term exposure to mercury vapour," *Br. J. Ind. Med.,* 46, 370, 1989.

64. Feldman, R. G., Niles, C. A., Kelly-Hayes, M., Sax, D. S., Dixon, W. J., Thompson, D. J., and Landau, E., "Peripheral neuropathy in arsenic smelter workers," *Neurology,* 29, 939, 1979.

65. Wennberg, A., Hagman, M, and Johansson, L., "Preclinical neurophysiological signs of Parkinsonism in occupational manganese exposure," *Neurotoxicology,* 13, 271, 1992.

66. Ruijten, M. W., Salle, H. J., Verberk, M. M., and Smink, M., "Effect of chronic mixed pesticide exposure on peripheral and autonomic nerve function," *Arch. Environ. Health,* 49, 188, 1994.

67. Misra, U. K., Prasda, M., and Pandey, C. M., "A study of cognitive functions and event related potentials following organophosphate exposure," *Electromyogr. Clin. Neurophysiol.,* 34, 197, 1994.

68. Seppäläinen, A. M., Vuojolahti, P., and Elo, O., "Reversible nerve lesions after accidental polychlorinated biphenyl exposure," *Scand. J. Work Environ. Health.,* 11, 91, 1985.

10 Role of Brain Imaging Techniques in Occupational Neurotoxicology

Gerhard Triebig

CONTENTS

1 INTRODUCTION

During past decades occupational neurotoxicology has become an important topic within the field of occupational medicine.

This is especially relevant for workers exposed to chemicals such as organic solvents, metals, and pesticides, which play an important role in both the industrialized and developing countries.

It has been estimated that in Germany (West) about 2 to 3 million employees have occasional or permanent contact with products containing organic solvents.[1] The number of persons exposed to neurotoxins at workplace in the U.S.A. is more than nine million.[2]

In order to examine toxic effects of chemicals in the human central nervous system, a multidisciplinary approach is necessary. The diagnostic procedures involve knowledge of several disciplines, e.g., occupational medicine, neurology, psychiatry, neuropsychology, and can be supported by special investigations such as those based

0-8493-9231-4/98/$0.00+$.50
© 1998 by CRC Press LLC

on brain imaging techniques (BIT). The latter includes X-ray computed tomography (CAT), magnetic resonance imaging (MRI), single photon emission computed tomography (SPECT) and positron emission tomography (PET).[3-5]

In this chapter, BIT studies are discussed in connection with exposure to neurotoxic substances that occur in occupational settings. The survey is not intended to cover all publications on the topic, especially with regard to technical, biochemical and radiological aspects.

2   BRAIN IMAGING TECHNIQUES

Neuroimaging methods like CAT, MRI, PET and SPECT differ in several important aspects, such as form of applied energy, diagnostic value, and cost of examination. In this section, some characteristics of these techniques are summarized. For detailed or specific technical information see Alavi and Hirsh;[6] Hartmann and Hoyer;[7] Verhoeff et al.;[8] Huk et al.;[9] Mazziotta and Gilman.[10]

Computed tomography (CAT) of the brain is the result of digital reproduction of computer-analyzed X-ray refraction in biological materials, such as bone, brain tissue, and body fluids. Because X-rays are used, there is a direct relation between the Hounsfield unit (HU) upon which the image is based and the CAT density of the substance imaged (range, -1000 to +1000 HU: -1000 for air, 0 for water, +1000 for bone).

Modern scanners have within-plane resolution of 4 to 5 mm when regions differ by less than 5 HUs in density, but less than 1 mm when the contrast (percentage difference between the object of interest and surrounding tissue) is maximum.

Limitations caused by bone-hardening artifacts, the effect of adjacent tissues of different density, window setting, and partial volume averaging of tissue within the width of the CAT slice, have been reviewed extensively (see DeCarli et al.).[11]

Magnetic resonance imaging (MRI) or nuclear magnetic resonance (NMR) is based on the absorption of radio-frequency energy by the magnetic moments of atomic nuclei in sample places in a strong magnetic field. So-called "T2-weighted" spin echoes provide information of the gray matter of the brain superior to CAT scan.[12]

Positron emission tomography (PET) is the only technique affording the quantitative three-dimensional imaging of various aspect of brain functions. It provides a reliable method to evaluate the metabolism, perfusion, and pharmacological responses in human organs in vivo.[13] A tracer tagged with a short-lived positron-emitting isotope is generated by a cyclotron and is administered intravenously, or by inhalation, to the subject.

Quantitative tomographic images of regional cerebral function can be generated using established mathematical models from the scan of regional cerebral uptake with the knowledge of the arterial plasma tracer activity.

Measuring the oxygen and glucose metabolism has been widely used to investigate cerebral functions and disease during the 10 to 15 years that PET has been used in humans. These substrates are used in brain as the sole providers of metabolic energy, and hence the consumption of glucose or oxygen, or both, reflects total neuronal function.

At present the clinical application of PET is limited. This technique is difficult due to the use of short-lived radionuclides. It is also expensive because a cyclotron or a positron emitter generator system is necessary.[14]

Investigation of cerebral glucose metabolism by means of ^{18}F-2-fluoro-2-deoxy-D-glucose (FDG) is particularly suitable for analysis of neurotoxic effects. The local cerebral glucose metabolism is closely related to the synapse density under normal resting conditions and rises with physiological activity of neurons. In degenerative diseases in which losses of functional synapses and neurones occur, there is a reduction of glucose turnover.[15]

In single photon emission computed tomography (SPECT) commercially available radionuclides that are capable of measuring cerebral function are used. One of these tracers, N-isopropyl-P-iodoamphetamine (IMP), labeled with iodine-123, is distributed in the brain in proportion to blood flow, and when scanned tomographically can provide three-dimensional information about regional cerebral blood flow (rCBF). Another tracer is hexamethylpropyleneamineoxin (HMPAO) marked with technetium-99m, which is characterized by rapid and intensive binding to substrates. SPECT has several advantages such as low cost and feasibility, but also an important limitation as most systems do not provide quantitative measurements but relative values.[16-17] The overall results depend therefore on the blood flow in the region of interest and do not necessarily indicate disturbed brain metabolism, rather than altered perfusion.[18-19]

In summary, while CAT and MRI give us anatomical and pathological images, PET and SPECT supply functional or biochemical information which cannot be obtained with other methods in a noninvasive manner. The imaging of the brain can provide a dynamic reflection of brain chemistry. The inaccessibility of brain tissue will necessitate various indirect methods of studying biochemical processes, in both normal and pathological states.

3   METALS

Brain imaging studies in metal exposed workers or in patients with metal intoxication are rare. Table 1 lists the main findings described in cases of intoxication with lead, mercury and manganese.

Cortical and cerebellar calcifications have been noted as the main imaging features in several lead intoxicated patients. The MRI abnormalities reported by Schröter et al. (1991)[25] have been suggested to reflect damage of small vessels or increased glial matter. Since systematic studies in occupationally lead-exposed workers are not available, it is not possible to relate the findings to metal concentrations or to specific biochemical or local structural abnormalities induced by metal toxicity.

In patients with manganese intoxication, MRI abnormalities in the region of interest were described. In a 48-year-old patient who had been exposed to mercury at his job in a thermometer factory, MRI showed mild central and cortical atrophy. Punctiform foci were noted in both frontal regions underlying the precentral gyri and in the subcortical myelin.[70]

TABLE 1
Brain Imaging Findings in Metal Intoxication

Year	Author(s)	Ref.	Metal/Cases	Method	Main Results
1977	Hungeford et al.	20	lead/1 male patient moonshine liquor consumption	CAT	Reduced density and lateralization
1983	Swartz et al.	21	lead/1 patient with renal disease	CAT	Calcific deposits in brain cortex
1986	Reyes et al.	22	lead/3 smelters, 2 had chronic renal disease and hypertension	CAT	Cerebral and cerebellar calcification
1988	Brückmann et al.	23	lead/2 smelters	CAT	Subcortical calcification
1989	Wolters et al.	24	manganese/4 smelters with mild Parkinsonism	MRI, PET with 6-FDG	No brain atrophy. In one patient high MR signal. PET with 6 FDG was normal. Reduced cerebral metabolic rates for glucose (FDG) in all patients.
1991	Schröter et al.	25	lead/1 potter production	CAT, MRI	Symmetrical and cerebellar calcifications, increased intensities (T2) in basal ganglia, thalamus and pons.
1992	Petsas et al.	26	lead/1 battery worker	CAT	Multiple subcortical calcification
1993	Nelson et al.	27	manganese/1 arc welder	CAT	No brain atrophy, hyperdense signals (T1) in basal ganglia and midbrain
1993	White et al.	70	inorganic mercury/1 thermometer factory worker	MRI	Mild central and cortical atrophy

4   ORGANIC SOLVENTS

4.1   SOLVENT ABUSE

Table 2 gives a summary of the brain imaging findings described in cases of solvent abuse, mainly by sniffing or huffing. Diffuse brain atrophy was observed in most of the cases with chronic solvent abuse.

4.2   TOXIC ENCEPHALOPATHY

In Table 3, literature data with the neuroimaging findings in patients with a suspected or confirmed solvent-associated toxic encephalopathy are summarized.

TABLE 2
Brain Imaging Findings in Solvent Abusers

Year	Author(s)	Ref.	Solvent/Cases	Method	Main Results
1977	Boor and Hurtig	28	toluene/1 optician, 1 glue sniffer	CAT	Normal CAT scan in optician. Diffuse widening of the cortical and cerebellar sulci
1983	Fornazzari *et al.*	29	cement cleaner/24 abusers (21 men, 3 women)	CAT	Diffuse general brain atrophy, cerebellar changes
1983	Lazar *et al.*	30	toluene/4 abusers	CAT	3 cases studied Diffuse atrophy in 2 cases
1988	Rosenberg *et al.*	68	clear spray lacquer (consisting mainly of toluene)/11 abusers	MRI	In 3 cases diffuse atrophy, loss of gray/white differentiation, an increased periventricular signal intensity
1990	Filley *et al.*	31	toluene/14 abusers (2–10 years history)	MRI	White matter changes
1990	Ikeda and Tsukagoshi	32	toluene/1 man with 10 years history of abuse	MRI	Diffuse brain atrophy
1994	Uitti *et al.*	33	lacquer thinner/ 1 woman with exposure over 9 months	CAT PET	Parkinsonism symptoms, normal CAT, PET indicating altered striatal dopaminergic function
1994	Goodheart and Dune	34	petrol/25 sniffers	CAT	Cerebellar atrophy (1), bilateral infarction of basal ganglia (1). Normal findings in the remaining 8 cases studied.

Before CAT was available, pneumoencephalography (PEG) was used to assess brain atrophy. The PEG allows imaging mainly of the inner ventricular system. In comparison to CAT the diagnostic sensitivity is lower. Therefore, PEG results are not further considered.

In patients with clinical overt dementia, brain atrophy of various degrees is not an uncommon finding. Brain atrophy may be confirmed by SPECT. Actual results of MRI examination correspond to the CAT findings, assuming an association between chronic solvent intoxication and mild brain atrophy. An unusually high number of cases of brain atrophy (12 out of 13 patients) was reported by Lorenz *et al.*[44] but this was probably due to a misclassification.[51-53]

4.3 SOLVENT EXPOSED WORKERS

The main results of BIT-examinations in solvent exposed workers not presenting clinical findings of a toxic encephalopathy are shown in Table 4.

TABLE 3
Survey of Neuroimaging Findings in Patients with Toxic Encephalopathy

Year	Author(s)	Ref.	Study Population	Method	Main Results
1979	Arlien-Soborg *et al.*	35	50 painters with suspected solvent intoxication	PEG (N = 12) CAT (N = 38)	Brain atrophy in 25 subjects (PEG = 12, CAT = 13)
1980	Juntunen *et al.*	36	37 patients with suspected solvent intoxication	PEG	Cerebellar atrophy and cerebral atrophy in 5 and 23 cases respectively
1980	Aquilonius *et al.*	37	6 cases of methanol intoxication	CAT	In 4 cases, necrotic changes in putamen area
1987	Orbaek *et al.*	38	32 patients	CAT	No cerebral atrophy
1987	Gregersen *et al.*	39	4 patients	CAT, PEG	Mild atrophy in 3 cases
1987	Anderson *et al.*	40	3 cases of metanol intoxication	CAT	Leukoencephalopathy and hemorrhage
1988	Skjodt *et al.*	41	181 patients with dementia possibly caused by solvents	CAT	No consistent findings
1989	Hagstadius *et al.*	42	Follow up of 28 patients with mild toxic encephalopathy	CAT	Cerebral blood flow reduced, predominantly in frontotemporal areas. No marked difference at follow up (2 to 7 years)
1989	Greve	43	44 men and 15 women with suspected solvent induced dementia	CAT SPECT	CBF findings indicating cerebral atrophy
1990	Lorenz *et al.*	44	13 patients with suspected tetrachloroethylene intoxication	MRI	Mild to moderate cerebral atrophy in 12 cases
1991	Lorenz *et al.*	45	8 men an 1 woman with suspected encephalopathy; exposure to solvent mixtures (dichloro-methane, trichlorofluoro-methane, 1,1,1-trichloroethane, gasoline) and isocyanates	CAT (2) MRI (7)	Cortical brain atrophy (7) with additional subcortical atrophy (4) or cerebellar atrophy (2)

TABLE 3 (continued)
Survey of Neuroimaging Findings in Patients with Toxic Encephalopathy

Year	Author(s)	Ref.	Study Population	Method	Main Results
1992	Heipertz et al.	46	17 patients with suspected solvent induced encephalopathy	CAT MRI SPECT	No consistent findings
1993	Deschamps et al.	47	50 patients with suspected solvent induced psycho-organic syndrome	SPECT	No difference for hemispheric cerebral blood flow
1993	Callender et al.	48	13 workers with acute exposure to mixtures of solvents, 8 with chronic exposure to mixtures of hazardous waste	CAT MRI SPECT	Various abnormal SPECT findings
1993	Ellingsen et al.	49	85 patients with suspected solvent induced encephalopathy	CAT	No significant correlation between duration of exposure and atrophy related parameters
1994	Orbaek et al.	50	24 patients with mild toxic encephalopathy	SPECT	Reduction of CBF, especially in fronto-temporal areas
1993	White et al.	70	1 patient with intoxication by 2,6-dimethyl-4-heptanone	MRI	Multiple small foci in the white matter and pons

In most studies, brain atrophy has not been observed. In Danish painters, a dose-effect-relationship between solvent exposure and CAT variables was reported.[57] However, the number and frequency of abnormal CAT findings in relation to exposure were not presented. The authors of these studies conclude that "solvent exposure was associated with both cortical and central CAT variables and the results suggest that a diffuse cerebral atrophy may be caused by occupational solvent exposure, high alcohol consumption, and age."[58] This conclusion is in accordance with the general experience from studies on the aging brain[61-63] and the influence of alcohol.[64]

With respect to other neuroimaging methods, only SPECT findings are available in small numbers of solvent-exposed workers. No studies using PET in solvent-exposed workers or in patients with suspected solvent intoxication have been published to date. Recently Heuser et al.[69] reported SPECT findings in a total of 41 patients exposed to neurotoxic chemicals, including pesticides, glues and solvents. However, no details were given regarding the duration and degree of exposures.

TABLE 4
Survey of Neuroimaging Findings in Solvent Exposed Workers

Year	Author(s)	Ref.	Study population	Method	Main Results
1980	Elofsson et al.	54	73 car painters	CAT	Remarkable brain atrophy
1982	Arlien-Soborg et al.	55	9 painters	CAT CBF	No or minimal cerebral atrophy, no differences in CBF
1983	Risberg and Hagstadius	56	50 paint-factory-workers, 50 control subjects	CBF	Slightly reduced mean CBF (4%, p<0.05) in exposed group. Largest differences in fronto-temporal areas
1986	Triebig	57	86 painters, 39 control subjects	CAT	No increased prevalence of atrophy in painters. No exposure-effect-relationship
1988	Mikkelsen et al.	58	46 painters, 34 bricklayers	CAT	Correlation between solvent exposure and cortical and central CAT variables
1989	Lang and Erbguth	59	82 painters, 42 control subjects	CAT	Cella-Media-Index higher in spray-painters. No significant associations between solvent exposure and CAT-variables
1990	Aaserud et al.	60	16 viscose rayon workers with CS_2-exposure	CAT SPECT	In 13 workers, cortical and/or central atrophy

The overall impression from these data is that some alterations of cerebral blood flow may be related to organic solvent exposure.

It should be noted that, according to current knowledge, there is no general biochemical or biophysical mechanism that can explain all aspects of acute and chronic neurotoxicity of organic solvents.[65]

With respect to manifestations of chronic neurotoxicity, Lang (1991)[66] emphazises that neuroimaging and mental abilities are conceptually different and empirically no comparable parameters. Therefore, a distinction has to be made between structural and functional techniques. The latter are, in principle, related to psychometric variables.[66] In this respect, more promising results may be obtained by PET examinations.[67]

5 CARBON MONOXIDE POISONING

Imaging techniques have been used in patients with carbon monoxide intoxication. On CT, white matter and globus pallidus lesions have been described.[71]

6 CONCLUSIONS

Modern brain imaging techniques can contribute to diagnosis of toxic encephalopathy. However, the existing literature does not offer consistent information to compare the various brain imaging techniques as a tool to assess the action of neuro-toxicants in individual subjects or groups.

Although no specific brain imaging alteration has been proven in cases of solvent-induced neurotoxicity, BIT findings can permit exclusion of other brain disease and may be useful for accurate follow-up observations. The possibility of correlating, at least in part, findings and symptoms is an important advantage associated with the "function-imaging" techniques, such as SPECT and PET. However, the results are influenced by several subjective and environmental factors.

In single cases, a synopsis of all information and findings, including a comprehensive occupational history, quantitative data of chemical exposures, neuropsychiatric evaluation based on psychodiagnostic tests and brain imaging techniques are necessary to confirm the diagnosis of toxic encephalopathy.

In perspective, studies are needed to examine:

(a) the outcome of cases presenting abnormal brain imaging findings

(b) the correlations between results of BIT and clinical, neurophysiological and neuropsychological parameters

(c) possible association of morphological, biochemical and functional findings particularly in subjects with confirmed solvent-induced encephalopathy.

7 REFERENCES

1. Triebig, G., "Neurotoxic risks from occupational exposure to chemical substances in Germany," in *Proc. 13rd Asian Conference on Occupational Health*, 1991,137.
2. Anger, W. K., "Workplace exposure," in *Neurobehavioral Toxicology*, Annau, Z., Ed., E. Arnold Ltd., London, 1987, 331.
3. Triebig, G. and Lang, C., "Brain imaging techniques applied to chronically solvent-exposed workers: current results and clinical evaluation," *Environ. Res.*, 61, 239, 1993.
4. Triebig, G., "Summary of workshop: neuroimaging methods," *Environ. Res.*, 60, 76, 1993.
5. Heipertx, W., "Moderne bildgebende Verfahren in der Diagnostik neurotoxischer Erkrankungen des Gerhirns," *Arbeitsmed. Sozialmed. Präventivmed.*, 10, 407, 1992.
6. Alavi, A. and Hirsch, L. J., "Studies of central nervous system disorders with single photon emission computed tomography and positron emission tomography. Evaluation over the past 2 decades," *Semin. Nucl. Med.*, 21, 58, 1991.
7. Hartmann, A. and Hoyer, S., Eds., *Cerebral Blood Flow and Metabolism Measurement*, Springer Verlag, Berlin, 1985.
8. Verhoeff, N., "Basics and recommendations for brain SPECT," *NuklearMedizin*, 31, 114, 1992.
9. Huk, W. J., Gademann, G. F. E., and Friedmann, G., *Magnetic Resonance Imaging of Central Nervous System Diseases*, Springer Verlag, Berlin, 1988.
10. Mazziotta, J. C. and Gilman, S., Eds., *Clinical Brain Imaging: Principles and Applications*, Davis, Philadelphia, 1992.

11. De Carli, C., Kaye, J. A., Horwitz, B., and Rapoport, S. I., "Critical analysis of the use of computer assisted transverse axial tomography to study human brain in aging and dementia of the Alzheimer type," *Neurology*, 40, 872, 1990.

12. Moonen, C. T. W., Zijl, P. C. M., Franck, J. A., Le Bihan, D., and Becker, E. D., "Functional magnetic resonance in medicine and physiology," *Science*, 250, 53, 1990.

13. Brooks, D. J., "PET: Its clinical role in neurology," *J. Neurol. Neurosurg. Psychiatry*, 54, 1, 1991.

14. Frackowiak, R. S. J. and Jones, T., "PET scanning," *Br. Med. J.*, 298, 693, 1989.

15. Herholz, K., "Möglichkeiten und Grenzen der Positronen-Emissionstomographie (PET) bei neurotoxischer Schädigung," *Arbeitsmed. Sozialmed. Präveantivmed.*, 27, 321, 1992.

16. English, R. and Brown, S. E., *SPECT. Single-photon emission computed tomography: A primer*, Society of Nuclear Medicine, New York, 1986.

17. Rootwelt, K., Dybevold, S., Nyberg-Hrensen, R., and Russel, D., "Measurement of cerebral blood flow with ^{133}Xe inhalation and dynamic single photon emission computer tomography: normal values," *Scand. J. Clin. Lab. Invest.* (Suppl.), 46, 97, 1986.

18. Deisenhammer, E., Reisecker, F., Leblhuber, F., Höll, K., Markut, H., Trenkler, S., and Schneider, I., "Single-Photon-Emissions-Computertomographie bei der Differentialdiagnose der Demenz," *Dtsch. Med. Wochenschr.*, 114, 1639, 1989.

19. Heiss, W. D., Herholz, K., Pawlik, G., and Szelies, B., "Beitrag der Positronen-Emissions-Tomographie zur Diagnose der Demenz," *Dtsch. Med. Wochenschr.*, 113, 1362, 1988.

20. Hungerford, G. D., Ross, P., Robertson, H. J., "Computed tomography in lead encephalopathy: A case report," *Radiology*, 123, 91, 1977.

21. Swartz, J. D., Faerbe, E. N., Singh, N., Polinsky, M. S., "CT demonstration of cerebral cortical calcification," *J. Comput. Assist.*, 7, 476, 1983.

22. Reyes, P. F., Gonzalez, C. F., Zalewska, M. K., Besarab, A., "Intracranial calcification in adults with chronic lead exposure," *AJR*, 146, 267, 1986.

23. Brückmann, H., Krieger, D., Zeumer, H., "Intrazerebrale subkortikale Verkalkungen nach langjähriger Bleiexposition," *Fortschr. Röntgenstr.*, 148, 95, 1988.

24. Wolters, E. Ch., Huang, C. C., Clark, C., Peppard, R. F., Okada, J., Chu, N. S., Adam, M. J., Ruth, T.J ., Li, D., and Calne-D. B., "Positron emission tomography in manganese intoxication," *Ann. Neurol.*, 26, 647, 1989.

25. Schröter, C., Schrote, H., and Huffmann, G., "Neurologische und psychiatrische Manifestationen der Bleiintoxication bei Erwachesenen (Fallbericht und Literaturübersicht)," *Fortschr. Neurol. Psychiat.*, 59, 413, 1991.

26. Petsas, Th., Fezoulidis, I., Ziogas, D., Kremydas, N., and Kelekis, D., "Gehirnverkalking bei chronischer Bleivergiftung," *Fortschr. Röntgenstr.*, 157, 192, 1992.

27. Nelson, K., Golnick, J., and Korn, T., "Manganese encephalopathy: utility of early magnetic resonance imaging," *Br. J. Ind. Med.*, 50, 510, 1993.

28. Boor, J. W. and Hurtig, H. I., "Persistent cerebellar ataxia after exposure to toluene," *Ann. Neurol.*, 2, 440, 1977.

29. Fornazzari, L., Wilkinson, D. A., Kapur, B. M., and Carlen, P. L., "Cerebellar, cortical and functional impairment in toluene abusers," *Acta Neurol. Scand.*, 67, 319, 1983.

30. Lazar, R. B., Ho, S. U., Melen, O., and Daghestani, A. N., "Multifocal central nervous system damage caused by toluene abuse," *Neurology*, 33, 1337, 1983.

31. Filley, C. H., Heaton, R. K., and Rosenberg, N. L., "White matter dementia in chronic toluene abuse," *Neurology*, 40, 532, 1990.

32. Ikeda, M. and Tsukagoshi, H., "Encephalopathy due to toluene sniffing," *Eur. Neurol.*, 30, 347, 1990.

33. Uitti, R. J., Snow, B. J., Shinotoh, H., Vingerhoets, F. J., Hayward, M ., Hashimoto, S., Richmond, J., Markey, S. P., Markey, C. J., and Calne, D. B., "Parkinsonism induced by solvent abuse," *Ann. Neurol.*, 35, 616, 1994.
34. Goodheart, R. S. and Dunne, J. W., "Petrol sniffer's encephalopathy, A study of 25 patients," *Med. J. Aust.*, 160, 178, 1994.
35. Arlien-Soborg, P., Bruhn, P., Gyldensted, C., and Melgaard, B., "Chronic painters syndrome," *Acta Neurol. Scand.*, 60, 149, 1979.
36. Juntunen, J., Hernberg, S., Eistola, P., and Hupli, V., "Exposure to industrial solvents and brain atrophy," *Eur. Neurol.*, 19, 366, 1980.
37. Aquilonius, St. M., Bergstrom, K., Enoksson, P., Hedstrand, U., Lundberg, P. O., Mostrom, U., and Olsson, Y., "Cerebral computed tomography in methanol intoxication," *J. Comput. Assist. Tomogr.*, 4, 425, 1980.
38. Orbaek, P., Lindfren, M., Olivecrona, H., and Haeger-Aronsen, B., "Computed tomography and psychometric test performances in patients with solvent induced chronic toxic encephalopathy and healthy controls," *Br. J. Ind. Med.*, 44, 175, 1987.
39. Gregersen, P., Klausen, H., and Elsnab, C. U., "Chronic toxic encephalopathy in solvent-exposed painters in Denmark 1976–1980: Clinical cases and social consequences after a 5-year follow-up," *Am. J. Ind. Med.*, 11, 399, 1987.
40. Anderson, T. J., Shuaib, A., and Becke, W. J., "Neurologic sequelae of methanol poisoning," *Can. Med. Assoc. J.*, 136, 1177, 1987.
41. Skjodt, T., Torfing, K. F., and Teisen, H., "Computed tomography in patients with dementia probably due to toxic encephalopathy," *Acta Radiol.*, 29, 495, 1988.
42. Hagstadius, S., Orbaek, P., Risberg, J., and Lindgren, M., "Regional cerebral blood flow at the time of diagnosis of chronic toxic encephalopathy induced by organic-solvent exposure and after the cessation of exposure," *Scand. J. Work. Environ. Health*, 15, 130, 1989.
43. Greve, E., "Computerized tomography findings among workers with chronic intoxication of the brain," *Scand. J. Soc. Med.*, 17, 147, 1989.
44. Lorenz, H., Omlor, A., Walter, G., Haas, A., Steigerwald, F., and Buchter, A., "Nachweis von Hirnschädigungen durch Tetrachlorethen," *Zbl. Arbeitsmed.*, 40, 355, 1990.
45. Lorenz, H., "Schädigung des zentralen Nervensystem durch heterogene Lösungsmittelgemische," *Zbl. Arbeitsmed.*, 41, 311, 1991.
46. Heipertz, W., "Die 'Single-photon-emission-computer-tomography' des Gehirns (Hirn-SPECT) in der arbeitsmedizinischen Begutachtung," in *Bericht über die 32. Jahrestagung der Deutschen Gesellschaft für Arbeitsmedizin e.V.*, Kreutz, R., and Piekarski, C., Eds., Gentner Verlag, Stuttgart, 1992, 533.
47. Deschamps, D., Garnier, R., Lille, F., Tran, Dinh. Y., Bertaux, L., Reygagne, A., and Dally, S., "Evoked potentials and cerebral blood flow in solvent induced psycho-organic syndrome," *Br. J. Ind. Med.*, 50, 325, 1993.
48. Callender, T. J., Morrow, L., and Subramanian, K., "Three-dimensional brain metabolic imaging in patients with toxic encephalopathy," *Environ. Res.*, 60, 295, 1993.
49. Ellingsen, D. G., Bekken, M., Kolsaker, L., and Langard, S., "Patients with suspected solvent-induced encephalopathy examined with cerebral computed tomography," *J. Occup. Med.*, 155, 1993.
50. Orbaek, P., Karison, B., and Risberg, J., "Reduced regional cerebral blood flow and cognitive dysfunction in organic solvent exposed subjects," *Proc. XIV Asian Conference on Occupational Health*, Beijing, China, 1994, 145.
51. Möllhoff, G., "Interdisziplinäres Kolloquium 'Möglichkeiten und Grenzen moderner bildgebender Verfahren bei neurotoxikologischen Fragestellungen in der Arbeits- und Umweltmedizin'," *Arbeitsmed. Sozialmed. Präventivmed.*, 26, 491, 1991.

52. Heipertz, W., "Moderne bildgebende Verfahren in der Diagnostik neurotoxischer Erkrankungen des Gehirns," *Arbeitsmed. Sozialmed. Präventivmed.*, 27, 407, 1992.

53. Triebig, G., "Kommentar zum Beitrag 'Schädigung des zentralen Nervensystems durch heterogene Lösungsmittel-gemische' von H. Lorenz *et al.*," *Zbl. Arbeitsmed.*, 42, 160, 1992.

54. Elofsson, S. A., Knave, B., Lydahl, E., Mindus, P., Persson, H. E., Philipson, B., Steby, M., Sruwe, G., Söderman, E., Gamberale, F., Hingmarsh, T., Iregren, A., Ikaksson, A., Johnsson, I., Wennberg, A., and Widen, L., "Exposure to organic solvents," *Scand. J. Work. Environ. Health*, 6, 239, 1980.

55. Ariien-Soborg, P., Henriksen, L., Gade, A., Gyldensted, C., and Paulson, O. B., "Cerebral blood flow in chronic toxic encephalopathy in house painters exposed to organic solvents," *Acta Neurol. Scand.*, 66, 34, 1982.

56. Risberg, J. and Hagstadius, S., "Effects on the regional cerebral blood flow of long-term exposure to organic solvents," *Acta Psychiat. Scand.*, 67, suppl. 303, 92, 1983.

57. Triebeg, G., "Erlanger Malerstudie, Multidisziplinäre Querschnittsuntersuchung zur Neurotoxizität von Lösemittelm in Farben und Lacken," *Arbeitsmed. Sozialmed. Präventivmed.*, Sonderfeft 9, 1986.

58. Mikkelsen, S., Jorgensen, M., Browne, E., and Gyldensted, C., "Mixed solvent exposure and organic brain damage," *Acta Neurol. Scand.*, Suppl. 78, 118, 1988.

59. Lang, C. and Erbguth, F., "Neuroradiologische Untersuchung," in *Arbeitsmed. Sozialmed. Präventivmed.*, Sonderfeft 13, Triebig, G., Ed., Gentner Verlag, Stuttgart, 1989, 43.

60. Aaserud, O., Hommeren, OJ., Tvedt, B., Nakstadt, P., Mowe, G., Efskind, J., Russell, D., Jörgensen, E. B., Nyber-Hansen, R., Rootwelt, K., and Gjerstad, L., "Carbon disulfide exposure and neurotoxic sequelae among viscose rayon workers," *Am. J. Ind. Med.*, 18, 25, 1990.

61. Gyldensten, C., "Measurements of the normal ventricular system and hemispheric sulci of 100 adults with computed tomography," *Neuroradiology*, 14, 183, 1977.

62. Nagatam, K., Basugi, N., Fukushima, T., Tango, T., Suzuki, I., Kaminuma, T., and Kurashina, S., "A quantitative study of physiological cerebral atrophy with aging," *Neuroradiology*, 29, 327, 1987.

63. Coffey, C. E., Wilkinson, W. E., Parashos, I. A., Soady, S. A., Sullivan, R. J., Patterson, L. J., Figiel, G. S., Webb, M. C., Spritzer, C. E., Djang, W. T., "Quantitative cerebral anatomy of the aging human brain: A cross-sectional study using magnetic resonance imaging," *Neurology*, 42, 527, 1992.

64. Ron, M. A., Acker, W., Shaw, G. K., and Lishman, W. A., "Computerized tomography of the brain in chronic alcoholism," *Brain*, 105, 497, 1982.

65. Arlien-Soborg, P., Hansen, L., Ladefoged, O., and Simonsen, L., "Report on a Conference on organic solvents and the nervous system," *Neurotoxicol. Teratol.*, 14, 81, 1992.

66. Lang, C., "Zur Interrelation bildgebender und psychometrischer Verfahren bei neurotoxikologischen Fragestellungen in der Arbeits- und Umwelt-Medizin,"*Arbeitsmed. Sozialmed. Präventivmed.*, 26, 433, 1991.

67. Herholz, K., "Möglichkeiten und Grenzen der Positronen-Emissionstomographie (PET) bei neurotoxischer Schädigung," *Arbeitsmed. Sozialmed. Präventivmed.*, 27, 321, 1992.

68. Rosenberg, N. L., Spitz, M. C., Filley, C. M., Davis, K., and Schaumburg, H.H., "Central nervous system effects of chronic toluene abuse. Clinical, brainstem evoked response and magnetic resonance imaging studies," *Neurotoxicol. Teratol.*, 10, 489, 1988.

69. Heuser, G., Mena, I., and Alamos, F., "Neurospect findings in patients exposed to neurotoxic chemicals," *Toxicol. Ind. Health*, 10, 561, 1994.
70. White, F. R., Feldman, R. G., Moss, M. B., and Proctor, S. P., "Magnetic resonance imaging (MRI), neurobehavioral testing, and toxic encephalopathy: two cases," *Environ. Res.*, 61, 117, 1993.
71. Prockop, L. D., "Neuroimaging in neurotoxicology," in *Neurotoxicology*, Chang, L. W. and Slikker, W., Jr., Eds., Academic Press, San Diego, 1995, 753.

11 Computer-Assisted Testing

Anders Iregren

CONTENTS

1 INTRODUCTION

This chapter describes some benefits gained from the automation of test procedures, as well as some drawbacks with the use of computers in psychological assessment. A brief history of the development of computerized testing is also presented. This history includes a listing of some conferences and symposia, which have been instrumental to the development of computerized testing within occupational neurotoxicology. Following this general introduction, seven of the test systems developed for studies of the neurotoxic impact from occupational and environmental exposures are briefly described. Some information on the validation available for these systems is also provided. In the last part of the chapter, efforts at the standardization of computerized test procedures and the use of behavioral data in the process of regulating environmental exposures are discussed.

2   DRAWBACKS WITH THE AUTOMATION OF TESTS

Benefits and drawbacks introduced by the automation of psychological tests and test batteries have been presented in several recent reviews of computerized testing.[1-4] For this reason, only some brief comments on these issues will be presented here.

The most important drawback with the use of computer-administered testing in research situations is that testing may be performed essentially with only one subject at the time. Even with fully automated systems and the present relatively low cost of computers, this is definitely a problem compared to the traditional testing situation.

Some concern has been expressed regarding the potentially threatening situation created by the computer. However, studies of computer use with populations as different as the elderly and demented,[5] the physically handicapped,[6] psychiatric patients,[7] and many more, have shown that computer-assisted testing is mostly perceived in a neutral way, or even as stimulating, rather than threatening.

Definite drawbacks with the use of present forms of microcomputers are the limitations with respect to response media that are easily available, and, to some extent, the limitations with respect to the stimulus material that can be presented. Although computers have been used to control various technical devices also for auditory or other stimuli, the vast majority of computerized tests and test batteries still rely almost exclusively on visual presentations on the computer screen. Similarly, the input mode is generally restricted to manual handling of keyboards and joy-sticks, or different kinds of pointing devices. Some of these devices put relatively high demands on motor skills. The only way to fully by-pass these limitations, which is presently available, is to use custom made equipment, thus severely limiting the possibilities to provide access to the tests for others than the developers. It is important, however, to point out that the drawbacks introduced by the technical limitations of the present computer generation will naturally diminish over time due to the development of more competent computer systems.

3   BENEFITS FROM THE AUTOMATION OF TESTS

In spite of the above mentioned limitations with respect to the modes of stimulation and response easily available using computers, the present author believes that one advantage with the computer is that the importance of various aspects of the tasks may be controlled. Thus, it is easier to reduce or increase the relative importance of perception, cognitive processing, and motor responding, for different tests by changes in the computer programs than to change traditional test procedures.

Another important advantage with computerized testing is the increase in reliability, and thus potentially also validity, that is gained by the automation. The reason for the reliability increase is of course the standardization provided by the computer presentation. Computers very seldom fail to consistently present in a standardized manner what they are programmed to, nor do they inadvertently cue or prompt the subject by changes in phrasing, gesture, or posture. Furthermore, automatic recording of responses and response latencies can be accurately performed using computers, complex scoring schemes may be easily included in a test program, analyses of response patterns can be performed in a simple manner, and results may be accessed almost instantaneously. In short, test administration may be strictly standardized and performed in an accurate way by the computer, while simultaneously freeing the psychologist from some of the most tedious parts of psychological assessment.

Still further advantages may be the reduction of the time required for testing, and the possibilities for so called adaptive testing. In a study of psychiatric patients[8] testing times were reduced by up to 60% with the use of computerized versions of traditional, clinical tests. By the use of tailored or adaptive testing, the time needed for assessment may be further reduced, as the subject is quickly run through the parts of the test which he finds easy, while concentrating more on the difficult items.[9,10] However, adaptive testing has not been used to any large extent in occupational neurotoxicology.

4   A BRIEF HISTORY OF COMPUTERIZED TESTING

Computerized tests and questionnaires have been used for more than three decades by now. The very first computers, in the 1950s and early 1960s, were available only to highly specialized technicians, and computer time was extremely expensive. For reasons like these, computerized tests and/or questionnaires were not developed until computer power was decentralized through the use of multiple terminals, and, later on, with the advent of personal computers.

The very first efforts at computerization of psychological assessment procedures were concentrated to tests and questionnaires that did not put high demands on timing, as the time sharing systems used with mainframe computers, and subsequent minicomputer systems, did not allow for exact timing of responses. Furthermore, the majority of early publications on computerized testing discuss the question of comparability of data gathered using both traditional and computerized forms of the same tests. Most early efforts were thus spent on the computerization of existing tests, and attempts to establish the validity of the computerized forms. Therefore, the potential of the computers was used only to a limited extent.

During the last decade, the rapid technical development of personal computers has substantially increased the computer power accessible for testing purposes. These developments include memory capacity, graphic resolution of computer screens, new media for data storage, devices for auditory presentations, etc. Another important reason for the availability of computer power is of course the decreased cost for computers. The psychometric implications of this technical development are increasingly being explored by test developers.

A list of scientific meeting and symposia that have been important to stimulate the computerization of tests within occupational neurotoxicology is presented in Table 1. As can be seen from the table, every three years since 1982, international symposia on neurobehavioral methods (lately including also effects) have been organized. The organization behind these symposia is the Scientific Committee on Neurotoxicology and Psychophysiology of the International Commission of Occupational Health, ICOH. A main goal of these symposia is to exchange state-of-the-art information on the development of neurobehavioral methods in occupational and environmental health.

TABLE 1
**Scientific Meetings that Have Been Instrumental to the Development of
Computerized Test Methods in the Field of Occupational Neurotoxicology**

Year	Title	Reference	Location
1982	1st International Symposium on Neurobehavioral Methods in Occupational Health	52	Como, Italy
1983	Meeting on Neurobehavioral Testing	11	North Carolina, USA
1985	2nd International Symposium on Neurobehavioral Methods in Occupational Health	53	Copenhagen, Denmark
1988	Symposium on "Behavioral Measures of Neurotoxicity"	54	Canberra, Australia
1988	3rd International Symposium on Neurobehavioral Methods in Occupational Health	55	Washington, DC, USA
1991	4th International Symposium on Neurobehavioral Methods and Effects in Occupational and Environmental Health	56	Tokyo, Japan
1993	Further Development of Computer-Based Neurobehavioral Testing in Environmental and Occupational Health		Oslo, Norway
1994	5th International Symposium on Neurobehavioral Methods and Effects in Occupational and Environmental Health		Cairo, Egypt
1995	Computerized Behavioral Testing of Humans in Neurotoxicology Research		Portland, USA
1997	6th International Symposium on Neurobehavioral Methods and Effects in Occupational and Environmental Health		Shanghai, China

Apart from this series of symposia, several single meetings with more specific aims have been organized, such as the 1983 meeting in Washington, DC. At this meeting, sponsored by the World Health Organization (WHO) and the U.S. National Institute of Occupational Safety and Health (NIOSH), the basic agreement concerning the WHO Neurobehavioral Core Test Battery (WHO-NCTB)[11] was reached by an international expert group.

Meetings specifically aimed at discussions of the development of computerized testing have been very few, and the only ones not organized solely for a specific test system user group are a meeting held in Oslo, Norway, in 1993, and a meeting that took place in Portland, Oregon, in 1995.

5   EXAMPLES OF TEST SYSTEMS

A review of computerized tests used in occupational neurotoxicology will probably never be exhaustive, since over the years numerous different tests and test systems have been employed in evaluations of toxic effects from occupational exposures. For this reason, only those computer-assisted test systems that were developed specifically for use in occupational neurotoxicology, and that are documented in the peer reviewed scientific literature, have been included in this review. Table 2 lists some basic characteristics of the test systems presented in this chapter. Descriptions will be given of the Swedish Performance Evaluation System (SPES),[12] the Information Processing and Performance Test Battery (IPPTB),[13] the Neurobehavioral Evaluation System (NES),[14] the Cognitive Function Scanner*,[15] the Milan Automated Neurobehavioral System (MANS),[16] the MicroTox system,[17] and the Behavioral Assessment and Research System (BARS).[18]

Some of the systems listed here are test batteries in the true sense of the word, i.e., sets of tests where the full set should be administered on each testing occasion. Other systems, like the SPES and the NES, are sets of research tools where the scientist makes his choice of tests suitable for use with each specific hypothesis and investigation.

These systems differ also in some other respects. One difference is that the systems have a more or less pronounced theoretical background. In some systems, tests have been chosen primarily to allow assessment of specific abilities which, from a theoretical point of view, were considered important. In other systems, empirical considerations have been at least equally important as a basis for the choice of tests; an important criterion has been proven sensitivity to neurotoxic insult.

The validation available varies widely between the test systems. Since validity is an extremely important aspect of a system, the internationally published studies on effects from neurotoxic exposures are referenced for each test system presented.

Another aspect is the degree to which test administration is automated. Some systems have (at least tried) fully automated testing, i.e., also computerized instructions, while several systems have not automated the instructional phase of testing, but merely item administration, response recording, calculations, and data storage.

Other differences between systems are with regard to the type of studies or measurements the systems were developed for, and the degree to which they use non-standard equipment. Knowledge on all these aspects is important for the understanding of the existing differences between the systems, and thus for the decision making when you have to choose among them. Therefore, information on these differences is given for each of the seven systems described.

5.1 THE SWEDISH PERFORMANCE EVALUATION SYSTEM

5.1.1 Background

The SPES was developed over the past 25 years at the Department of Psychophysiology, Swedish National Institute of Occupational Health. Originally, the development was stimulated by the need for tests sensitive to the narcotic effects from short term solvent exposure. Thus, the starting point for the SPES was experimental psychology and arousal theory. Several tests of simple and choice reaction time, as well as short term memory, were developed. They were applied in a series of experiments regarding the neurotoxic effects from short-term exposure to different solvents.[9-25]

Following these experimental uses, the SPES also was applied in several cross-sectional epidemiological studies of neurotoxic effects from long-term exposures to jet fuel,[26] toluene,[27] solvent mixtures,[27-28] manganese,[29-32] and aluminum.[32] Tests

TABLE 2
Test Systems

System Name	Reference	Computers Used	External Equipment	Number of		Administer All the Tests?	Validation
				Tests	Rating Scales		
The Swedish Performance Evaluation System	12	IBM compatibles	Some	18	5	No	Extensive
The Information Processing and Performance Test Battery	13	IBM compatibles	A lot	8	0	Yes	Sparse
The Neurobehavioral Evaluation System	14	IBM compatibles	Some	17	1	No	Extensive
The Cognitive Function Scanner	15	IBM compatibles	A lot	8	0	Yes	Sparse
The Milan Automated Neurobehavioral System	16	IBM compatibles	Some	7	0	Yes	Sparse
The MicroTox System	17	AppleII	Some	17	0	No	Sparse
The Behavioral Assessment and Research System	18	Macintosh	Some	4	0	No	Sparse

measuring skills like verbal and logical ability were included in these studies, since they were needed to verify the comparability between study groups with respect to intellectual capacity. More recently, tests sensitive to effects from exposure also to other factors than toxic chemicals have been included. Thus, tests useful in studies of exposures such as cold, heat, noise, vibration, and electromagnetic fields have been added to the SPES during later years.[33-37]

The technical equipment for the administration of the SPES tests has varied a lot over time. In the beginning, electronic apparatuses built specifically for the purpose of test administration were used. With the advent of the microcomputers in the late 1970s, the SPES was computerized. Initially the SPES programs were written in Basic for a Swedish computer brand using the Z80 chip, and the first study using only computerized tests from the SPES was an experimental evaluation performed in 1982 of the possible effects from exposure to toluene, xylene, and their combination.[38]

5.1.2 Equipment

More recent versions of the SPES were programmed for IBM compatible computers. The present version 5 is programmed in Pascal and runs in protected mode on IBM compatibles with an 80286 processor or later, a minimum of 2 Mb RAM, and at least EGA color graphics. Printer use is optional. In addition, a hardware lock, provided by the developers, and a joystick, are needed to run the system.

5.1.3 Tests

Version 5 of the SPES is a menu driven DOS program, capable of presenting some 20 different performance tests and rating scales for mood and symptoms. The test list, which is presented in Table 3, includes four tests of simple, choice, and complex reaction time, search and memory, digit span, symbol digit, three tests of logical reasoning, two tests of vocabulary, additions, digit classification, serial digit addition, two tests of finger tapping, and a continuous tracking task. Rating scales include three symptom scales, a mood scale, and a scale for the rating of performance. From this list of tests and rating scales, the investigator makes his specific choice for each hypothesis and study at hand. Instructions for the SPES tests are presented orally by the investigator, with written support on the computer screen. For most of the SPES tests (exceptions are the two verbal tests and the category test) standardization is independent of verbal material. What is needed for use in other languages is therefore only a straightforward translation of the instructions, and the text files containing instructions are easily edited. Thus, the SPES can be easily translated into almost any language. Presently there are Swedish, Danish, English, Finnish, French, German, Hebrew, and Italian versions in use.

5.1.4 Validation

As mentioned above, validation for the SPES includes studies of experimental as well as long-term exposures to various solvents, several studies of effects from occupational exposure to neurotoxic metals like aluminium and manganese, and also studies of effects from exposures to physical factors like cold, heat, noise, and electromagnetic fields. In total, more than 40 experimental and epidemiological

TABLE 3
Tests Available in the Swedish Performance Evaluation System
with Approximate Testing Time, Reliability Coefficients,
and References to Studies Where the Tests Were Applied

Performance test	Testing Time	r_{tt}	References
Simple Reaction Time	6	0.85	19-33,35-38,57-77
Choice Reaction Time	10	0.88	31,37,38,62-64,68
Color Word Vigilance	8	0.90	33,36,38,62-64,68,78
Color Word Stress	7		33,35,78
Search and Memory	10	0.75	68,78
Symbol Digit	5	0.74	29,30,32,36,64,68
Digit Span	10	0.70	29-32,36,64,68
Additions	8	0.94	22-30,65,66,68-70
Digit Classification	5		33,78
Digit Addition	7		33
Grammatical Reasoning	8	0.76	35,64,68,72,74,76-79
Logical Series	5		
Logical Categories	5		
Vocabulary	8		29,32
Word Reaction Time	5		
Finger Tapping Speed	5	0.82	29-32,68,71
Finger Tapping Endurance	3		29,32
Continuous Tracking Test	6		32
Performance	1		
Mood	3		29,32,36,65,70,73
Acute Symptoms	4		29,32,62,65,70
Long-term Symptoms	5		29,32,62,65
EURONEST Symptoms Scale	10		

r_{tt} is the test-retest reliability in a group of 60 employees at the National Institute of Occupational Health

retested after about 100 days. In August 1995, a similar reliability study was being performed with 10 of the SPES tasks.

studies using the SPES have been reported in the peer reviewed scientific literature. Table 3 lists for each test references to the studies where it was used.

5.2   THE INFORMATION PROCESSING AND PERFORMANCE TEST BATTERY

5.2.1   Background

The IPPTB was developed in Sydney, Australia, at the National Institute of Occupational Health and Safety.[13] A leading idea behind the development of this test battery was to provide a reasonably comprehensive coverage of the major areas of known neurobehavioral functions, while at the same time providing an integrated framework for the interpretation of results. Therefore, the tests were selected using an information processing model of human performance generation.[39] This model was used to choose tests with a stepwise increase in complexity with respect to the psychological functions tested, advancing from the measurement of simple perception to the assessment of complex cognitive functions like learning and memory.

5.2.2   Equipment

The IPPTB is a computer aided battery in that not all the tests are computerized. Stimulus and response panels are attached as peripheral equipment to an IBM compatible computer. In effect, the subject tested does not interact directly with the computer at all, since stimuli are presented on custom made peripheral equipment, and responses are likewise given without the use of the standard computer peripherals.

5.2.3   Tests

The tests included in the battery are critical flicker fusion, vigilance, hand steadiness, simple reaction time, visual pursuit, Sternberg memory test, paired associates short-term memory test, and paired associates long-term memory test. The IPPTB is a true test battery, where all tests should be administered on each test occasion, and instructions are provided by the investigator.

5.2.4   Validation

Validation for the IPPTB is relatively sparse. The small number of validation studies is most probably due to the fact that the many custom made peripheral units which are needed to run the system make the distribution of the system to other research groups quite difficult. Thus, validation is limited to the studies performed by the developers. Epidemiological studies of the effects of long-term exposure to lead, [40] mercury, [41] and underwater environments[42] have been reported. Furthermore, a longitudinal follow up of painter apprentices is running since several years. [43]

5.3   THE NEUROBEHAVIORAL EVALUATION SYSTEM

5.3.1   Background

The Neurobehavioral Evaluation System (NES) was developed "… to facilitate the conduct of epidemiological studies of populations at risk for or suffering from central nervous dysfunction due to environmental agents." [14] The NES was developed in the early 1980s, and runs on IBM compatible personal computers. Following the initial version of the NES that was distributed widely, a NES2, which is protected against unauthorized distribution, was developed. In this second release, the known bugs were corrected, and a refined algorithm for the timing of responses was introduced.

5.3.2   Equipment

The NES runs on standard IBM clones. Additional equipment used is an overlay to the keyboard, a joy-stick, and a hardware lock to prevent unauthorized use of the system. This additional equipment is provided by the developers, together with the program and manual.

5.3.3   Tests

The present version of NES2 includes some 20 different tests and rating scales. Psychomotor tests like finger tapping, hand–eye coordination, simple reaction time, continuous performance test, pattern comparison, and symbol digit substitution are available. Tests of pattern memory, digit span, serial digit learning, associate learning, associate recall, and memory scanning may be used to assess memory function. Cognitive tests include verbal ability, horizontal addition, switching attention, grammatical reasoning, color word, and color naming tests, and affect can be assessed with the use of mood scales. The tests are generally relatively short, with testing times of three–six minutes. Initially, the NES was fully automated, i.e., without interaction between the investigator and the subject during testing. In the NES2, this has been revised, and the investigator provides oral instructions for the subject.

5.3.4   Validation

The NES2 has a thorough validation in the scientific literature, with more than 40 publications describing mainly cross-sectional epidemiological investigations. These studies investigate effects from agents like ethylene oxide, lead, mercury, methylene chloride, naphta, and organophosphates, and worker groups like car, house and shipyard painters, as well as some patient groups and the elderly. The NES2 has been applied also in a few experimental investigations, for the measurement of effects from, e.g., ethanol, nitrous oxide, perchloroethylene, toluene, volatile organic mixtures, and some drugs. For reviews of the NES2 validation, see References 44, 45.

5.4   THE COGNITIVE FUNCTION SCANNER

5.4.1   Background

The purpose of the Cognitive Function Scanner[15] is "... to be sensitive to mild brain dysfunction and suitable for modern microcomputer administration." It was developed in a clinical, neuropsychological tradition, and it is commercially available through Cognitive Research Scandinavia.

5.4.2   Equipment

The Cognitive Function Scanner runs on IBM compatible computers with at least a 33 MHz 80486D processor, high resolution graphics (VGA or multisync), attached to a laser printer, and requires a digitizer, adapter card, hand switch, video adapter, headphones, and a client keyboard.

5.4.3   Tests

Tests included in the Cognitive Function Scanner are a face recognition test, number learning, figure drawing, a pen-to-point test, a parallelogram test, a continuous graphics test, a Bourdon-Wiersma test, and a continuous reaction time test. Standardized instructions are given orally by the investigator.

5.4.4 Validation

Although the Cognitive Function Scanner is supported by one of the largest standardization studies (using a sample of 1026 persons),[15] validation studies regarding neurotoxic effects published in the peer reviewed literature are still lacking.

5.5 THE MILAN AUTOMATED NEUROBEHAVIORAL SYSTEM

5.5.1 Background

The Milan Automated Neurobehavioral System (MANS) was developed in an "... attempt to convert six out of seven tests of the WHO-NCTB into an automated form.."[16] The authors state that in doing so they use the assembled expertise of the international group gathered by the WHO and the NIOSH in Cincinnati, OH 1983. Furthermore, they state that the reliability of the tests should be increased, in comparison with the standard forms, by the computerization, and figures for the "equivalence" of traditional and computerized version are given. These correlations vary between 0.73 for the Symbol Digit test to 0.79 for the Benton memory test. Still, these figures allow for 38–47% unique variance for the different forms. This means, e.g., that half of what is measured by the computerized Symbol Digit test is something else from what is measured by the traditional Digit Symbol test.

5.5.2 Equipment

The MANS runs on IBM compatible personal computers. A color screen is used, as well as a graphic tablet. The programming language used for this test battery is Pascal.

5.5.3 Tests

Tasks included in the MANS are computerized analogues to six of the tests in the WHO-NCTB and the POMS mood scale. The six tests are simple reaction time, digit span, serial digit learning, digit symbol, visual recognition, and aiming pursuit.

5.5.4 Validation

Judging from the number of publications in the scientific literature, validation for the MANS is not impressive. One study performed in the U.S.A. reports on the effects of the potential confounders age, sex, and education differences.[46] In another publication the authors of the battery present data from seven different groups exposed to various neurotoxic agents.[16] However, no references to original studies in the international literature are given. Furthermore, as pointed out by Laursen[15] "... this computerized version seems of little use, because the intention of the Battery (*i.e., the WHO-NCTB*) was the opposite, i.e., to select a set of tests applicable at places where only limited resources are available — as, for instance, in the developing countries."

5.6 THE MICROTOX SYSTEM

5.6.1 Background

The MicroTox system[17] was developed "... for the following reasons; standardizing procedures of administration; making measurements more objective, reliable, and easily scored; and orienting staffing recruitment and training more to the needs of the tested population rather than to the needs of follow-up litigation." The MicroTox represents an interesting approach, as the choice of tests was based not only on sensitivity to neurotoxic insult, but based also on a theory regarding cognitive factors and elementary cognitive tasks.[47]

5.6.2 Equipment

The MicroTox system was developed for the Apple II computer. Additional peripheral equipment needed was an external clock card, and a custom made keyboard.

5.6.3 Tests

The MicroTox system is fully automated, and capable of presenting 17 different tasks, related to six of the eight "paradigms" described by the theory.[47] The interested reader is referred to references 17 and 47 for more details on the tests.

5.6.4 Validation

As the computer company discontinued the production of the Apple II computer, the MicroTox system never became widely spread. Thus, validation available is limited to a couple of pilot studies, using alcohol and carbon monoxide exposures, performed by the authors of the system.

5.7 THE BEHAVIORAL ASSESSMENT AND RESEARCH SYSTEM

5.7.1 Background

The Behavioral Assessment and Research System (BARS)[18] is an interesting approach that is presently being developed. The authors of the system are implementing tasks with demonstrated sensitivity to neurotoxic insult on Macintosh computers. The novel idea implemented with the BARS is to use shaping procedures for the instructions instead of the more common written and/or oral verbal instructions. The BARS system was presented at the recent symposium on computerized testing in Portland, Oregon.[48]

5.7.2 Equipment

The BARS is administered on Macintosh computers, using the standard keyboard with an overlay covering redundant keys.

5.7.3   Tests

As yet, the BARS comprises only a few tests. The symbol digit test, simple reaction time, a selective attention task, and a vigilance attention task are included so far in the system.[48]

5.7.4   Validation

Validation of the BARS[18] is naturally limited, since the development was started recently. Still, there is an international publication presenting experimental data regarding the novel way of presenting instructions. The data obtained indicate that shaping procedures may be effectively used to present test instructions. Such procedures may have a potential for efficiently testing people who are illiterate, or at least have poor language skills. Thus, shaping procedures may be useful to overcome at least some of the bias in performance testing caused by language and other inter-cultural differences.

6   STANDARDIZATION EFFORTS

6.1   ONE STANDARDIZED TEST BATTERY

The world-wide use of "a standard test battery" in all studies on the effects from exposure to neurotoxic agents in the workplace certainly would enhance the possibilities to compare results from one study to the other, and from one country to the other. Such a development would allow for a much faster growth of knowledge within this area of research. However, history has shown that efforts at achieving standardization of methodology are often more or less fruitless. To illustrate this problem, one need only search the literature for neurotoxicological studies, reported during recent years, that are not using the WHO-NCTB.

One important reason for this is that most researchers and clinicians are reluctant, and rightly so, to give up the extensive experience gained with the methods he or she has been using for a long time. Starting to use another, maybe totally new, test battery is likely to reduce one's productivity for a considerable period of time.

The present lack of knowledge regarding the normal functioning of the central nervous system, as well as the possible behavioral expressions of neurotoxic insult, is most probably another important reason for the low degree of acceptance of a single, standard test battery. At present, we simply do not have the information needed to establish an exhaustive test battery.

6.2   A MINIMAL SET OF COMMON TESTS

In a recent review of computerized testing in neurobehavioral toxicology, a "Minimal Common Core Computerized Battery, MCCCB," to be used as an addition to other tests, was proposed by Iregren and Letz.[2] They stated that "It is recognized that the probability of acceptance of a MCCCB is inversely proportional to the number of tests included. Therefore, we suggest that the MCCCB should consist of three tests, at least for the present." The tests proposed for inclusion in the MCCCB were simple reaction time, symbol-digit substitution, and finger tapping. Reaction time and finger tapping were to be measured the same way as in the SPES system, and the symbol digit test should be identical to the NES version.

To our knowledge, even the proposal of the addition of this small "test battery" to other tests used has had little or no effect during the three years passed since its presentation. Still, it is our sincere hope that some standardization may be reached within computer-assisted neurotoxicological testing. There is definitely a point to be made gaining possibilities to express effects of one toxic agent in direct comparison with the effects of another, or to compare the effects caused by an exposure with the effects from, say, aging a certain number of years.

7 BEHAVIORAL DATA AND ENVIRONMENTAL REGULATION

Agencies in many different countries summarize the evidence for health effects from numerous compounds in documents serving as a basis for the regulation of the general environment, or specifically the work environment. Such documents are prepared, e.g., by Dutch, Swedish, U.S., and other agencies.[49] There is considerable variation as to the extent to which data from behavioral studies are taken into account in these reviews. However, during recent years, an increased willingness to acknowledge the importance of behavioral effects in the setting of exposure limits may be noted. This is be seen e.g., within the European Community, where countries like the Federal Republic of Germany and the United Kingdom are presently changing their policies towards the acceptance of behavioral evidence of neurotoxicity. The regulating authorities in Sweden are asking more and more not only for data relating to behavioral measures such as performance, but even to data regarding the perception of discomfort.

The process of actually arriving at an occupational exposure limit (OEL), following the compilation of such "criteria documents," varies considerably between countries, and it has been described elsewhere.[50]

In Sweden, the establishment of new or revised occupational exposure limits is performed during a process of discussion, where officials of the government work closely together with representatives of both employer's and employee's organizations.[51] Toxicological data, as well as technical and economical aspects are taken into consideration, and nowadays quite extensive documents with "consequence descriptions" are used during these discussions. For this reason, it is possible for the latest revisions of Swedish OELs to follow in detail the considerations which constitute the basis for the limit values established. From these documents, it is obvious that behavioral performance data, and even data concerning perceived comfort, are increasingly considered important information for the establishment of Swedish exposure limits.

8 CONCLUSIONS

The following conclusions regarding the use of computer-assisted behavioral testing in occupational neurotoxicology may be drawn:

1. There are obvious advantages gained from using computer-assisted testing as compared to traditional testing formats.
2. There is considerable validation available in the scientific literature for at least some of the test systems presented here.
3. There are some efforts at standardization of computerized test methods for use in occupational neurotoxicology.
4. There are continuous efforts at development within single systems, i.e., none of the prominent test systems used in occupational neurotoxicology is static.
5. There is evidence that data collected using behavioral methods increasingly are being included in the scientific bases for occupational exposure limits.

9 REFERENCES

1. Bartram, D., "Computer-based assessment," *Int. Rev. Ind. Organ. Psychol.*, 9, 31, 1994.
2. Iregren, A. and Letz, R., "Computerized testing in neurobehavioural toxicology,"*Appl. Psychol. Int. Rev.*, 41, 247, 1992.
3. Wilson, S. and McMillan, T., "Computer-based assessment in neuropsychology." In *A Handbook of Neuropsychological Assessment*, Crawford, J. R., Parker, D. M. and McKinlay, W. W., Eds., Lawrence Erlbaum Associates LTD, Hove, UK, 1992.
4. Kane, R. L. and Kay, G. G., "Computerized assessment in neuropsychology: A review of tests and test batteries," *Neuropsych. Rev.*, 3, 1, 1992.
5. Almkvist, O. and Bäckman, L., "Detection and staging of early clinical dementia," *Acta Neurol.Scand.*, 88, 10, 1993.
6. Wilson, S. L. and McMillan, T. M., "Finding able minds in disabled bodies," *Lancet*, 8521/22, 1444, 1986.
7. French, C. C. and Beaumont, J. G., "The reaction of psychiatric patients to computerized assessment," *Brit. J. Clin. Psychol.*, 26, 267, 1987.
8. Beaumont, J. G. and French, C. C., "A clinical field study of eight automated psychometric procedures: the Leicester/DHSS project," *Int. J. Man-Machine Stud.*, 26, 661, 1987.
9. Watts, K., Baddeley, A., and Williams, M., "Automated tailored testing using Raven's Matrices and the Mill Hill Vocabulary tests: a comparison with manual administration," *Int. J. Man-Machine Stud.*, 17, 331, 1982.
10. Weiss, D. J., "Adaptive testing by computer," *J. Consult. Clin. Psychol.*, 53, 774, 1985.
11. Johnson, B., Baker, E., Batawi, M., Gilioli, R., Hänninen, H., Seppäläinen, A., and Xintaras, C., Eds., *Prevention of Neurotoxic Illness in Working Populations*, John Wiley & Sons, New York, 1987.
12. Gamberale, F., Iregren, A., and Kjellberg, A. "Computerized performance testing in neurotoxicology. Why, what, how, and whereto? The SPES example." In *Behavioral Measures of Neurotoxicity*, Russel, R. W., Flattau, P. E., and Pope, A. M., Eds., National Academy Press, Washington, DC, 1990, 359.
13. Williamson, A. M., "The development of a neurobehavioral test battery for use in hazard evaluations in occupational settings," *Neurotoxicol. Teratol.*, 12, 509, 1990.
14. Baker, E. L., Letz, R., and Fidler, A., "A computer-administered neurobehavioral evaluation system for occupational and environmental epidemiology," *J. Occup. Med.*, 27, 206, 1985.

15. Laursen, P., "A computer-aided technique for testing cognitive functions validated on a sample of Danes 30 to 60 years of age," *Acta Neurol. Scand.*, 82, 82, 1990.
16. Cassitto, M. G., Gilioli, R., and Camerino, D., "Experiences with the Milan Automated Neurobehavioral System (MANS) in occupational neurotoxic exposure," *Neurotoxicol. Teratol.*, 11, 571, 1989.
17. Eckerman, D. A., Carroll, J. B., Foree, D., Gullion, C. M., Lansman, M., Long, E. R., Waller, M. B., and Wallsten, T. S., "An approach to brief field testing for neurotoxicity," *Neurobehav. Toxicol. Teratol.*, 7, 387, 1985.
18. Anger, W. K., Rohlman, D. S., and Sizemore, O. J., "A comparison of instruction formats for administering a computerized behavioral test," *Behav. Res. Method. Instrum. Comp.*, 26, 209, 1994.
19. Gamberale, F. and Hultengren, M., "Toluene exposure II. Psychophysiological functions," *Work Environ. Health*, 9, 131, 1972.
20. Gamberale, F. and Hultengren, M., "Methylchloroform exposure: II. Psychophysiological functions," *Work Environ. Health*, 10, 82, 1973.
21. Gamberale, F. and Hultengren, M., "Exposure to styrene. II. Psychological functions," *Work Environ. Health*, 11, 86, 1974.
22. Gamberale, F., Annwall, G., and Hultengren, M., "Exposure to methylene chloride II. Psychological functions," *Scand. J. Work Environ. Health*, 1, 95, 1975.
23. Gamberale, F., Annwall, G., and Hultengren, M., "Exposure to white spirit II. Psychological functions," *Scand. J. Work Environ. Health*, 1, 31, 1975.
24. Gamberale, F., Annwall, G., and Anshelm Olson, B., "Exposure to trichloroethylene II. Psychological functions," *Scand. J. Work Environ. Health*, 4, 220, 1976.
25. Gamberale, F., Annwall, G., and Hultengren, M., "Exposure to xylene and ethylbenzene III. Effects on central nervous functions," *Scand. J. Work Environ. Health*, 4, 204, 1978.
26. Knave, B., Anshelm Olson, B., Elofsson, S. A., Gamberale, F., Isaksson, A., Mindus, P., Persson, H. E., Struwe, G., Wennberg, A., and Westerholm, P., "Long-term exposure to jet fuel. A cross-sectional epidemiologic investigation on occupationally exposed workers with special reference to the nervous system," *Scand. J. Work Environ. Health*, 4, 19, 1978.
27. Iregren, A., "Effects on psychological test performance of workers exposed to a single solvent (toluene) — A comparison with effects of exposure to a mixture of organic solvents," *Neurobehav. Toxicol. Teratol.*, 4, 695, 1982.
28. Elofsson, S. A., Gamberale, F., Hindmarsh, T., Iregren, A., Isaksson, A., Johnsson, I., Knave, B., Lydahl, E., Mindus, P., Persson, H. E., Philipson, B., Steby, M., Struwe, G., Söderman, E., Wennberg, A., and Widén, L., "Exposure to organic solvents. A cross-sectional epidemiologic investigation on occupationally exposed car and industrial spray painters with special reference to the nervous system," *Scand. J. Work Environ. Health*, 6, 239, 1980.
29. Iregren, A., "Psychological test performance among foundry workers exposed to low levels of manganese," *Neurotoxicol. Teratol.*, 12, 673, 1990.
30. Lucchini, R., Selis, L., Folli, D., Apostoli, P., Mutti, A., Vanoni, O., Iregren, A., and Alessio, L., "Neurobehavioral effects of manganese in workers from a ferroalloy plant after temporary cessation of exposure," *Scand. J. Work Environ. Health*, 21, 143, 1995.
31. Mergler, D., Huel, G., Bowler, R., Iregren, A., Bélanger, S., Baldwin, M., Tardif, R., Smarigiassi, A., and Martin, L., "Nervous system dysfunction among workers with long-term exposure to manganese," *Environ. Res.*, 64, 151 — 180, 1994.

32. Sjögren, B., Iregren, A., Frech, W., Hagman, M., Johansson, L., Tesarz, M., and Wennberg, A., "Effects on the Nervous System in Welders Exposed to Aluminium or Manganese" (in Swedish), *Arbete och Hälsa*, 1994:27, National Institute of Occupational Health, S-17184 Solna, Sweden.

33. Enander, A., "Effects of moderate cold on performance of cognitive and psycho-motor tasks," *Ergonomics,* 30, 1431, 1987.

34. Kjellberg, A., Sköldström, B., Andersson, P., and Lindberg, L., "Fatigue Effects of Noise among Airplane Mechanics," *Internoise 91*, 1991, Sydney.

35. Kjellberg, A. and Sköldström, B., "Noise annoyance during performance of different non-auditory tasks," *Percept. Mot. Skills,* 73, 39, 1991.

36. Gamberale, F., Anshelm Olson, B., Eneroth, P., Lindh, T., and Wennberg, A., "Acute effects of ELF electromagnetic fields. A field study of linesmen working with 400kV power lines," *Scand. J.Work Environ. Health,* 46, 729, 1989.

37. Razmjou, S. and Kjellberg, A., "Sustained attention and serial responding in heat: Mental effort in the control of performance," *Aviat. Space Environ. Med.,* 63, 594–601, 1992.

38. Anshelm Olson, B., Gamberale, F., and Iregren, A., "Coexposure to toluene and *p*-xylene in man: central nervous functions," *Br. J. Ind. Med.,* 42, 117, 1985.

39. Wickens, C. D. *Engineering Psychology and Human Performance*, Charles Merrill, Columbus, OH, 1984.

40. Williamson, A. M. and Teo, R. K. C., "Neurobehavioural effects of occupational exposure to lead," *Br. J. Ind. Med.,* 43, 374, 1986.

41. Williamson, A. M., Teo, R. K. C., and Sanderson, J., "Occupational mercury exposure and its consequences for behaviour," *Int. Arch. Occup. Environ. Health,* 50, 273, 1982.

42. Williamson, A. M., Clarke, B., and Edmonds, C., "Neurobehavioral effects of professional abalone diving," *Br. J. Ind. Med.,* 44, 459, 1987.

43. Williamson, A. M. and Winder, C., "A prospective cohort study of the chronic effects of solvent exposure," *Environ. Res.,* 62, 256, 1993.

44. Letz, R., "Use of computerized test batteries for quantifying neurobehavioral outcomes," *Environ. Health Perspect.,* 90, 195, 1991.

45. Letz, R., "Covariates of computerized neurobehavioral test performance in epidemiologic investigations," *Environ. Res.,* 61, 124, 1993.

46. Fittro, K. P., Bolla, K. I., Heller, J. R., and Meyd, C. J., "The Milan Automated Neurobehavioral System: Age, sex, and education differences," *J. Occup. Med.,* 34, 918, 1992.

47. Carroll, J. B., "Individual Difference Relations in Psychometric and Experimental Cognitive Tasks," The L L Thurstone Psychometric Laboratory, NTIS Document AD-A086 057, The University of North Carolina, Chapel Hill, NC, 1980.

48. Anger, W. K., Rohlman, D., Sizemore, O. J., and Kovera, C., "Behavioral Assessment and Research System; Computerized Behavioral Testing of Humans in Neurotoxicology Research," 1995, Portland, OR.

49. Beije, B. and Lundberg, P., Eds., "Occupational Exposure Limits — Health Based Values or Administrative Norms?" *Arbete och Hälsa 1993*,15, National Institute of Occupational Health, S-171 84 Solna, Sweden.

50. Holmberg, B. and Lundberg, P., "Assessment and management of occupational risks in the Nordic (Scandinavian) countries," *Am. J. Ind. Med.,* 15, 615, 1989.

51. Lundberg, P. and Holmberg, B., "Occupational standard setting in Sweden — procedure and criteria," *Ann. Em. Conf. Ind. Hyg.,* 12, 249–252, 1985.

52. Gilioli, R., Cassitto, M. G., and Foá, V., Eds, *Neurobehavioral Methods in Occupational Health, Advances in the Biosciences*, Vol. 46, Pergamon Press, Oxford, UK, 1983.

53. WHO *Neurobehavioral Methods in Occupational and Environmental Health: Extended Abstracts from the Second International Symposium, Environmental Health,* Vol. 3, WHO Regional Office for Europe, Copenhagen, 1985.

54. Russell, R. W., Flattau, P. E., and Pope, A. M., Eds., *Behavioral Measures of Neurotoxicity;* National Academy Press, Washington, DC, 1990.

55. Johnson, B. L., Anger, W. K., Durao, A., and Xintaras, C., Eds., *Advances in Neurobehavioral Toxicology: Applications in Environmental and Occupational Health,* Lewis, Chelsea, MI, 1990.

56. Araki, S., Ed., *Neurobehavioral Methods and Effects in Occupational and Environmental Health,* Academic Press, San Diego, 1994, pp 1020.

57. Lisper, H. O. and Kjellberg, A., "Effects of 24-hour sleep deprivation on rate of decrement in a 10-minute auditory reaction time task," *J. Exp. Psychol.,* 96, 287, 1972.

58. Anshelm Olson, B., Gamberale, F., and Grönqvist, B., "Reaction time changes among steel workers exposed to solvent vapors. A longitudinal study," *Int. Arch. Occup. Environ. Health,* 48, 211, 1981.

59. Anshelm Olson, B., "Effects of organic solvents on behavioral performance of workers in the paint industry," *Neurobehav. Toxicol. Teratol.,* 4, 703, 1982.

60. Gamberale, F. and Svensson, G., "The effect of anaesthetic gases on the psychomotor and perceptual functions of anaesthetic nurses," *Work Environ. Health,* 11, 108, 1974.

61. Gamberale, F., Lisper, H. O., and Anshelm Olson, B. "The effect of styrene vapor on the reaction time of workers in the plastic boat industry." In *Adverse Effects of Environmental Chemicals and Psychtropic Drugs,* Horvath, M., Ed., Elsevier, Amsterdam, 1976, Vol. 2, p. 135.

62. Iregren, A., "Subjective and objective signs of organic solvent toxicity among occupationally exposed workers: An experimental evaluation," *Scand. J. Work Environ Health,* 12, 469, 1986.

63. Iregren, A., Åkerstedt, T., Anshelm Olson, B. and Gamberale, F., "Experimental exposure to toluene in combination with ethanol intake," *Scand. J. Work Environ Health,* 12, 128, 1986.

64. Iregren, A., Almkvist, O., Klevegård, M., and Åslund, U., "A clinical validation of six computerized tests for diagnosing solvent caused occupational illness" (in Swedish), *Arbete och Hälsa 1987:*13, National Board of Occupational Safety and Health, S-171 84 Solna, Sweden.

65. Iregren, A., Tesarz, M., and Wigaeus-Hjelm, E., "Human experimental MIBK exposure: Effects on heart rate, performance, and symptoms," *Environ. Res.,* 63, 101, 1993.

66. Kjellberg, A., Wigaeus, E., Engström, J., Åstrand, I., and Ljungquist, E., "Long-term effects from styrene exposure at a plastic boatyard" (in Swedish), *Arbete och Hälsa 1979:*18, National Board of Occupational Safety and Health, S-171 84 Solna, Sweden.

67. Kjellberg, A. and Strandberg, M., "The effects of anaestethic gases on reaction time of anaesthetic nurses" (in Swedish), *Undersökningsrapport 1979:*11, National Board of Occupational Safety and Health, S-171 84 Solna, Sweden.

68. Kjellberg, A. and Wisung, H., "Metrical properties in a computer administered test battery for investigations in behavioral toxicology" (in Swedish), *Undersökningsrapport 1987:*1, National Board of Occupational Safety and Health, S-171 84 Solna, Sweden.

69. Knave, B., Gamberale, F., Bergström, S., Birke, E., Iregren, A., Kolmodin-Hedman, B., and Wennberg, A., "Long-term exposure to electric fields. A cross-sectional epidemiologic investigation of occupationally exposed workers in high-voltage substations," *Scand. J. Work Environ. Health,* 2, 115, 1979.

70. Wigaeus Hjelm, E., Hagberg, M., Iregren, A., and Löf, A., "Exposure to methyl isobutyl ketone: Toxicokinetics and occurrence of irritative and CNS symptoms in man," *Int. Arch. Occup. Environ. Health*, 62, 19, 1990.

71. Hänninen, H., Matikainen, E., Kovala, T., Valkonen, S., and Riihimäki, V., "Internal load of aluminium and the central nervous system function of aluminium welders," *Scand. J. Work Environ. Health*, 20, 279, 1994.

72. Kjellberg, A., Sköldström, B., Tesarz, M., and Dallner, M., "Facial EMG responses to noise," *Work Stress*, 79, 1203, 1994.

73 Kjellberg, A., Sköldström, B., Andersson, P., and Lindberg, L., "Fatigue effects of noise among airplane mechanics," *Work Stress*, in press.

74. Landström, U., Byström, M., and Kjellberg, A., "Annoyance and performance related to frequency of noise exposure and type of work." In *Proc. Nordic Acoustical Meeting 90*, Sundbäck, U., Ed., Luleå University of Technology, Luleå, Sweden, 1990, p 177.

75. Landström, U., Kjellberg, A., and Lundström, R., "Combined effects of exposures to noise and whole-body vibration in dumpers, helicopters and railway engines," *J. Low Frequation Noise Vib.*, 12, 75, 1993.

76. Landström, U., Kjellberg, A., and Byström, M., "Acceptable levels of tonal and broadband repetitive and continuous sounds during the performance of non-auditory tasks," *Percept. Mot. Skills*, in press

77. Landström, U., Kjellberg, A., and Byström, M., "Acceptable levels of sounds with different spectral characteristics during the performance of a simple and a complex non-auditory task," *J. Sound Vib.*, 160, 533, 1993.

78. Gamberale, F., Kjellberg, A., and Razmjou, S., *Effects of moderate cold and heat on performance of sensory-motor and cognitive tasks*, Report No 692, Department of Psychology, University of Stockholm, Sweden, 1989.

79. Kjellberg, A. and Wide, P., "Effects of simulated ventilation noise on performance of a grammatical reasoning task." In *Proc. 5th Int. Congr. Noise as a Public Health Problem. Vol. 3*, Berglund, B., Berglund, U., Karlsson, J. and Lindvall, T., Eds., Swedish Council for Building Research, Stockholm, 1988, p 31.

12 Computer Systems in Prevention and Diagnosis of Occupational Neurotoxicity

Sara Lloyd, Alan H. Hall, and Barry H. Rumack

CONTENTS

0-8493-9231-4/98/$0.00+$.50
© 1998 by CRC Press LLC

1   INTRODUCTION

Computerized information systems and integrated medical surveillance/biological monitoring tracking databases are two generic types of computer systems useful in identifying potential neurotoxicants, other hazards, and location/process-specific neurotoxic risks. The following discussion provides an overview of some existing systems and gives practical guidance regarding selection, implementation, and use of these tools.

2   COMPUTERIZED INFORMATION SOURCES OVERVIEW

A wide variety of computerized information systems are available. These systems can be categorized as:

1. databases which contain specific, limited data sets
2. special purpose systems
3. bibliographic databases and abstracting services
4. multipurpose, comprehensive toxicological and medical information bases.

Some systems have features of more than one of the above categories. A second major type of computer systems are medical surveillance/biological monitoring tools.

Major uses of the above information systems are to identify potential health and safety hazards, preventive or treatment measures, and strategies for safe storage and handling of causative agents. A thorough risk assessment requires use of more than one information system, or use of a comprehensive information base providing access to multiple databases.

2.2   Specific Data Set Systems

Examples include the CHEMID and CHEMLINE® systems, available through the U.S. National Library of Medicine (NLM) MEDLARS® online service. These databases are useful for determining unique chemical identifiers, such as Chemical Abstracts Service (CAS) numbers (including "old" CAS numbers), other identifiers such as UN/NA numbers, RTECS (US National Institute for Occupational Safety and Health Registry of Toxic Effects of Chemical Substances) numbers, RCRA hazardous waste numbers, molecular formulae, synonyms, international synonyms and designations, and other specific chemical identifiers.

2.2.1   Applications

Retrieving correct identifiers for the chemical(s) of interest is necessary for searching for toxicological/medical information in bibliographic citation or data-bases. It also familiarizes the investigator with terms found in the literature pertaining to the specific chemical(s) of concern, as well as to similar chemicals or chemical groups which may have common toxicological or hazardous properties.

The utility of these systems is particularly apparent when evaluating substances within similar chemical groups, but which are quite different toxicologically. For example, while many arsenic compounds are neurotoxic (causing peripheral axonopathies), trimethylated organic arsenical compounds, such as arsenobetaine and arsenocholine ("fish arsenic"), found in seafood are essentially nontoxic.[1] CHEMID, CHEMLINE® or similar specific data set systems enable researchers to quickly obtain needed chemical toxicity or hazard information.

Using such systems allows identification of common synonyms. Users may thus be able to recognize that "aethylmethylketon" (in the German literature) pertains to methyl ethyl ketone (MEK; 2-butanone), while the abbreviation TCE refers to trichloroethylene and not to tetrachloroethylene (PCE).

The NLM and other similar specific data set systems list available databases containing information about the chemical(s) in question, and also provide specific terms used in these databases for information retrieval. A user thus efficiently enters appropriate terms and limit searches to databases known to contain desired chemical information. This reduces online search time and costs.

Other uses of specific data set systems include identification of the toxicity and hazards of structurally similar chemicals (which may be helpful for evaluating substances for which a hazard evaluation is needed, but about which there is no or very limited available information). CHEMID and CHEMLINE® allow searching by partial or complete structural formulas. Structure-activity relationships may thus be considered by the investigator. Performing such searches are more complex than conducting a simple search on a chemical name or identifier. NLM provides training for conducting such searches. It may be helpful to obtain guidance from a trained research librarian.

2.2.2 Choosing Systems

Specific data set systems may not be necessary for information retrieval on specific chemicals which have terms and unique identifiers widely understood and universally applied on a consistent basis (e.g., benzene; nickel). Thorough information, however, is quite often only obtained with initial searches in specific data set systems. Features and functions which should be considered when selecting a specific data set system are summarized in Table 1.

CHEMLINE® provides a very comprehensive data set for most chemicals, but the cost is generally greater than that of other databases, such as CHEMID. Although CHEMID generally provides less extensive information than CHEMLINE®, the information obtained from the former system is often sufficient for a comprehensive hazard evaluation. Online connect times for either CHEMLINE® or CHEMID are usually minimal, even when there are numerous entries for chemically related substances.

Performing a search in either of these databases usually takes less time and results in better and more accurate results than manually searching through reference texts for unique chemical identifiers and key synonyms. When many chemicals or unfamiliar chemicals are to be evaluated for potential health hazards, using a specific data set system such as CHEMID or CHEMLINE® is highly recommended.

TABLE 1
Considerations in Selecting Various
Computerized Systems

Specific Data Set Systems
- Breadth of Information
- Specific Features
  — Structural formulae
  — Foreign synonyms
  — Identifies hits in databases

Bibliographic and Abstracting Services
- Can Customize Search
- CAS# Searching
- Journal Coverage
- Adverse Effects Categorization
- Abstract and Key Word Accuracy

Special Purpose Systems
- Integration with Existing Systems
- Functionality
- Technical Support
- Enhancements
- Scientifically Valid

Comprehensive, Multipurpose Systems
- Breadth
- Quality of Information
- Peer Reviewed
- Updating Frequency
- Searchable
  — CAS #
  — Specific topics
  — Detailed adverse effects
  — Absence of effects

2.3 BIBLIOGRAPHIC/ABSTRACTING SERVICES

Bibliographic and abstracting services available for either online or CD-ROM access enable users to search and retrieve literature citations and abstracts pertaining to key words or other identifiers. Examples of these systems are NIOSHTIC®, CDPlus®, Current Contents®, BIOSIS® and the NLM MEDLARS® system (MED-LINE®, TOXLINE®, and other TOXNET® bibliographic citation or fact databases). A public domain system providing references to over 32,000 publications concerning poisonous plants is offered by the FDA on the INTERNET.[2,3]

2.3.1   Benefits

Currently available systems are much more user friendly than was the case in the past. Modern menu-driven or graphical user interfaces (GUI's) guide users through the steps necessary to conduct a search, provide on-screen help, and recommended MeSH or other searchable terms. CD-ROM systems (for either stand-alone personal computers or multi-user networks) provide unlimited searching for a fixed annual cost. Such systems provide up-to-date literature citation and abstract access.

Abstract retrieval aids users in limiting costly library or literature service full-text article retrieval. Some literature citation and abstracting services, such as the Genuine Article® offered with the Current Contents® product, allow users to directly order hard-copy publications. This is useful for obtaining international literature when local libraries do not have ready access, or when the user's location is remote from university or other scientific/medical libraries.

2.3.2   Limitations

Bibliographic citation systems do have certain limitations. Abstracts and key words provided may or may not have been written by the original author(s) (i.e., in MEDLINE® and TOXLINE®, abstracts are those written by the paper's authors, while in the NIOSHTIC database, an independent reviewer's abstract is provided). Regardless of the abstract's source, the content may not always accurately reflect the main body of the paper. Abstracts are often not available, particularly for older journal articles, or may not be available in the user's native language.

Key words and unique identifiers may be inaccurate, regardless of whether or not the author(s) provided this information. Some systems do not allow for searching by CAS number (one of the most specific identifiers for individual chemicals). Search terms may also be of questionable value in some systems: a search on the neurotoxic metal "lead" in one system results in many false retrievals of information pertaining to "EKG leads," for example.

Journal article ordering services and online searching can be relatively costly. Users should have search strategies pre-determined and at hand before logging-on to online services, and may first wish to identify local sources in order to reduce costs.

User-friendly systems, such as Grateful Med, CDPLUS®, and Current Contents®, cannot always guarantee complete chemical toxicology and hazard information retrieval. Failure of complete retrieval may be due to inadequate search approaches, as some systems do not allow viewing of results at each step. Some bibliographic information retrieval systems do not allow users to revise or customize searches based on interim retrieval results.

Specific system-provided searchable categories do not always sufficiently differentiate adverse effects, and may not provide search options for chemical-induced neurotoxicity. Either excessive or very limited retrieval of literature citations may thus result.

Journals cited by bibliographic and abstracting services may not cover critical chemical toxicology or hazard information, particularly if the focus is on medical journals and does not also include those publications addressing basic and applied toxicology and hazard research. For obscure chemicals, environmental and life sciences journals may need to be included.

2.3.3 Recommendations

Basic aspects which one should consider when selecting bibliographic and abstracting services are summarized in Table 1. Users should understand the limitations of chosen bibliographic citation systems in terms of ability to customize searches, to limit retrieval to the specific chemical(s), chemical group(s), or effect(s), and breadth of journal coverage (including publication dates as much useful human chemical exposure data are found in citations older than those included in some bibliographic information systems).

Hazard identification focusing primarily on human data may be well served by systems such as MEDLINE®. However, hazard identification based on human, animal, and *in vitro* data, or for less common chemicals, should utilize literature citation services covering a broader range of toxicological, basic science(s), and medical journals. TOXNET®, Pol-Tox®, and NIOSHTIC® are examples of literature citation services which may have broader coverage.

Abstracts and key terms (including CAS numbers) listed for citations can be helpful, but should not be relied upon exclusively, unless they have been confirmed through review of primary journal articles. A knowledge of the manner in which various systems develop provided abstracts, key words, and entries for each citation is useful.

Users require a basic understanding of Boolean search logic (including the concepts of "and," "or," and "and not") to appreciate inherent limitations of bibliographic citation databases. In some cases, a search strategy may be "logged-in" to a regularly-updated bibliographic citation database, such that any new information posted is automatically transmitted to the user. Automatic generation of NLM database update searches through STORESEARCH and SDI are examples.

Sources of free or low cost information should be considered. University libraries may have CD-ROM or other locally-available batabases available on a "first-come, first-served" basis. In the U.S., certain regional libraries in the Depository Library System provide free access to public domain U.S. government publications (hard copy and electronic versions), and some offer use of proprietary toxicological databases on CD-ROM.[4] Interlibrary loan services (through a University or other library) may be less costly than ordering full-text journal articles from a citation or abstracting service.

Grateful MED and similar searching tools are inexpensive resources which may be sufficient for users requiring basic literature searches. However, subscription to CD-ROM or network versions of bibliographic citation services which cover a wide range of journal types is likely to be more cost-effective and practical for other users who require frequent and comprehensive information retrieval on many chemical substances. Professionals who must respond quickly to community or workplace concerns are also likely to benefit most by subscribing to citation services.

2.4 Special Purpose Systems

Special purpose systems have specific, limited functions. The Electronic NIOSH pocket Guide to Chemical Hazards (Industrial Hygiene Services, Inc.), the ACGIH TLVs and other Occupational Exposure Values on CD-ROM, and Risk Assessment programs, such as Cristobal®, Risk* Assistant™, and physiologically-based pharmacokinetic (PBPK) modeling programs are examples of systems designed for retrieving exposure limits information or calculating exposure and risk probability. Risk assessment programs (such as Risk* Assistant™) provide education and training about risk assessment processes, in addition to generating risk estimates.[5]

2.4.1 Recommendations

Key consideration for selection of special purpose systems are determination of specific data requirements, updating frequency, technical support availability, and other factors as summarized in Table 1.

Risk assessment programs vary greatly in utility, ease of use, and ability to be used with various hardware. Users of such systems should have access to the formulae, assumptions, and defaults used in risk assessment tools (consideration should also be given to a contract which provides the user with the source code for such systems, in the event that the system providee ceases business). Understanding on-going developments and uncertainties in the risk assessment process (such as presented by the U.S. National Research Council, 1994), is essential to determine the scientific validity and, thus, the utility of a computerized risk assessment tool.[6]

2.5 Comprehensive Multipurpose Systems

Growing demand for a wide range of information in a single computerized system has resulted in availability of comprehensive multipurpose systems providing toxicological, medical, industrial hygiene, safety, and regulatory information. Such systems vary in scope and formats, but all attempt to provide access to multiple databases through single systems, with consistent search and retrieval strategies for ease of use.

Examples of available multipurpose systems include: the Canadian CCOHS CcinfoDisc series of bibliographic citation and fact databases (online and CD-ROM access), the Silver Platter® series of toxicology databases, and the Micromedex, Inc. TOMES Plus® and related Environmental Health and Safety products. Other publishers are providing portions of such information on CD-ROM, such as the U.S. Agency for Toxic Substances and Disease Registry (ATSDR) Toxicological Profiles and Sax's Dangerous Properties of Industrial Materials.

2.5.1 Cautions

Regardless of the system(s) utilized, users should be aware that chemical substances may be discussed as part of a chemically or toxicologically similar group, with little or no discussion of unique toxic effects associated with the specific chemical of interest. Toxic effects of trivalent chromium, for example, might be discussed with those of hexavalent chromium, without differentiation of which effects are associated with which form of the metal.

2.5.2 Selecting multipurpose systems

Given the variety of available multipurpose systems, selection of the most useful system(s) may be challenging. Basic features which should be considered are summarized in Table 1. The types and quantity of information needed, ease of use, searching capabilities, technical support, and frequency of file updating should be considered during the selection process. The quality, quantity, and timeliness of provided information should be carefully evaluated to determine if the system meets the user's needs.

3   MEDICAL SURVEILLANCE/BIOLOGICAL MONITORING SYSTEMS

Computerized medical surveillance/biological monitoring systems can be a solution to the problem of maintaining, tracking, integrating, and using site- or process-specific monitoring data in injury/illness prevention efforts. There are three basic options available for instituting medical surveillance/biological monitoring tracking systems:

1. develop the system from resources available "in-house"
2. hire a consultant or consulting firm to develop the system
3. purchase an existing commercialized system.

A number of frustrations or problems may be encountered in this process, regardless of which of the above alternatives is utilized. The chosen system(s) may not be able to fulfil initial expectations, and may also not be flexible enough to respond to changing, ongoing demands for information retrieval, customized reports, etc. Technical support and source codes are often not available beyond the date of system purchase, making necessary changes and improvements next to impossible.

Lack of significant commitment from upper management and persons using the system, or unacceptability to workers, can defeat system development and applications. Benefits provided by computerization also may not justify the time, resources, and financial outlay. In order to avoid these problems, careful needs analysis should be performed before proceeding.

3.1 FORMAL/INFORMAL NEEDS ANALYSIS

Formal prolem oriented approaches,[7] or more informal needs analysis to system development can aid in determining whether or not a computerized system is needed and of appropriate cost-benefit. A basic approach begins with careful evaluation of the benefits and limitations of the currently used information retrieval and tracking systems. Problems inherent in paper-based surveillance systems (e.g., inaccurate or biased record-keeping) will be transferred into a computer-based system; therefore, methods of decision making and data entry must also be scrutinized.

If it is determined that current information systems are deficient, consider the type of information to be collected, use(s) of such information, logistics of data collection and interpretation, and what actions will be taken based on data retrieval and/or evaluation.

Basic questions which should be addressed include:

Data collection
— what information should be collected and by whom?
— how will the data collection and entry be integrated?
Data interpretation
— who reviews data and makes medical and risk management decisions?
— how are the data interpreted and action levels established?
— what are the action levels?
Decisions based on data collection and interpretation
— environmental redesign and controls?
— personal protective equipment?
— job reassignment or medical follow-up and treatment?
— no action?
Additional considerations
• Time/resource allocation?
• Acceptance by workers, data collectors, and users?
• Can the system be updated or customized?
• Technical support readily available?
• Practical system uses:
    — can generate medical examination schedules, protocols and reports?
    — can be used in epidemiological research?
    — enables record keeping and analysis of accident and compensation claims?

3.2 SURVEILLANCE/MONITORING USES AND INTEGRATION

Computerized medical surveillance/biological monitoring systems should be designed to provide data or analyses which could not be accomplished by existing manual-paper data systems. The computerized systems should focus on prevention, with provisions for early detection and diagnosis of neurotoxicity and other adverse effects. Systems which provide only for identification of disease(s)/injury(ies), are generally not acceptable.

Common goals of such computerized systems are to:

1. provide consistency in provision of medical care and examinations, workplace monitoring, analysis and reporting of results for all workers
2. gather data identifying individuals at risk of excessive exposure or adverse effects

3. maintain a chronological record of each employee's work category, work location, and exposure level(s)
4. identify individuals or groups with adverse health effects
5. document medical treatment and follow-up
6. document health, safety, and Workers' Compensation claims

A variety of data are collected and integrated to meet these goals. The processes and specific chemical/physical exposures by work site/process, or job title must be documented. Industrial hygiene monitoring of work area and individual worker breathing zones are also required. Results of biological monitoring, medical examinations, and histories concerning past work and medical problems complete the data sets needed for the basic surveillance/monitoring system. These data are generally entered into personal computers, with integration by a centralized processing unit.

3.3   CHALLENGES: DATA COLLECTION FOR NEUROTOXICITY ASSESSMENT

Data collection for evaluating specific neurotoxic effects is problematic. Although there have been many advances concerning biological markers and electrophysiological techniques for identifying neurotoxicant exposure and effects, many surveillance systems have not yet incorporated these measurements. Practical issues, including the expense and invasiveness of some procedures, the requirements for baseline and repeated measures, questions regarding data interpretation and use, and the wide variability in normal ranges for some tests, have not yet been resolved.

However, several examples do exist in which medical surveillance systems have been used to identify exposure or effects of neurotoxins. For example, carpal tunnel syndrome has been confirmed through the use of EMG and Mayo Clinic diagnostic criteria.[8] Red blood cell acetylcholinesterase and blood lead levels have proven useful for monitoring potential neurotoxic anticholinesterase pesticide and lead exposures.

3.4   CURRENT STATUS: SURVEILLANCE/MONITORING SYSTEMS

Computerized medical surveillance/biological monitoring systems are generally only available at larger industries which have extensive fiscal and human resources. The full benefits of such systems often have not been realized. Many of the large companies originally developed aspects of the computerized surveillance systems separately and did not foresee the need for future integration or upgrading of the systems. The original systems were based on main-frame technology, and have been difficult to adapt to newer technology which largely relies on personal computers linked to a central processing unit.

All too commonly separate computerized data collection systems are in place, without cross-referencing capabilities or integration. Electronic automatic scheduling of exposure-specific or health effect(s)-specific examinations for workers can be performed by sophisticated medical surveillance/biological monitoring systems. Many users of the computerized systems, however, have not incorporated this function, or have had difficulties due to an inability to override the automated functions.[9]

TABLE 2
Considerations in Selecting and Developing Integrated Surveillance Systems

Needs
- Current System Limitations
- Medical/Toxicological
- Industrial Hygiene
- Safety
- Record-Keeping
- Regulatory Compliance

Functionality
- Cost
- Ease of Use
- Accepted
- Compatible with Other Systems
- Meets Specific Needs
- Adaptable
- Allows User Input

Health Impact
- Prevention
- Reduce Adverse Effects
- Early Recognition and Treatment

Although there is interest in and perceived value of surveillance systems, there is little published information concerning the efficacy and cost-effectiveness of these systems. Survey-based study of the effects of the more focused occupational medical surveillance programs has been inherently flawed, making it difficult to draw conclusions regarding system benefits.[10] Well-conducted prospective or retrospective studies are needed in order to validate the comprehensive computerized systems.

3.5 RECOMMENDATIONS

Basic factors which should be addressed before selecting or developing a computerized medical surveillance/biological monitoring system are presented in Table 2. Critical to the success of the system are adequate resources, management support, acceptance of data collection methods and intent, and system adaptability, with data integration and analysis which results in positive impacts on worker health.

4　REFERENCES

1. ATSDR, *ATSDR Case Studies in Environmental Medicine: Arsenic Toxicity,* Agency for Toxic Substances and Disease Registry, Atlanta, GA, 1990.
2. Wagstaff, D. G., personal communication, 1995.
3. Wagstaff, D. J. and Chambers, T. C., *Poisonous plant bibliographic file in internet,* 34th Annual Meeting of the Society of Toxicology, Baltimore, MD, March 5–9, 1995 (Abstract 277).
4. Beleu, S., personal communication, 1995.
5. Muir, W. R., Young, J. S., and Benes, C. M., *Use of Risk* Assistant personal computer software to teach toxicological risk assessment and exposure assessment,* 34th Annual Meeting of the Society of Toxicology, Baltimore, MD, March 5–9, 1995 (Abstract 280b).
6. U.S. National Research Council, *Science and Judgment in Risk Assessment,* National Academy Press, Washington, DC, 1994.
7. Willson, R. D., "Systems analysis techniques," in *Computerized Occupational Health Record Systems,* The American Conference of Governmental Industrial Hygienists, Cincinnati, OH, 1993, 13-20.
8. Kirschberg, G. J., Fillingim, R., and Davies, V. P., "Carpal tunnel syndrome: classic clinical symptoms and electrodiagnostic studies in poultry workers with hand, wrist, and forearm pain," *South. Med. J.,* 87, 328, 1994.
9. Timlin, E. L., Hillman, G., and Peterson, K. W., "IBM's experience with its environmental, chemical, and occupational evaluation systems, ECHOES," in *Computerized Occupational Health Record Systems,* The American Conference of Governmental Industrial Hygienists, Cincinnati, OH, 1983, 31.
10. Hathaway, J. A., "Medical surveillance: extent and effectiveness?" *J. Occup. Med.,* 35, 698, 1983.

5   ELECTRONIC RESOURCES

ACGIH TLVS™ and other Occupational Exposure Values:
ACGIH 1330 Kemper Meadow Drive, Cincinnati, OH, 45240-1634
Phone: 513-742-2020   Fax: 513-742-3355   E-mail: ACGIH_mem@pol.com

ATSDR'S Toxicological Profiles on CD-ROM:
Lewis Publishers, 2000 Corporate Blvd., N.W., Boca Raton, FL, Florida 33431-9868
Phone: 800-272-7737   Fax: 800-374-3401

CDPLUS®/OVID TECHNOLOGIES: Phone: 800-950-2035   E-mail: cdplus@cdplus.com

CURRENT CONTENTS® and THE GENUINE ARTICLE®:
Institute for Scientific Information, 3501 Market Street, Philadelphia, PA 19104
Phone: 215-386-0100   Fax: 215-386-2911
Phone: 800-336-4474   E-mail: custserv@isinet.com

NATIONAL LIBRARY OF MEDICINE (NLM), 8600 Rockville Pike, Bethesda, MD 20894
General Information Phone: 800-272-4787   E-mail: publicifo@occshost.nlm.nih.gov

RISK* ASSISTANT™:
Thistle Publishing, P.O. Box 1327, Alexandria, Virginia 22313-1327
Phone: 703-684-5203   Fax: 703-684-7704

SAX'S DANGEROUS PROPERTIES OF INDUSTRIAL MATERIALS on CD-ROM:
Richard J. Lewis, Sr., (Available from ACGIH)
ACGIH Phone: 513-742-2020   Fax: 513-742-3355   E-mail: ACGIH_pubs@pol.com

SILVER PLATTER® INFORMATION, Ltd. (POL-TOX® and other databases):
100 River Ridge Drive, Norwood, Massachusetts 02062-5026
Phone: 617-769-2599   Phone 800-343-0064   Fax: 617-769-8763

TOMES PLUS® (RTECS®, HSDB, REPROTEXT®/REPROTOX™, NIOSH Pocket Guide and
other Knowledge Bases for Healthcare, Safety and the Environment):
MICROMEDEX, INC., 6200 S. Syracuse Way, Suite 300, Englewood, Colorado 80111
Phone: 303-486-6400   International Phone: 1-303-486-6444
Phone: 800-525-9083   Fax: 303-486-6464   E-mail: micromedex@mdx.com

U.S. FOOD AND DRUG ADMINISTRATION, POISONOUS PLANT BIBLIOGRAPHIC
   FILE, PLANTOX
D. Jesse Wagstaff, Phone: 202-205-8705   E-mail: DJW@FDACF.SSW.DHHS.GOV
Access Internet file by: Anonymous FTP at VAX8.CFSAN.FDA.GOV and get
ANON.FILES.PUBLIC.PLANT-TOX PLANTOX.REF. (also on World Wide Web)

U.S. GOVERNMENT PRINTING OFFICE:
General Information Phone: 202-512-0000 Library Programs Phone: 202-512-1114
Federal Bulletin Board: 202-512-1387 (set modem to 891 full, duplex)
U.S. Federal Bulletin Board Internet e-mail address: fed.bbs.gpo.gov

13 Risk Assessment for Occupational Neurotoxicants

William Slikker, Jr.

CONTENTS

1 INTRODUCTION

Occupational risk assessment refers to the use of toxicological data to define the relationship between an occupational exposure and adverse health outcome. The goal is to develop a quantitative estimate of the probability that a given health outcome results from a specific exposure experienced at one's workplace, for a particular length of time, by a particular route.[1] Neurotoxicity may be defined as any adverse effect on the structure or function of the central and/or peripheral nervous system by a biological, chemical, or physical agent. Neurotoxic effects may be permanent or reversible, produced by neuropharmacological or neurodegenerative properties of a neurotoxicant, or the result of direct or indirect actions on the nervous system. Adverse effects can include both unwanted effects and any alteration from baseline that diminishes the ability of an organism to survive, reproduce, or adapt to its environment.[2,3] So to put them together, occupational neurotoxicity risk assessment seeks to define the relationship between adverse health outcome and an occupational exposure to an agent that affects the structure or function of the nervous system.

Examples of occupational neurotoxicity are plentiful and therefore the need for assessment and prevention are evident. More than 35 different agents have been reported to produce occupational neurotoxicity.[4] The symptoms resulting from these occupational exposures are diverse and the agents thought to be responsible are equally varied and include gases, metals, organic solvents and pesticides (see Chapter 1). Therefore, the need for comprehensive and multidisciplinary risk assessment techniques is evident.

Although many occupational neurotoxicants have been identified via human case reports and epidemiological studies, they may frequently involve extrapolation based on data obtained from toxicity testing in animals.[1] Therefore, the need to extrapolate toxicity data obtained in animal studies to humans is evident. Considering that inherent species differences including maximal achievable life span, abundance and activity of endogenous chemical metabolizing enzymes and behavioral response indices are prevalent, it is not surprising that the uncertainty of the cross-species extrapolation assumption is large. One method to reduce uncertainty of risk assessment assumptions is to develop a comprehensive database to serve as the foundation for the risk assessment of a particular agent. Such a database may contain multispecies information on multiple endpoints of toxicity, pharmacokinetic data, structure activity relationships and mechanistic data. Concordance of animal and human toxicity data may well be enhanced with the use of databases, although development of such comprehensive databases are frequently expensive and time consuming. Additional approaches such as *in vitro* test methods and predictive tests systems may reduce resource utilization.

Other assumptions inherent in the risk assessment of occupational neurotoxicants involve high to low dose extrapolation. Because the determination of the exposure level or dose is so critical to both the national health and economy, dose-response models and quantitative risk assessment procedures for high to low dose extrapolation are critical. Several methods such as the benchmark dose approach and quantitative methods for continuous data have recently been published and are gaining acceptance.[5]

This chapter will review the approaches and methods used in the risk assessment of occupational neurotoxicants and stress the more recent advances in predictive neurotoxicology.

2 OCCURRENCE OF OCCUPATIONAL NEUROTOXICANTS

The occurrence of neurotoxicants in the workplace is widespread. It has been estimated that nearly 25% of all chemicals that produce toxicity in the workplace have an effect on the nervous system.[6] Approximations from the current literature suggest that more than 25% of chemicals for which threshold limit values have been set had their threshold limit determined solely or in part on the basis of direct nervous system effects.[7]

TABLE 1
Selected Neurotoxicants Known to Affect the Nervous System

Organic Solvents	Pesticides	Metals	Gases	Other
Acetone	Aldrin	Alkyltins	Carbon Monoxide	Acrylamide
Carbon disulfide	Carbamates	Aluminum	Ethylene Oxide	β-Dimethylamino-propionitrile
Ethyl Benzene	Chlordane	Arsenic	Formaldehyde	Triorthocresyl-phosphate
Methanol	Chlordecone	Lead	Hydrogen Sulfide	
Methyl chloroform	2,4-Dichlorphen-oxyacetic acid	Manganese	Nitrous Oxide	
Methyl n-Butyl ketone Ketone	2,4,5-Trichloro-phenoxyacetic acid	Mercury	Methyl Chloride	
Methyl Ethyl Ketone	Lindane	Thallium		
Methylene Chloride	Methyl Bromide			
n-Hexane	Organophosphate Compounds			
Styrene	Pyrethroids			
Toluene				
Trichloroethylene				
Xylene				
Mixture of Solvents				

Modified from Bleecker.[4]

The chemical structures of agents producing neurotoxicity are diverse. A listing of selected neurotoxicants exemplify the many chemical classes known to affect the nervous system (Table 1; see also chapter 1). This list does not include another set of chemicals comprised of more than 20 agents (e.g., dimethylbenzanthracene, ethyl methanesulphonate and methylnitrosourea) reported to produce brain tumors in laboratory animals.[8] Taken together, this diverse set of chemicals produce a wide variety of nervous system dysfunctions. Many of these human neurotoxic syndromes have been recently reviewed[4,8,9] and include alterations of both the central and peripheral nervous system. In order to evaluate such a diverse set of chemicals that produces a wide array of symptoms, a comprehensive assessment approach is necessary.

3   RISK ASSESSMENT STRATEGIES

A multidisciplinary approach is necessary for the assessment of neurotoxicity because of the complexity and diverse functions of the nervous system. Many effects relevant to the neurotoxicologist can be measured directly by neurochemical, neurophysiological, and neuropathological techniques, whereas, others must be inferred from observed behavior (Figure 1).

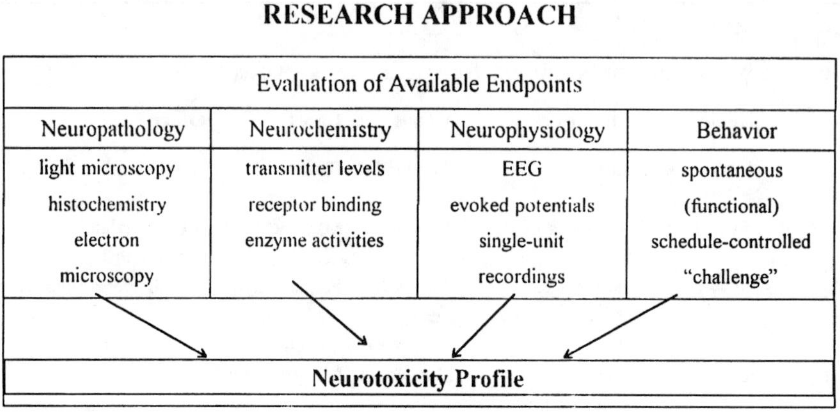

RESEARCH APPROACH

Evaluation of Available Endpoints			
Neuropathology	Neurochemistry	Neurophysiology	Behavior
light microscopy	transmitter levels	EEG	spontaneous
histochemistry	receptor binding	evoked potentials	(functional)
electron	enzyme activities	single-unit	schedule-controlled
microscopy		recordings	"challenge"
Neurotoxicity Profile			

FIGURE 1   Description of research approach including endpoints that may be evaluated to generate a neurotoxicity profile for a given agent.

The research approach consists of the following three steps:

1. gathering information from all available endpoints or disciplines and using it to generate a neurotoxicity profile
2. correlating structural, neurochemical and neurophysiological data with overt behavioral manifestations of neurotoxicity
3. developing a pharmacokinetic/metabolic basis to aid in data interpretation and for interspecies extrapolation. Future strategies for neurotoxicity assessments may include additional disciplines and approaches such as molecular biology and psychology (Figure 2). To allow for successful interspecies extrapolation, physiologically-based pharmacokinetic modeling (PBPk) may be used to predict target tissue exposure to an agent in several species including humans.[5]

3.1   ASSESSING HUMAN NEUROTOXICITY

Because of technical and ethical considerations, neurotoxicity in humans is primarily measured by noninvasive neurophysiological and neurobehavioral methods. As illustrated in other chapters in this book, clinical neurology and neuropsychology approaches have been used extensively to evaluate neurological diseases, and these same methods have been used to assess patients suspected of having neurotoxic insults.[2] In addition, many neurobehavioral methods have been proposed or used to assess neurobehavioral function in humans.[2,10] In recent years, imaging techniques including magnetic resonance imaging (MRI) and computerized tomography (CT), and computerized brain electrical activity mapping, as well as operant behavioral assessments have provided noninvasive and quantitative methods for human neurotoxicity assessment. Epidemiological approaches, including retrospective and prospective studies also provide the means for evaluating the effects of neurotoxic substances in human populations.[2]

Discipline - Continuum Approach

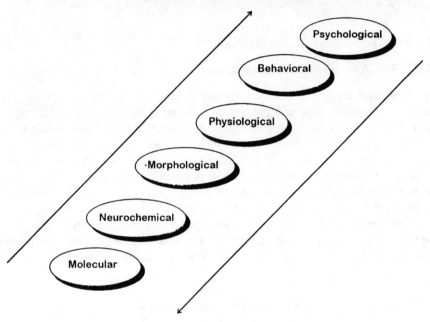

FIGURE 2 Description of the discipline-continuum approach for the evaluation of agents. Initial studies of an agent begin within any discipline on the continuum and additional studies may follow within other disciplines. Data generated from the various disciplines or from different species may be integrated with the use of pharmacokinetics.

3.2 ANIMAL MODELS

While human data are available from case reports and epidemiological studies concerning occupational neurotoxicants, studies to determine dose, pharmacokinetic and safety parameters are often performed in animal models. *In vivo* animal studies currently serve as the principal approach for detecting and characterizing neurotoxic hazards and helping to identify factors affecting susceptibility to neurotoxicity. *In vitro* tests have been proposed as a means of complementing whole animal tests and, when properly developed, may be less time consuming and more cost effective than *in vivo* assessments.[2] The currently used *in vitro* tests, however, have certain limitations including the inability to model neurobehavioral effects (e.g., memory, learning or sensory dysfunction). Validation of animal models, whether *in vivo* or *in vitro*, is of paramount importance and may include measures of construct, criterion and predictive validity. Despite the biological similarity of humans and many animal models, differential susceptibility to toxicants is well documented between species. Predictive capability or concordance between human and animal models after exposure to human neurotoxicants has frequently been reported.[2,11,12] These comparable outcomes provide a firm foundation for the use of animal models in neurotoxicity risk assessment. Descriptions of these electrophysiological, neuropathological, neurochemical and behavioral assessments have recently been published.[2,10]

4   RISK ASSESSMENT METHODS

4.1   RISK ASSESSMENT/RISK MANAGEMENT

Risk assessment is an empirically-based process used to estimate the risk that exposure of an individual or population to a chemical, physical, or biological agent will be associated with an adverse or abnormal effect. The risk assessment process usually involves four steps: hazard identification; dose-response assessment; exposure assessment; and risk characterization.[13] Risk assessment should not be confused with risk management, a process whereby information concerning the benefits as well as the risks of exposure are considered to determine whether, or at what exposure level, an individual chemical should be used. Most often, risk assessments are performed by researchers, whereas risk management is often conducted by regulatory agencies as directed by legislative bodies via legislation, or by physician/patient interactions focused on risk/benefit considerations for the individual in need of treatment.

4.2   CURRENTLY USED RISK ASSESSMENT APPROACHES

The most frequently used approach for risk assessment of neurotoxicants and other noncancer endpoints is the uncertainty or safety-factor approach.[14,15] Within the Environmental Protection Agency (EPA), for example, this approach involves the determination of reference doses (RfDs) by dividing a no observed adverse effect level (NOAEL) by uncertainty factors that presumably account for interspecies and intraspecies differences in sensitivity.[14] Generally, an uncertainty factor of 10 is used to allow for the presumed greater sensitivity in humans than in animals and another uncertainty factor of 10 is used to allow for variability in sensitivity among humans. In this case, the RfD is equal to the NOAEL divided by 100. If the NOAEL cannot be established, it is replaced by the lowest observed adverse effect level (LOAEL) in the RfD calculation and an additional uncertainty factor of 10 is generally introduced (i.e., the RfD equals the LOAEL divided by 1000).

4.3   LIMITATIONS AND ASSUMPTIONS

There are several features of this RfD or safety factor approach which deserve consideration. First, the method assumes a theoretical threshold dose below which no biological effects of any type are observed in a heterogeneous population. The theoretical bases of a threshold dose may be questioned. If, due to normal variation in cellular function an adverse effect can occur in untreated control subjects, then endogenous or exogenous factors may already be supplying a stimulus which is equivalent to a dose above the threshold dose. If exposure to an agent augments this stimulus, then an additional risk is expected and no threshold dose exists for that agent.[16] Second, the determination of a threshold dose is influenced by the sensitivity of the analytical methods employed. Unfortunately, less sensitive experiments can result in higher RfD's. Third, the magnitudes of the safety factors used to determine RfDs [interspecies extrapolation (10) and intraspecies extrapolation (10)] are based more on best estimates than actual data.[2,10]

The RfD approach relies on a single experimental observation (the NOAEL or LOAEL), instead of using complete dose-response curve data in the calculation of risk estimates. Because chemical interactions with biological systems are often specific, stereoselective, and/or saturable, a chemical's dose-response curve may not be linear. Examples include enzyme-substrate binding leading to substrate metabolism, transport, and receptor-binding, any or all of which may be a requirement for an agent's effect or toxicity. The certainty of low-dose extrapolation has been determined to be markedly affected by the shape of the dose-response curve.[17] Therefore, the use of dose-response curve data should enhance the certainty of risk estimations when thresholds are not assumed or determined.

4.4 DOSE-RESPONSE/QUANTITATIVE MODELS

Dose-response models have generated considerable interest and are seen by many to be more appropriate (and quantitative) than the safety-factor approach to risk assessment. Rather than routinely applying a "fixed" safety factor to the NOAEL (which is based on a single dose) to estimate a "safe" dose, these approaches use data from the entire dose-response curve.[2]

Two fundamentally different approaches in the use of dose-response data on quantitative neurotoxic effects to estimate risk have been developed. Dews and coworkers[18,19] and Crump[20] demonstrated an approach in which they used information on the shape of the dose-response curve to estimate levels of exposure associated with relatively small effects (i.e., a 1, 5, or 10 percent change in a biological endpoint). Both Dews and Crump fit a mathematical function to the data and provided an estimate of the variability in exposure levels associated with a relatively small effect.

In another approach developed by Gaylor and Slikker,[16,21] a mathematical relationship is first established between the average biological effect and the dose of a given chemical. A second step determines the distribution (variability) of individual measurements of biological effects about the dose-response curve. The third step statistically defines an adverse or "abnormal" level of a biological effect in an untreated population. The fourth step estimates the probability of an adverse or abnormal level as a function of dose utilizing the information from the first three steps. The advantages of these dose-response models are that they encourage the generation and use of data needed to define a complete dose-response curve and provide an estimate of risk and/or changes in the average response as a function of dose.

4.5 BIOLOGICALLY-BASED RISK ASSESSMENT

The development of quantitative risk assessment approaches depends, in part, on the availability of information on the mechanism of action and pharmacokinetics of the agent in question.[5] In the development of a biologically-based, dose-response model for the psychoactive agent, methylenedioxymethamphetamine (MDMA), Slikker and Gaylor[21] considered several factors, including the pharmacokinetics of the parent chemical, the target tissue concentrations of the parent chemical or its bioactivated proximate toxicant, the uptake kinetics of the parent chemical or metabolite into the target cell and membrane interactions, and the interaction of the chemical or metabolite with presumed receptor site(s). Because these theoretical factors contain a saturable step due to limited amounts of required enzyme, reuptake, or receptor site(s), a nonlinear, saturable dose-response curve was predicted. In this case of neurochemical effects of MDMA in the rodent, saturation mechanisms were hypothesized and indeed saturation curves provided relatively good fits to the experimental results. Some of the advantages of the biologically-based, quantitative approaches over the currently used RfD risk assessment procedures include the ability to (1) utilize continuous data, (2) utilize all of the dose-response curve data, (3) incorporate biological information into the dose-response model, and (4) provide an actual risk of exposure to a given dose.[3] The conclusion was that use of dose-response models based on plausible biological mechanisms provide more validity to prediction than purely empirical models.

A single risk assessment model may not be adequate for all conditions of exposure, for all endpoints, or for all agents. Risk assessment models of the future may well include biomarkers of both effect and exposure as well as biologically-based mechanistic and pharmacokinetic considerations derived from both epidemiologic and experimental test system data.[5,22]

5 PREDICTIVE VALUES OF HUMAN RISK

The ultimate goal of risk assessment of occupational neurotoxicants is to predict the exposure level of a workplace agent that results in human neurotoxicity. Because of the many hundreds of chemical, biological and physical agents to which a worker may be exposed, it is not likely that all these agents can be evaluated individually. Strategies have been developed to aid in the risk assessment process in an attempt to improve the efficiency of risk assessment.

5.1 DATABASE VS. KNOWLEDGE BASE

A database is a collection of data in a computer, organized so that it can be expanded, deleted, and retrieved using a database management system. In contrast, a knowledge base is a collection of "knowledge," organized so that it can be expanded, updated, deleted, retrieved, and most important of all, it can be used as a basis for prediction and classification using a knowledge-based system (KBS). By "knowledge" we mean facts, inferences, and procedures.[23] For a database to act like a knowledge base some operation must be added and/or performed on the data in the database to acquire knowledge. These operations are collectively known as "knowledge acquisition systems."[24]

Some of the knowledge acquisition systems try to induce knowledge from a database (machine learning systems, neural networks, rough sets-based systems, and others) but often these systems try to acquire knowledge directly from the expert.

The acquisition of knowledge through the elicitation from an expert is done by a "knowledge engineer." Knowledge representation and knowledge validation are two other essential tasks performed by the knowledge engineer.

5.2 KNOWLEDGE BASE APPROACH

Development of a knowledge base is an integral component of a more comprehensive system called a knowledge-based system (KBS). We expect that the KBS uses its knowledge base to infer other knowledge (predictive values) that is not explicitly stored but can be logically deduced using one or more reasoning methodologies (e.g., reasoning by induction, reasoning by deduction, probabilistic reasoning, commonsense reasoning, etc.).[25] The foundations of the knowledge base consist of biological endpoints (e.g., neuropathological, neurophysiological, neurochemical, molecular biological and behavioral), mechanisms, structure activity relationships (SAR), target tissue concentrations, and physical/chemical properties of the agent. Hence, the prediction of human risk can be derived from the working model (Figure 3) by assembling information in an ascending order of complexity from method-, agent-, or concept-driven research to strategies for prediction (e.g., SAR and quantitative risk assessment procedures) to databases. A complete database can be envisioned as the product of an interactive and iterative process between the several foundation components (e.g., endpoints and mechanisms). In the process of developing knowledge bases from various data sources, deficits will be identified in existing data that will determine directions of new research priorities. Subsequent studies can then be conducted to fill previously identified data gaps in a knowledge base.

6 CONCLUSIONS

Although workplace exposures to potentially toxic agents may well occur at higher levels than in the general environment, absolute exposure levels often remain uncertain. The uncertainty of workplace exposures, especially for chronic/semichronic, low dose situations, poses a special problem for the assessment of human risk. As the need for quantitative risk assessment has increased so as to protect both the national economy and human health, newer approaches to risk assessment have evolved. The occurrence of neurotoxic exposures in the workplace and animal model studies provide an opportunity to develop and validate these more quantitative and predictive approaches.

Quantitative approaches for neurotoxicity risk assessment are generally based on dose response data from either human or animal model studies. Recent examples include the benchmark approach[13] and the quantitative abnormal-range approach for continuous data.[5,16,26] Either one of these approaches can provide an actual risk of exposure to a given dose (e.g., 1 in 100,000 risk at dose X). In order to increase the efficiency of using the complex databases needed for quantitative and predictive assessment, the concept of a knowledge base has been developed. Predictive capabilities are achieved through the application of artificial intelligence programs to database information. Future quantitative risk assessments for occupational neuro-

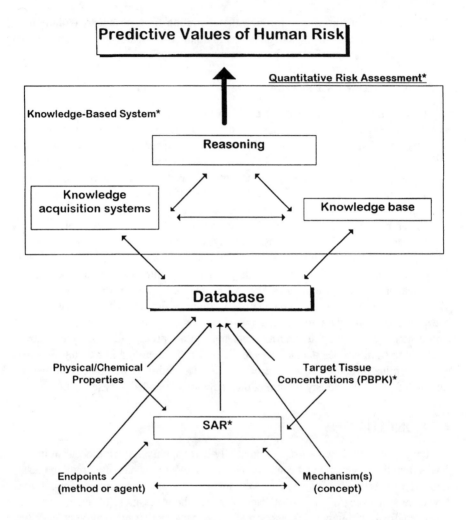

FIGURE 3 Integrative strategy for determination of predictive values of human risk. The foundations of the approach include information from endpoints and mechanisms. This information can be integrated with the use of strategies for prediction such as SAR (structure activity relationship) and PBPK (physiologically based pharmacokinetics). This information is assembled in a database. The building of a database may require multiple or iterative interaction of the foundation building blocks, i.e., physical/chemical properties of a chemical, target tissue concentrations, proposed mechanism(s) and endpoints. The knowledge base is derived from a database through the application of artificial intelligence methodologies such as neural networks, machine learning, experts systems, or other approaches. The knowledge base serves to generate predictive values of human risk and this process can be enhanced with the use of quantitative risk assessment methods.

toxicants may well include the application of quantitative risk assessment procedures to knowledge base generated information.

7 ACKNOWLEDGMENTS

I sincerely thank Drs. Bernard Schwetz and Ray Hashemi for their definitions and helpful discussions concerning knowledge bases. I also thank Ms. Barbara Jacks and Loetta Bradford for their clerical and editorial assistance.

8 REFERENCES

1. Kilbourne, E. M., "Overview of Environmental Medicine," in *Textbook of Clinical Occupational Environmental Medicine*, Rosenstack, L. and Cullen, M. R., Eds., W. B. Saunders, Philadelphia, 1994, 41.
2. Reiter, L. W., Tilson, H. A., Dougherty, J., Harry, G. J., Jones, C. J., McMaster, S., Slikker, W., Jr., and Sabotka, T. J. "Final Report: Principles of neurotoxicity risk assessment," *Fed. Reg.* 59(158), 42360, 1994.
3. Slikker, W., Jr., and Gaylor, D.W., "Risk Assessment Strategies for Neuroprotective Agents, in Neuroprotective Agents: Clinical and Experimental Aspects," Trembly, B. and Slikker, W., Jr., Eds., *Annals of the New York Academy of Sciences*, Vol 765, New York, 198, 1995.
4. Bleecker, M. L., "Clinical Presentation of Selected Neurotoxic Compounds," in *Occupational Neurology and Clinical Neurotoxicology*, Bleecker, M.L., Ed., Williams & Wilkins, Baltimore, MD, 1994, 207.
5. Slikker, W., Jr., Crump, K. S., Andersen, M.E., and Bellinger, D., "Biologically-based, quantitative risk assessment of neurotoxicants," *Fundam. Appl. Toxicol*, 29, 18, 1996.
6. OTA (Office of Technology Assessment), *Impact of Neuroscience: A Background Paper*, OTA-BP-BA-24, U.S. Government Printing Office, Washington, DC, 1984.
7. Anger, W. K., "Neurobehavioral testing of chemicals: Impact on recommended standards," *Neurobehav. Toxicol. Teratol.*, 6, 147, 1984.
8. Stacey, N. H., Haschek, M., and Winder, C., *Occupational Toxicology*, Stacey, N. H., Ed., Taylor & Francis, London, 1993, 50.
9. Baker, E. L., *Neurologic Disorders, in Environmental and Occupational Medicine*, Rom, W. N., Ed., Little, Brown & Co., Boston, 1992, 561.
10. Slikker, W., Jr., and Gaylor, D. W., "Concepts on quantitative risk assessment of neurotoxicants." In *Neurotoxicology: Approaches and Methods*. L. W. Chang and W. Slikker, Jr., Eds., Academic Press, San Diego, 771, 1995.
11. Slikker, W., Jr., "Principles of developmental neurotoxicology." *Neurotoxicology*, 15(1), 11, 1994.
12. Slikker, W., Jr., "Placental transfer and pharmacokinetics of developmental neurotoxicants." In *Principles of Neurotoxicology*, L. Chang, Ed., Marcel Dekker, Inc., New York, NY, 659, 1994.
13. National Research Council (NRC), *Risk Assessment in the Federal Government*, National Academy Press, Washington, DC, 1983.
14. Barnes, D. G. and Dourson, M. "Reference dose (RfD): Description and use in health risk assessments," *Reg. Toxicol. Pharmacol.* 8, 471, 1988.

15. Kimmel, C. A., "Quantitative approaches to human risk assessment for noncancer health effects," *Neurotoxicology*, 11, 189, 1990.
16. Gaylor, D. W. and Slikker, W., Jr., "Risk assessment for neurotoxic effects," *Neurotoxicology*, 11, 211, 1990.
17. FDA (Food and Drug Administration Advisory Committee on Protocols for Safety Evaluation), "Panel on carcinogenesis report on cancer testing in the safety evaluation of food additives and pesticides," *Toxicol. Appl. Pharmacol.*, 20, 419, 1971.
18. Dews, P. B., "On the assessment of risk." In *Developmental Behavioral Pharmacology*, N. Krasnegor, J. Gray and T. Thompson, Eds., Lawrence Erlbaum Associates, Hillsdale, NJ, 53, 1986.
19. Glowa, J. R. and Dews, P. B., "Behavioral toxicology of volatile organic solvents. IV. Comparison of the behavioral effects of acetone, methyl ethyl ketone, ethyl acetate, carbon disulfide, and toluene on the responding of mice." *J. Am. Coll. Toxicol.* 6, 461, 1987.
20. Crump, K. S., "A new method for determining allowable daily intake," *Fund. Appl. Toxicol.*, 4, 854, 1984.
21. Slikker, W., Jr., and Gaylor, D.W., "Biologically based dose-response model for neurotoxicity risk assessment," *Korean J. Toxicol.*, 6, 204, 1990.
22. Slikker, W., Jr., "Biomarkers of neurotoxicity: An overview. Recent advances on biomarker research," *Biomed. Environ. Sci.*, 4, 192, 1991.
23. Barr, W. and Feigenbaum, E. A., *The Handbook of Artificial Intelligence*, 2, 34, Addison Wesley, 1981.
24. Shaw, M. L. G. and Gaines, B. R., "Techniques for knowledge acquisition and transfer," *Proceedings of the first AAAI Knowledge Acquisition for Knowledge-based Systems Workshop*, J. H. Boose, and B. R. Gaines Eds., 39:0-13, 1986.
25. Bible, W., "Methods of automated reasoning," W. Bible and P. Jorrand, Eds., *Fundamentals of Artificial Intelligence — An advanced Course*, pp. 173-222, Springer, LNCS 232, Berlin.
26. Kodell, R. L., Chen J. J., and Gaylor D. W., "Neurotoxicity modeling for risk assessment." *Reg. Toxicol. Pharmacol.*, 22:24-29, 1995.

14 Occupational Neurotoxic Diseases in Developing Countries

Fengsheng He

CONTENTS

1 INTRODUCTION

The total working population accounts for 45% of the world's population which was estimated to be 5559 million in 1993.[1] In the global work force, 75% work in the developing countries, and almost eight out of 10 workers will work in the developing world by the year 2000.

In the occupational settings of many developing nations, there are more opportunities of exposure to various toxic chemicals and, for several reasons, higher risks of occupational poisonings. These include a growing use of agrochemicals in agricultural workers under poor working conditions and with no appropriate personal protection, as well as an increasing use of basic industrial chemicals in small scale industries under uncontrolled environment. Moreover, many of the most hazardous chemicals or pesticides which have been prohibited for use in industrialized countries, are however, still commonly used in developing countries.

In recent years, the shift of hazardous industries from highly regulated countries to less developed countries has particularly been of concern. In some instances, individuals receive a massive, or even fatal exposure through a chemical disaster in the workplace.[2-4]

On the other hand, there is a general lack of health, safety, and welfare services in developing countries. Due to inadequate coverage of occupational health services, many of the occupational disease reporting systems in the developing nations are either incomplete or inaccurate. Although there is scarce information concerning epidemiological surveys on occupational poisonings in developing countries, available data show that the number of cases of chemical poisonings stands the highest or the second among the occupational diseases in many developing countries,[3,5] and the most common occupational poisonings are neurotoxic diseases.[6]

The nervous system is particularly vulnerable to many chemical insults. Different chemicals may affect different parts of the nervous system at certain dosages and may induce neurotoxic diseases by different mechanisms. This chapter describes some neurotoxic diseases associated with occupational exposure which are more prevalent in the developing world.

2 AGROCHEMICAL EXPOSURE

Developing countries are steadily increasing their demand for agrochemicals, many of which are pesticides. Recent world-wide estimates suggest that there are about 3 million acute pesticide poisonings and 220,000 deaths each year.[7] Most of the poisonings and 99% of the deaths are believed to occur in developing countries.[8]

In the Asian region, a survey of acute pesticide poisoning among agricultural workers revealed that occupation-related accidental pesticide poisonings were about 1.9% of cases in Indonesia and as high as 31.9% in Sri Lanka, while suicides accounted for 62.6% of cases in Indonesia, 67.9% in Malaysia, 36.2% in Sri Lanka and 61.4% in Thailand.[9] In China, there were 52287 cases of acute pesticide poisoning and 6281 deaths reported from 27 provinces in 1993. Among total cases, occupational pesticide poisoning accounted for 17.8% with a fatality at 0.9%, and the suicidal cases were 82.2%, with a fatality rate at 14.2%. However, in Latin America, it appears that most cases of pesticide poisoning are occupational. For example, in Costa Rica, 67.8% were work-related compared with 6.4% that were suicidal;[10] in Nicaragua, 91% of the reported cases of pesticide poisoning were occupational, 8% involved other accidents, and 1% were suicide attempts.[11]

The main classes of pesticides responsible for acute poisoning in many developing countries have been identified to be insecticides which are mostly neurotoxic. The causal factors contributing to acute occupational pesticides poisoning include sloppy handling during preparing and spraying pesticides, using highly toxic pesticides or high concentrations of pesticide, spraying every row and direct contact with sprayed crop, going forward with the wind during spraying, lack of personal protection, and poor personal hygiene.[12,13]

2.1   ORGANOPHOSPHATES POISONING

Acute organophosphate poisoning accounts for 78.8% of total pesticide poisonings in China, 69.1% in Sri Lanka, and 53.6% in Malaysia.[9]

2.1.1   Acute Poisoning

Organophosphates (OPs) cause acute toxic effects by inhibiting the enzyme acetylcholinesterase (AChE) leading to the accumulation of acetylcholine at all cholinergic transmission sites in the peripheral and central nervous system. Clinical manifestations constitute a typical syndrome composed of muscarinic, nicotinic, and central nervous signs.[14] Symptoms and signs resulting from mild poisoning include dizziness, headache, nausea, vomiting, miosis, excessive sweating and tracheobroncheal and salivary secretions. Muscular fasciculations and shortness of breath appear in cases of moderate poisoning, while severe poisoning causes coma, pulmonary edema and respiratory depression. The cholinergic effects of acute OPs poisoning usually correlate well with blood AChE inhibition at the initial stage.[15]

Acute respiratory failure is the main cause of death. In addition to rapid removal of toxic compounds, early treatment with atropine and oxime cholinesterase reactivators at repeated and sufficient dosages are necessary. A comparative study on the efficacy of four oxime cholinesterase reactivators, e.g., pyridine aldoxime methiodide (PAM), pralidoxime (pyridine aldoxime methylchloride, PAC), trimedoxime (TMB4), and obidoxime (toxogonin, LuH6) in humans revealed that all of them proved capable of restoring erythrocyte cholinesterase activity and relieving symptoms and signs of organophosphate insecticide poisoning, such as parathion, demeton, and thiodiphosphoric acid tetraethyl ether poisoning. However, these antidotes are less effective against dipterex and dichlorvos poisoning; and least effective against dimethoate and fenitrothion poisoning. All severely intoxicated patients need the synergistic action of atropine. TMB4 and DMO4 show higher potency among the four oxime cholinesterase reactivators, however, they are not recommended for routine treatment because of their side effects.[16]

2.1.2   Organophosphate-Induced Delayed Polyneuropathy

The organophosphate-induced delayed polyneuropathy (OPIDP) resulting from a single or repeated exposure to an organophophorous compound is independent of AChE inhibition. The neuropathy develops following a latent period of two–four weeks after the cholinergic crisis. The main symptoms are distal weakness of the feet and hands, calf pain preceding the weakness and in some cases paraesthesia in the distal parts of the limbs. Wasting of distal muscles is often observed. In severe cases, pyramidal signs (spasticity, increased tendon reflexes, and pathological reflexes) appear after several weeks or months. Treatments are mainly symptomatic and recovery from OPIDP is variable.

There were several outbreaks of OPIDP in Morocco, Durban, Fiji, Vietnam, and Sri Lanka due to contamination of cooking oil with mineral oil which contained tri-ortho-cresyl phosphate (TOCP).[17] Two recent outbreaks of OPIDP involving about 200 patients occurred in China in 1990 and 1995 resulting from leakage of lubricant oil containing TOCP on the flour milling machine (unpublished data). There have been several reports of OPIDP with insecticides, mipafox, leptophos, trichlorphon, trichlornat, chlorpyrifos and methamidophos.[16] In China, there were 218 cases of OPIDP reported in the medical journals during 1960–1990 (published in Chinese), mainly induced by exposure to methamidophos (134 cases of ingestive poisoning and 11 cases of occupational poisoning), and suicidal attempts by dichlorvos (35 cases), trichlorphon (15 cases), dimethoate (seven cases), isocarbophos (four cases), and parathion (three cases). A follow-up study in 110 patients of acute methamidophos poisoning indicated a prevalence of OPIDP at 9.1%, while the prevalence of OPIDP caused by other OPs was found to be 4.2% in severe patients (published in Chinese).

The "neuropathy target esterase" (NTE) is thought to be the molecular target of OPIDP and neuropathy to be initiated with a two-step mechanism: progressive inhibition of NTE and aging of the phosphorylated enzyme.[18] However, recent evidence indicates that aging may not always be essential in causing neuropathy.[18] Other recent hypotheses include the interaction of OP compounds with Ca/calmodulin kinase II.[19]

2.1.3   Intermediate Syndrome

The "Intermediate Syndrome (IMS)" in OP poisoning is a syndrome of muscle paralysis which occurs after recovery from the cholinergic crisis, but before the expected onset of the delayed polyneuropathy.[20] The syndrome is of acute onset, often seen 24–96 hours after poisoning, affecting conscious patients without fasciculations or other cholinergic manifestations. The cardinal clinical feature is muscle weakness affecting predominantly the proximal limb muscles, neck flexors, muscles innervated by motor cranial nerves III-VII and X in different combination, and respiratory muscles. Respiratory insufficiency usually draws attention to the onset of the IMS and is the cause of death if not recognized early and treated with artificial ventilation adequately. Tetanic stimulation can reveal a marked fade at 20 and 50 Hz.[20]

The term "Intermediate Syndrome" first was proposed by Senanayake and Kalliedde in 1987,[20] and was previously described by Wadia as the Type 2 paralysis of organophosphorous poisoning in 1974.[21] The agents responsible were diasinonon, fenthion, dimethoate and monocrotophos. Since then, there have been more than 70 cases reported from the developing countries, including 39 cases from India,[21, 22] 29 cases from Sri Lanka,[17,20] and 11 cases from China resulting from exposure to dichlorvos, parathion, phoxim, omethoate and other compounds (published in Chinese).

The pathogenesis of IMS remains to be investigated. It seems that the IMS is likely to result from post-synaptic neuromuscular junction dysfunction.[20]

2.2 CARBAMATE INSECTICIDES

Insecticidal and nematocidal carbamates have a high acute toxicity.[23] In Nicaragua, almost half of the poisonings involved the carbamates insecticide,[11] while 15.9% in Thailand and 10.7% in Malaysia.[9] In China, thousands of cases of acute carbamates poisoning have been reported, mainly caused by carbofuran (more than 1500 cases) and fewer by carbaryl, MTMC and isoprocarb (MIPC).[24]

Carbamate insecticides produce a clinical picture of cholinergic excess similar to that of organophosphate toxicity. However, because of spontaneous hydrolysis of the carbamylated AChE enzyme, the symptoms are less severe and of shorter duration. The carbamates poorly penetrate the blood-brain barrier and therefore produce minimal effects on brain ChE activity and few CNS symptoms.[23,24]

The prognosis of carbamates poisoning is usually good. Fatal cases are all related to poisoning by ingestion. Atropine is the antidote of choice as in organophosphate poisoning, but the total amount of atropine required usually is less. Oxime cholinesterase reactivators are not recommended for the treatment of carbamate poisoning.[23,24]

2.3 PYRETHROID POISONING

Synthetic pyrethroids have a high insecticidal activity and a low toxicity in mammals. All the pyrethroids share the same mechanism of action on voltage-dependent Na^+ channels. Cyanopyrethroids prolong the Na^+ current to a much greater extent than noncyanopyrethroids.[25]

With good work practice and safety precautions, pyrethroids are unlikely to present acute toxicity for those occupationally exposed. However, cases of acute occupational pyrethroid poisoning were first reported in China in 1982, due to inappropriate handling, spraying with higher concentration than that recommended on the label, longer exposure duration, spraying against the wind, clearing blocked nozzles by mouth and hands, poor personal hygiene and lack of personal protection. During 1983–1988, there were 573 cases of acute pyrethroids poisoning reported in the Chinese medical literature.[26] Occupational pyrethroid poisoning accounted for one third of the total patients, and two third were of oral poisoning. The majority of cases resulted from exposure to deltamethrin, followed by fenvalerate, cypermethrin and other pyrethroids (cyfluthrin, fenpropathrin).[26] In an epidemiological survey of occupational poisoning in China, the prevalence of mild acute pyrethroid intoxication in 3113 spraymen was 0.38%.[27]

The initial symptoms of acute occupational pyrethroids poisoning are burning and itching sensations in the face or dizziness which usually develop 4–6 hours after exposure. The oral poisoning often starts with epigastric pain, nausea and vomiting within 10 minutes–1 hour after ingestion. The systemic symptoms include dizziness, headache, nausea, anorexia, fatigue, increased stomal secretion, and muscular fasciculation. Convulsive attacks, disturbances of consciousness (such as drowsiness, cloudiness, and coma), as well as dyspnea, cyanosis and moist rales indicating pulmonary edema occur mainly in severe cases of ingestive poisoning. Blood cholinesterase is usually normal except in cases with combined pyrethroid-organophosphate poisoning whose clinical manifestations are mainly of acute organophosphate poisoning. The acidic metabolite 3-(2,2-dichlorovinyl)-2,2'-dimethyl-cyclopropane carboxylic acid (Cl2CA) is excreted for both cypermethrin and permethrin and the bromo-analogue (Br2CA) for deltamethrin. These metabolites are excreted in the urine within a 24-hour period.[28] Pyrethroids prolong the inward sodium current across the nerve membrane shown as increased supernormal excitability of peripheral nerves. This action can be detected by a non-invasive percutaneous electroneuromyographic technique with pairs of stimuli on the median nerves, as demonstrated in deltamethrin spraymen.[29]

The vast majority of patients with acute pyrethroids poisoning recovers fully after symptomatic and supportive treatments for one–six days. The prognosis was proved to be good even in severely poisoned patients. Few fatal cases were reported due to misdiagnosis and inappropriate treatments, including those poisoned by pyrethroid-organophosphate mixtures and treated with insufficient dosage of atropine, as well as those with pure pyrethroid poisoning dying of atropine poisoning.[26]

3   INDUSTRIAL CHEMICAL EXPOSURE

3.1   CENTRAL NERVOUS SYSTEM INVOLVEMENT

3.1.1   Carbon Monoxide

Carbon monoxide (CO) poisoning is the leading cause of poisoning deaths in many developed and developing countries.[30-33] In Seoul, Korea, the incidence of CO intoxication was estimated at 306 per 10,000 persons in 1977, and the mortality rate at one per 10,000.[31] The annual rates of morbidity and mortality from acute CO poisoning are the highest among all acute occupational poisonings in China.[32,33] The statistics from the National Occupational Disease Reporting System in China showed that there were 853 cases of acute CO poisoning in 1990, accounting for more than one third of the total cases of acute occupational poisonings in that year.[32] Acute CO poisoning is also a common life threatening disorder for inhabitants of northern China, Korea, and the Bloemfontein area in Africa, because of indoor heating with coal fires and faulty ventilation.[31,32,34]

Acute CO poisoning is manifested by neurologic dysfunction resulting from cerebral hypoxia and brain edema, e.g., headache, unconsciousness, and convulsion. The confirmation of CO poisoning is easily obtained via assessment of carboxyhemoglobin (COHb), however, the blood sampling time should not be later than 8 hours after onset. Hyperbaric oxygen therapy is indicated for severe patients, in addition to appropriate supportive treatments and intensive care.[30]

Acute mortality is not the sole risk of CO poisoning. Of growing concern is the syndrome of delayed encephalopathy (DE), which occurs following a variable interval (usually 2–60 days after recovery from acute CO poisoning) and is clinically characterized by mental abnormalities, Parkinsonism and pyramidal signs. About 11–13% of the acute CO poisoning patients admitted to hospitals have developed DE.[30-32] This severe consequence of acute CO poisoning is at present unpredictable.

The cerebral CT scanning may not reveal any hypodensity at the subcortical white matter and globus pallidus until 7–14 days after the appearance of the symptoms and signs of DE. In a follow-up study of patients with CO poisoning, the somatosensory (SEP) or visual (VEP) evoked potentials in some patients showed recovery of initial abnormalities after regaining consciousness, but became abnormal again just three days prior to the onset of delayed encephalopathy, suggesting that both SEP and VEP may be helpful in the acute stage of intoxication to predict the occurrence of delayed encephalopathy.[33] Another study on 207 cases of acute CO poisoning followed for three months in China has led to characterization of aspects present in the acute stage that seem to be significant risk factors for the development of DE. These were: age >40 years, occupation that involved mental work, previous hypertension, coma at the acute stage lasting for two–three days, long persistence of dizziness and fatigue after regaining consciousness, and being mentally stimulated during the recovery stage of acute CO poisoning.[32]

3.3.2   Metals

While occupational lead exposure is adequately controlled in most developed countries to prevent the occurrence of overt lead poisoning, there is a higher risk of lead poisoning in workers of small scale industries in many developing countries.[35, 36] Lead encephalopathy is generally rare, yet occasionally occurs due to accidental exposure to lead.[37] There have been several studies showing that lead exposed workers manifest disorders of behavioral functions as blood lead reaches 45–50 μg/dl.[35,38]

An outbreak of manganese induced Parkinsonism was recently reported from Taiwan which resulted from exposure to high concentrations of manganese fumes (>28.8 mg/m^3) through the breakdown of a ventilation system in a ferromanganese smelter.[39] The patients showed a bradykinetic-rigid syndrome and had increased manganese concentrations in the blood, scalp and pubic hair.[40]

Exposure to high levels of mercury can occur in mercury refining and chloralkali industries. Significant association was noted between the urine mercury concentration and prevalence of symptoms, including weakness, tiredness, headache, dizziness and irritability.[35,41] Workers with heavy exposure to mercury may develop tremor and ataxia. Chelation therapy with sodium dimercaptosulphonate (unithiol) or sodium dimercaptosuccinate (Na -DMS) is evidently beneficial for reducing urine mercury and some symptoms, but not for neurological and stomal signs. However, the overall prognosis of chronic metallic mercury poisoning is encouraging after termination of mercury exposure for a longer period.[41]

3.2   PERIPHERAL NERVOUS SYSTEM

3.2.1   Acrylamide

Acrylamide monomer is the raw material used in the production of polyacrylamides which are useful as flocculates or grouting agents.[42] Since the 1950s, monometric acrylamide has been known as a neurotoxicant to man, and the total number of reported cases of acrylamide poisoning had been about 60 until 1977.[42,43] In the last ten years, several cases of occupational acrylamide poisoning have been reported in China and South Africa,[43-47] due to heavy exposure to acrylamide in the small scale industries with the prevalence of acrylamide poisoning at 32 to 73.2% in two factories.[43,44]

Skin peeling from the hands and feet, weak legs, sweating hands and numbness in hands and feet are the early symptoms of acrylamide poisoning. The early signs are impairments of vibration sensation in the toes and loss of ankle reflexes. Severe patients may have cerebellar involvement followed by polyneuropathy. Electro-neuromyographic changes, including a decrease in the sensory action potential amplitude, neurogenic abnormalities in electromyography, and prolongation of the ankle tendon reflex latency are of greater importance in the early detection of acrylamide neurotoxicity since they can precede the neuropathic symptoms and signs.[43] Vibration thresholds quantitatively determined with a Vibratron II device in acrylamide workers are higher than those in unexposed subjects.[45]

In the assessment of acrylamide exposure, biological monitoring is required because of the dermal route of absorption of acrylamide. Recently, the gas chroma-tography-mass spectrometry (GC-MS) determination of acrylamide hemoglobin adducts has been developed as a marker of the blood dose accumulated over the preceding four months in exposed workers.[47]

3.2.2 Allyl chloride

The chlorinated hydrocarbon, allyl chloride is a liquid agent readily volatile at room temperature. It is used as one of the raw materials for manufacturing a variety of products such as epichlorohydrin, epoxy resin, glycerin, pesticides and sodium allyl sulphonate which is one of the monomers of polyacrylonitrile. The neurotoxicity of allyl chloride was first reported from China when two outbreaks of polyneuropathy occurred in China in the early 1970s.[48,49] Epidemiological studies revealed weakness, paresthesia and numbness in extremities with sensory impairment in the glove-stocking distribution, as well as reduced ankle reflexes. Manifestations were more prevalent in workers exposed to allyl chloride at levels of 2.6–6650 mg/m^3 for 2.5 months to 6 years than in workers with milder exposure (0.2–25.13 mg/kg). The electroneuromyography (ENMG) showed neurogenic abnormalities in half of the exposed workers, even in those with few or very mild neuropathic symptoms. The recovery of neuropathic symptoms was shown to begin at three months after onset, however, the ENMG abnormalities did not recover until 11 months after cessation of exposure.[48]

Neuropathological studies on rabbits and mice have indicated that allyl chloride mainly induces damage to the peripheral nervous system causing a central-peripheral distal type of axonopathy.[49] Allyl chloride was found to induce accumulation of neurofilaments in the distal peripheral nerves, an effect similar to that seen in acrylamide neuropathy and 2,5-hexanedione axonopathy. However, a comparative study examining the modification and degradation of neurofilament proteins in rats subchronically treated with allyl chloride, acrylamide, or 2,5-hexanedione did not reveal any common mechanism.[50] The pathogenesis of allyl chloride neuropathy remains to be further investigated.

4   CONCLUSION

In developing countries, industrialization is a necessary feature of economic growth and development. Inevitably, rapid economic development has brought in its wake considerable occupational health and safety problems. The more frequent occurrence of occupational neurotoxic diseases in the developing world strongly indicates an increasing demand on safe working conditions, better occupational health services, and broader coverage of health education. While multi-disciplinary efforts are encouraged toward the research on the pathogenesis, diagnosis and treatment of neurotoxic disorders, multisectoral collaboration is strongly required for improving the prevention of occupational neurotoxic diseases world-wide.

5   REFERENCES

1. *The World Health Report 1995*, World Health Organization, Geneva, Switzerland, 1995, 113.
2. Forget, G., "Pesticides and the third world," *J. Toxicol. Environ. Health*, 32, 11, 1991.
3. Ong, C. N., Jeyaratnam, J., and Koh, D., "Factors influencing the assessment and control of occupational hazards in developing countries," *Environ. Res.*, 60, 112, 1993.
4. Jeyaratnam, J., "The transfer of hazardous industries," *J. Soc Occup Med.* 4,123, 1990.
5. Department of Health Inspection, Ministry of Public Health, *Annual Bulletin of Health Inspection (1994)*, Ministry of Public Health, The People's Republic of China, Beijing, 1995, 8.
6. Department of Health Inspection, Ministry of Public Health, *Annual Bulletin of Health Inspection (1991)*, Ministry of Public Health, The People's Republic of China, Beijing, 1992, 8.
7. Jeyaratnam, J., "Acute pesticide poisoning: a major global health problem," *World Health Stat. Q*, 43, 139, 1990.
8. WHO, *Public Health Impact of Pesticides Used in Agriculture*, World Health Organization, Geneva, Switzerland, 1990, 30.
9. Jeyaratnam, J., Lun, K. C., and Phoon, W. O., "Survey of acute pesticide poisoning among agricultural workers in four Asian countries," *Bull. World Health Organization*, 65, 521, 1987.
10. PAHO (Pan American Health Organization), "Pesticide and Health," *Hum. Ecol. Health*, 5, 2, 1986.
11. McConnell, R. and Hruska, A. J., "An epidemic of pesticide poisoning in Nicaragua: Implications for prevention in developing countries," *Am. J. Pub. Health*, 83, 1559, 1993.
12. Chen, S., Zhang, Z., He, F., Yao, P., Wu, Y., Sun, J., Liu, L., and Li, Q., "An epidemiological study on occupational acute pyrethroid poisoning in cotton farmers," *Br. J. Ind. Med.*, 48, 77, 1991.
13. Ambridge, E. M., Haines, I. H., and Lambert, M. R. K., "Operator contamination during pesticide application to tropical crops," *Med. Lav.*, 81, 457, 1990.
14. Jeyaratnam, J. and Maroni, M., "Organophosphorous compounds." *Toxicology* 91, 15, 1994.
15. He, F., "Biological monitoring of occupational pesticides exposure," *Int. Arch. Occup. Environ. Health* 65, S 69, 1993.

16. Xue, S. Z., Ding, X. J., and Ding, Y., "Clinical observation and comparison of the effectiveness of several oxime cholinesterase reactivators," *Scan. J. Work Environ. Health* 11, suppl 4, 46, 1985.

17. Karalliedde, L. and Senanayake, N., "Organophophorus insecticide poisoning," *Br. J. Anaesth.* 63, 736, 1989.

18. Lotti, M., Moretto, A., Capodicasa, E., Bertolazzi, M., Peraica, M., and Scapellato, M. L., "Interactions between neuropathy target esterase and its inhibitors and the development of polyneuropathy," *Toxicol. Appl. Pharmacol.* 122, 165, 1993.

19. Abou-Donia, M. B. and Lapadula, D. M., "Mechanisms of organophosphorus ester-induced delayed neurotoxicity: type 1 and type 2," *Annu. Rev. Pharmacol. Toxicol.* 30, 405, 1990.

20. Senanayake, N. and Karalliedde, L., "Neurotoxic effects of organophosphorus insecticides. An intermediate syndrome," *New. Eng. J. Med.* 316, 761, 1987.

21. Wadia, R. S., Sadagopan, C., Amin, R. B., and Sardesai, H. V., "Neurological manifestations of organophosphorous insecticide poisoning," *J. Neurol. Neurosurg. Psychiat.* 37, 841, 1947.

22. Mani, A., Thomas, M. S., and Abraham, A. P., "Type II paralysis or intermediate syndrome following organophosphorous poisoning," *J. Assoc. Physicians India* 40, 542, 1992.

23. Zhang, S., Huang, J., and Ding, M., "The mechanism and clinical features of carbamate insecticides poisoning," *Chi. J. Ind. Med.* 1 (1), 29, 1988.

24. Machmer, L.H. and Pickel, M., "Carbamate insecticides," *Toxicology* 91, 29, 1991.

25. Vijverberg, H. P. M. and van den Bercken, J., "Neurotoxicological effects and the mode of action of pyrethroid insecticides," *Crit. Rev. Toxicol.* 21, 2, 105, 1990.

26. He, F., Wang, S., Liu, L., Zhang, Z., and Sun, J., "Clinical manifestations and diagnosis of acute pyrethroid poisoning," *Arch. Toxicol.* 63, 54, 1989.

27. Chen, S., Zhang, Z., He, F., Yao, P., Wu, Y., Sun, J., Liu, L., and Li, Q., "An epidemiological study on occupational acute pyrethroid poisoning in cotton farmers," *Br. J. Ind. Med.* 48, 77, 1991.

28. He, F., "Synthetic pyrethroids," *Toxicology* 91, 43-49, 1994.

29. He, F., Deng, H., Ji, X., Zhang, Z., Sun, J., and Yao, P., "Changes of nerve excitability and urinary deltamethrin in sprayers," *Int. Arch. Environ. Health* 62, 587, 1991.

30. Thom, S. R. and Keim, L. W., "Carbon monoxide poisoning: A review. Epidemiology, pathophysiology, clinical findings, and treatment options including hyperbaric oxygen therapy," *Clin. Toxicol.* 27, 141, 1989.

31. Choi, I. S., "Delayed neurologic sequelae in carbon monoxide intoxication," *Arch. Neurol.* 40, 433, 1983.

32. He, F., Qin, J., Chen, S., Pan, X., Xu, G., Zhang, S., and Fang, G., "Risk factors of development of delayed encephalopathy following acute carbon monoxide poisoning," *Indoor Environ.* 1, 268, 1992.

33. He, F., Liu, X., Yang, S., Zhang, S., Xu, G., Fang, G., and Pan, X., "Evaluation of brain function in acute carbon monoxide poisoning with multimodality evoked potentials," *Environ. Res.* 60, 213, 1993.

34. Joubert, P. H., "Acute poisoning in developing countries," *Adverse Drug React. Acute Poisoning Rev.* 8, 165, 1989.

35. Chia, S. E., Phoon, W. H., Lee, H. S., Tan, K. T., and Jeyaratnam, J., "Exposure to neurotoxic metals among workers in Singapore: an overview," *Occup. Med.* 43, 18, 1993.

36. Matte, T. D., Figueroa, J. P., Ostrowski, S., Burr, G., Jackson-Hunt, L., Keenlyside, R. A., and Baker, E. L., "Lead poisoning among household members exposed to lead-acid battery repair shops in Kingston, Jamaica." *Int. J. Epidemiol.* 18, 874, 1989.

37. Eisenberg, A., Avni, A., Grauer, F., Weissenberg, E., Acker, C., Hamdallah, M., Shahin, S., Moreb, J., and Hershko, C., "Identification of lead poisoning in West Bank Arabs," *Arch. Intern. Med.* 145, 1848, 1985.

38. Wang, Y., "Industrial lead poisoning in China over the past 33 years." *Ecotoxicol. Environ. Safety* 8, 526, 1984.

39. Wang, J. D., Huang, C. C., Hwang, Y. H., Chiang, J. R., Lin, J. M., and Chen, J. S., "Manganese induced Parkinsonism: an outbreak due to an unpaired ventilation control system in a ferromanganese smelter." *Br. J. Ind. Med.* 46, 856, 1989.

40. Huang, C. C., Chu, N. S., Lu, C. S., Wang, J. D., Tsai, J. L., Tzeng, J. L., Wolters, E. C., and Calne, D. B., "Chronic manganese intoxication," *Arch. Neurol.* 46, 1104, 1989.

41. He, F., Zhow, X. R., Lin, B. X., Xiung, Y. P., Chen, S. Y., Zhang, S. L., Ru, J. Y., and Deng, M. H., "Prognosis of mercury poisoning in mercury refinery workers," *Ann. Acad. Med. Singapore* 13, Suppl, 389, 1984.

42. WHO Task Group, *Environmental Health Criteria 59. Acrylamide.* World Health Organization., Geneva, Switzerland, 1985.

43. He, F., Zhang, S., Wang, H., Li, G., Zhang, Z., Li, F., Dong, X., and Hu, F., "Neurological and electromyographic assessment of the adverse effects of acrylamide on occupationally exposed workers." *Scan. J. Work Environ. Health* 15, 125, 1989.

44. Myers, J., and Macun, I., "Acrylamide neuropathy in a South African factory: An epidemiologic investigation," *Am. J. Ind. Med.* 19, 487, 1991.

45. Bachmann, M., Myers, J., and Bezuidenhout, B. N., "Acrylamide monomer and peripheral neuropathy in chemical workers," *Am. J. Ind. Med.* 21, 217, 1992.

46. Deng, H., He, F., Zhang, S., Calleman, C. J., and Costa, L. G., "Quantitative measurement of vibration thresholds in healthy adults and acrylamide workers," *Int. Arch. Occup. Environ. Health* 65, 53, 1993.

47. Bergmark, E., Calleman, C. J., He, F., and Costa, L. G., "Determination of haemoglobin adducts in humans occupationally exposed to acrylamide," *Toxicol. Appl. Pharmacol.* 120, 45, 1993.

48. He, F. and Zhang, S., "Effects of allyl chloride on occupationally exposed subjects," *Scan. J. Work Environ. Health* 11, Suppl 4, 43, 1985.

49. He, F., Jacobs, J. M., and Scaravilli, F., "The pathology of allyl chloride neurotoxicity in mice," *Acta Neuropathol* 55, 125, 1980.

50. Nagano, M., Yamamoto, H., Harada, K., Miyamoto, E., Futatsuka, M., "Comparison study of modification and degradation of neurofilament proteins in rats subchronically treated with allyl chloride, acrylamide, or 2,5-hexanedione," *Environ. Res.* 63, 229, 1993.

Index

A

Acetone, 249
Acetylcholinesterase, 80, 84–85
  inhibitors, 4, 6–7, 63, 138, 261
  reactivators, 261
Acetylene, 41
2-Acetylfuran, 61
Acrylamide, 4, 83–84, 122
  axonopathy, 12, 83, 118
  adducts, 62, 83–84, 266
  conversion to glycidamide, 84
  neurophysiological investigations,
    193, 266
  poisoning, 265–266
Acrylonitrile, 14, 62, 82
Adaptive psychological
    testing, 215
Adrenoceptors in blood cells, 81
Agrochemicals, 4–5, 122, 260–264
ACGIH, *see* Americal Conference
    of Governmental Industrial Hygienists
Alcohol, *see* ethanol
Aldrin, 3, 40, 65, 249
Aliphatic alcohols, 137
Alkyl phosphates, 63, 87, 142
Alkyl tins, 3, *see also* individual tin
    compounds
Alloy production workers, 8
Allyl alcohol, 118
Allyl chloride, 4, 122, 266
Aluminum, 3, 249
  absorption, 50
  encephalopathy, 3, 148
  levels in body fluids, 50
  neurobehavioral studies, 217, 219
  and neurodegenerative disorders,
    106, 134, 140, 147–148
  neuroendocrine effects, 148
  occupational exposure, 148
Alzheimer's disease, 139, 146–149
Americal Conference of Governmental
    Industrial Hygienists (ACGIH),
    51, 56, 58–61, 63–66
δ-Aminolevulinic acid, 53–54

δ-Aminolevulinic acid dehydratase
  and lead toxicity, 36, 53, 78
  polymorphism, 35
Amyotrophic lateral sclerosis, 139, 149–152
Anesthetic gases, 3, 220
Animal models
  neurotoxicity assessment, 251–252
  visual toxicity in, 171–172, 175
Antidotes in organophosphate poisoning, 261
Anti-12-hydroxyendrin, 65
Arsenic, 3, 5, 249
  electrophysiological studies, 193
  encephalopath, 8
  levels in body fluids, 51
  methylation, 51
  peripheral neuropathy, 12–13, 118
Arsenobetaine, 51, 235
Arsenocholine, 235
Arsine, 51
Associate learning test, 222
Atropine, 261, 263–264
Auditory evoked potentials
  carbon disulfide, 187
  in shoe-workers, 191
  lead, 192
  mercury, 193
  solvent mixtures, 191
Autonomic system dysfunction,
    13, 118–119, 188, 192
Axonopathy, 11–12, 87, 90

B

BAEP, *see* auditory evoked potentials
BARS, *see* Behavioral Assessment
    and Research System
Battery making/recycling, 5, 122, 202
Behavioral Assessment and Research System
    (BARS), 217–218, 224–225
Benton visual recognition test, 103, 223
Bibliographic/abstracting services, 236–238
Biological monitoring, 49–67
  acrylamide, 62, 84–85
  aluminium, 50
  arsenic, 51